S0-AVN-726

HEAT TRANSFER AT LOW TEMPERATURES

THE INTERNATIONAL CRYOGENICS MONOGRAPH SERIES

H. J. Goldsmid
Thermoelectric Refrigeration, 1964

G. T. Meaden
Electrical Resistance of Metals, 1965

E. S. R. Gopal
Specific Heats at Low Temperatures, 1966

M. G. Zabetakis
Safety with Cryogenic Fluids, 1967

D. H. Parkinson and B. E. Mulhall
The Generation of High Magnetic Fields, 1967

W. E. Keller
Helium-3 and Helium-4, 1969

A. J. Croft
Cryogenic Laboratory Equipment, 1970

A. U. Smith
Current Trends in Cryobiology, 1970

C. A. Bailey
Advanced Cryogenics, 1971

D. A. Wigley
Mechanical Properties of Materials
at Low Temperatures, 1971

C. M. Hurd
The Hall Effect in Metals and Alloys, 1972

E. M. Savitskii, V. V. Baron, Yu. V. Efimov,
M. I. Bychkova, and L. F. Myzenkova
Superconducting Materials, 1973

W. Frost
Heat Transfer at Low Temperatures, 1975

HEAT TRANSFER AT LOW TEMPERATURES

Edited by Walter Frost
The University of Tennessee Space Institute
Tullahoma, Tennessee

PLENUM PRESS • NEW YORK AND LONDON

Library of Congress Cataloging in Publication Data

Frost, Walter, 1935-
Heat transfer at low temperatures.

(The International cryogenics monograph series)
"Many of the chapters were originally presented as lectures in a short course sponsored by the University of Tennessee Space Institute."
Includes bibliographical references and index.
1. Low temperature engineering. 2. Heat—Transmission. I. Title. II. Series.
TP482.F76 621.5 74-34014
ISBN 0-306-30575-5

A Division of Plenum Publishing Corporation
227 West 17th Street, New York, N.Y. 10011

United Kingdom edition published by Plenum Press, London
A Division of Plenum Publishing Company, Ltd.
Davis House (4th Floor), 8 Scrubs Lane, Harlesden, London, NW10 6SE, England

Printed in the United States of America

CONTRIBUTORS

D. L. Clements, *U.S. Army Ordnance Center and School, Aberdeen Proving Ground, Maryland*

F. J. Edeskuty, *Los Alamos Scientific Laboratory, Los Alamos, New Mexico*

H. K. Fauske, *Reactor Analysis and Safety Division, Argonne National Laboratory, Argonne, Illinois*

Walter Frost, *University of Tennessee Space Institute, Tullahoma, Tennessee*

R. W. Graham, *NASA, Lewis Research Center, Cleveland, Ohio*

M. A. Grolmes, *Argonne National Laboratory, Argonne, Illinois*

W. L. Harper, *University of Tennessee Space Institute, Tullahoma, Tennessee*

R. C. Hendricks, *NASA, Lewis Research Center, Cleveland, Ohio*

R. E. Henry, *Argonne National Laboratory, Argonne, Illinois*

R. K. Irey, *Department of Mechanical Engineering, University of Florida, Gainesville, Florida*

M. A. Lechtenberger, *Engineering Research Center, University of Colorado, Boulder, Colorado*

G. H. Nix, *President, Optimal Systems, Inc., Atlanta, Georgia*

R. D. von Reth, *Messerschmeitt-Bolkow-Blohm, Hamburg, Germany*

W. M. Rohsenow, *Heat Transfer Laboratory, Department of Mechanical Engineering, Massachusetts Institute of Technology, Cambridge, Massachusetts*

J. A. Roux, *Arnold Engineering Development Center, Arnold AFS, Tennessee*

R. J. Simoneau, *NASA, Lewis Research Center, Cleveland, Ohio*

K. E. Tempelmeyer, *University of Tennessee Space Institute, Tullahoma, Tennessee*

K. D. Timmerhaus, *Engineering Research Center, University of Colorado, Boulder, Colorado*

R. I. Vachon, *Auburn University, Auburn, Alabama*

K. D. Williamson, Jr., *Los Alamos Scientific Laboratory, Los Alamos, New Mexico*

PREFACE

The purposes of this book are to provide insight and to draw attention to problems peculiar to heat transfer at low temperatures. This does not imply that the theories of classical heat transfer fail at low temperatures, but rather that many of the approximations employed in standard solutions techniques are not valid in this regime. Physical properties, for example, have more pronounced variations at low temperatures and cannot, as is conventionally done, be held constant. Fluids readily become mixtures of two or more phases and their analysis is different from that for a single-phase fluid. These and other problems which occur more frequently at low temperatures than at standard conditions are discussed in this book.

Although the title specifies heat transfer, the book also contains a very comprehensive chapter on two-phase fluid flow and a partial chapter on the flow of fluids in the thermodynamically critical state. Emphasis is placed on those flow phenomena that occur at low temperatures. Flow analyses are, of course, a prerequisite to forced-convection heat transfer analyses, and thus these chapters add continuity to the text.

The book is primarily written for the design engineer, but does broach many topics which should prove interesting to the researcher. For the student and teacher the book will serve as a useful reference and possibly as a text for a special topics course in heat transfer.

The concept of the book is due to Dr. Timmerhaus, series editor of the International Cryogenic Monograph Series. Dr. Timmerhaus recognized the need for a treatise on heat transfer at cryogenic conditions and the potential for expanding a short course given at The University of Tennessee Space Institute into such an exposition. He provided much of the impetus to bring the book to print, and the editor expresses a deep appreciation for the association and the experience gained in working with him on the manuscript.

The cooperation of all the authors in preparing and condensing their material for the final draft and for presenting lectures at the short course from which the book sprung is gratefully acknowledged. Thanks also go to

Betty Spray, who typed many of the revised chapter manuscripts; this required efforts above her normal work load.

Finally, appreciation is expressed to Dr. Goethert, who's concept of an educational institute, such as the Space Institute, provides the environment which allows UTSI faculty close professional association with the many learned people who have written the chapters contained herein.

Walter Frost

CONTENTS

PART II. TWO-PHASE PHENOMENA

PART III. RADIATION AND HELIUM II HEAT TRANSPORT

INTRODUCTION 1

W. FROST

1.1 INTRODUCTION

The addition of this volume to the *International Cryogenic Monograph Series* is intended to enhance the knowledge of the thermal design engineer faced with solving heat transfer and fluid flow problems at low temperatures. Chapters authored by various experts in the field have been integrated into a consistent format such that the monograph, hopefully, has continuity and a central theme. Many of the chapters were originally presented as lectures at a short course sponsored by The University of Tennessee Space Institute.

The expression "low temperature" used here is not precise but is intended to connote temperatures ranging from those normally associated with common liquid refrigerants, which may be on the order of room temperature, 298°K, to those associated with liquid helium, 4K. Scott[1] lists 36 fluids with normal boiling points in this temperature range which he classifies as cryogenic fluids. Other authors[1–4] define cryogenic fluids as those fluids that have normal boiling points below 123K. Typical of these are the so-called permanent gases such as helium, hydrogen, neon, nitrogen, oxygen, and air.

A partial tabulation of physical properties of cryogenic liquids is given in Table 1-1. More complete physical property data are given in Refs. 1–6.

The transport and storage of cryogens at very low temperatures result in numerous thermal design problems related to all three classical mechanisms of heat transfer. Chapter 2 covers the mechanics of conduction heat transfer, emphasizing those effects that, although normally negligible at room

W. FROST The University of Tennessee Space Institute, Tullahoma, Tennessee.

TABLE 1-1
Thermodynamic Properties of Common Cryogens[7]
(Pressure = 1 atm)

Cryogenic substance	Boiling point, °F	Liquid density, lb_m/ft^3	Heat of vaporization		Volume of gas (ft^3) at STP per ft^3 of liquid
			Btu/lb_m	Btu/ft^3 liquid	
Oxygen	−297	71.3	91.6	6531	862
Nitrogen	−320	50.4	85.6	4319	696
Argon	−303	87.5	70.0	6126	846
Methane	−259	26.5	219.0	5808	637
Fluorine	−307	94.3	74.0	6975	959
Hydrogen	−423	4.4	193.0	849	844
Helium	−452	7.8	8.8	69	754

temperature, are very significant at cryogenic temperatures. Typical of these are analysis including variable thermal conductivity since the physical properties of many substances exhibit strong temperature dependence at low temperature. Cryodeposits formed at low temperatures frequently display anisotropic thermal conductivities which also require heat conduction analysis different from common design practice.

Moreover, substantial temperature differences are established between cryogenic storage tanks or pipe lines and ambient surroundings. Thus, special insulations and methods of designing structural supports and piping to avoid heat leaks require detailed conduction heat transfer analysis. Also, transient heat conduction associated with rapid cooldown of massive cryogenic systems becomes more important than ever.

Chapter 3 considers convection heat transfer. In general cryogenic fluids behave as "classical" fluids (with the important exception of helium II discussed in Chapter 15) in that they characteristically obey the laws of mechanics and thermodynamics such that many of the convective analyses, scaling laws, and experimental correlations based on dimensional analysis developed for common liquids and gases are applicable to cryogens. However, cryogenic systems frequently operate near the thermodynamic critical state, where fluid properties show strong variation with temperature and pressure. Thus, constant property solutions of the convective energy equation given in standard heat transfer texts cannot be used indiscriminantly. Chapter 3 covers heat transfer with near-critical fluid properties and presents single-phase forced and natural convection correlations recommended for low temperatures.

Boiling and two-phase flow readily occur in cryogenic fluids during storage in vessels and during transport through pipe lines. Generally, liquid-to-vapor phase change reduces the amount of liquid stored or transferred and represents an undesirable condition. In other applications, such as cooling superconducting electronic components and magnets or cooling panels in cryogenically cooled space simulators, the latent energy absorbed with phase change is desirable. Thus, methods of predicting when two-phase flow will occur and the heat transfer rate associated with it are needed.

Flow of mixed fluid phases occurs readily in transfer systems due to the ever-present problem of heat leaks. The presence of two-phase flow complicates the prediction of such parameters as pressure drop, allowable propellant loss, acceptable flow rates, cooldown requirements, dilution effects, and many others. With two-phase flow several flow regimes, depending on the distribution of vapor and liquid in the duct, may occur. Different combinations of turbulent and laminar flows governed by different physical laws result due to these regimes and different equations are required for engineering predictions. Additionally, these flow regimes change along the length of the duct due to changes in the mass fractions of vapor and liquid present and also change with time during cooldown periods. The various regimes of two-phase flow are discussed in Chapter 4 while methods of predicting pressure drops, pressure-wave propagation, critical flow, and inlet effects are described in Chapter 11. Chapter 13 covers some of the transient two-phase flow problems associated with cooldown, blowdown from high pressures, and quenching of hot solids. The nature of flow oscillations and instabilities frequently resulting in two-phase flow processes are also referred to in this chapter.

In storage vessels and other cryogenic equipment, boiling will begin on the walls at temperatures only slightly over the saturation value. With the inception of boiling, the heat transfer to the stored fluid becomes much more rapid, resulting in high rates of boiloff. The phenomenon of boiling in pools of fluid and prediction techniques for the rate of heat transfer and the inception of boiling under these conditions are discussed in Chapters 4–8.

Boiling also occurs in transfer lines under forced convection conditions. Again the mechanics of heat transfer are influenced by the two-phase flow regime that exists at that time or location in the duct. There are sufficient differences between boiling in a pool or under natural convection conditions and boiling under forced-flow conditions that Chapter 12 has been included.

An alternative to boiling in cryogenic systems is phase change due to condensation. Commonly, one encounters vapor-to-liquid or vapor-to-solid condensation processes. The former process frequently occurs in heat exchangers where a high-temperature vapor condenses on the tube, giving up its latent heat of vaporization and thus heating the fluid on the opposite side of the tube. Cryogens are not frequently used for heating; however, vapor–liquid condensation may readily occur in heat exchangers in liquefaction systems or refrigeration units and in gas storage vessels in space facilities or in petroleum plants. The heat transfer mechanism associated with vapor–liquid condensation is treated in Chapter 9.

The process of vapor–solid condensation readily occurs in cryopumping of vacuums where the gases remaining after mechanical pumping are frozen onto cryogenically cooled surfaces. The solids formed are called cryodeposits. Heat transfer considerations are important in the design of the cooling panels and the sizing of the refrigeration unit necessary for providing the coolant. Also, condensation occurs on cryogenic pipe lines and storage vessels and influences the heat leak from these components of the system. Chapter 10 discusses the vapor-to-solid condensation phenomenon.

Finally, Part III of the text deals with the mechanism of radiation heat transfer at low temperatures in Chapter 14 and with heat transfer in helium II in Chapter 15.

Radiation heat transfer at low temperatures differs from conventional theory[8,9] only through the effects of surface property changes. Emissivity, reflectivity, and absorptivity of a surface are altered by the formation of condensates (cryodeposits) or frosts on the solid surfaces of the system, and thin dielectric coatings have been observed to show increased emissivity at very low temperatures.[9] Thus, the exchange of radiant energy in a system may change because of these variations in surface properties. Information on these effects is available in Chapter 14.

Heat transfer in helium II is a very intriguing phenomenon which is strongly influenced by its peculiar superproperties. Much needs to be learned about the transport of energy and momentum in this fluid in order to harness the advantages which can be achieved from superconduction, frictionless flow, etc. Chapter 15 gives a detailed account of the heat transfer processes in helium II.

1.2 ORGANIZATION

Each chapter has its own bibliography and list of nomenclature. Although in general the nomenclature has been standardized, small differences occur due to the preference of the individual authors.

The reference numbers appear in numerical order starting with one in each chapter. Equation, table, and figure numbers are preceded by the chapter number.

1.3 REFERENCES

1. R. B. Scott, *Cryogenic Engineering*, D. Van Nostrand Co., New York (1959).
2. R. F. Barron, *Cryogenic Systems*, McGraw-Hill Book Co., New York (1966).
3. M. McClintock, *Cryogenics*, Reinhold Publishing Co., New York (1964).
4. R. W. Vance, ed., *Applied Cryogenic Engineering*, John Wiley and Sons, New York (1963).
5. V. J. Johnson, "A Compendium of the Properties of Materials at Low Temperature (Phase 1), Part 1, Properties of Fluids," PB171 616, Air Research and Development Command, Wright-Patterson Air Force Base, Ohio (1960).
6. K. D. Timmerhaus, "A Correlation of Thermodynamic Properties of Cryogenic Fluids," lecture notes for short course at University of Tennessee Space Institute, Tullahoma, Tennessee (1969).
7. J. A. Clark, in *Advances in Heat Transfer*, Vol. 5, Academic Press, New York (1968).
8. E. M. Sparrow and R. D. Cess, *Radiation Heat Transfer*, Brooks/Cole Publishing Co., California (1970).
9. R. Siegel and J. R. Howell, *Thermal Radiation Heat Transfer*, McGraw-Hill Book Co., New York (1972).

PART I

CONDUCTION AND CONVECTION HEAT TRANSFER

CONDUCTIVE HEAT TRANSFER 2

K. D. TIMMERHAUS

2.1 INTRODUCTION

As indicated earlier, the use of cryogenic fluids has introduced several unique problems in heat transfer. The handling and transporting of such fluids at low temperatures in the presence of an atmospheric ambient have required the innovation and development of specialized insulating methods and design techniques. As a result, new insulation systems, consisting of many thin layers of highly reflecting radiation shields separated by thin spacers or insulators, have been developed. These insulations, termed multilayer insulations or superinsulations, are very effective when subjected to a high vacuum (5×10^{-5} to 10^{-6} Torr) in reducing the heat leak into low-temperature systems. One type, which has as many as 50 layers of 0.00025-in.-thick, aluminum-coated polyester radiation shields, has an apparent thermal conductivity at these pressures that is only 1/2000 that of ambient air.[1] In addition to new insulations for minimizing heat losses, special attention has been given to the thermal conduction effects associated with the design of supporting structures of cryogenic containers.

At low temperatures the physical properties of many substances are significantly temperature dependent. This has made it necessary to consider the effects of variable properties in the analysis and calculation of conductive heat transfer processes. This has been particularly true in the case of the thermal conductivity and heat capacity properties.

K. D. TIMMERHAUS University of Colorado, Boulder, Colorado.

Although most of the material to be presented here can also be applied to noncryogenic behavior, the emphasis will be on applications valid for cryogenic operations. Where possible or available, cryogenic data will be cited and the uniqueness of low-temperature application will be underscored. Additional surveys of conductive heat transfer appropriate to low temperatures are given in the references cited at the conclusion of this chapter.

2.2 STEADY-STATE CONDUCTIVE HEAT TRANSFER

Heat transfer frequently is described as a rate process or transport phenomenon. It is energy in transit by virtue of a temperature difference or gradient in a system capable of thermal communication. Invariably it is the instantaneous time rate of heat flow or state of temperature distribution which is of interest in a heat transfer calculation. In conductive heat transfer, the energy is transmitted by direct molecular communication with appreciable displacement of the molecules. According to kinetic theory, the temperature of an element of a system is proportional to the mean kinetic energy of its constituent molecules. When molecules in one region acquire a mean kinetic energy greater than that of molecules in an adjacent region, as evidenced by a difference in temperature, the molecules possessing the greater energy will transmit part of their energy to the molecules in the lower-temperature region. The observable effect of conductive heat transfer is an equalization of temperature. If differences in temperatures are, however, maintained by the addition and removal of energy at different points in the system, a continuous flow of heat from the warmer to the cooler region will be established.

For an isotropic material the instantaneous rate of heat flow $q_x(x, y, z, \theta)$ across an orthogonal plane of area $A(x)$ is given by the relationship

$$Q_x(x, y, z, \theta) = -k(T)A(x)\,[\partial T(x, y, z, \theta)/\partial x] \tag{2-1}$$

The instantaneous rate of heat flow in this case is a directional quantity. Obviously, there are also heat flows in the y and z directions if temperature gradients exist in those directions. The minus sign satisfies the requirement that the direction of heat flow be from regions of high to those of low temperature and to identify the direction of the heat transfer in terms of the arbitrarily established directions of the coordinate system. Equation (2-1) describes the rate of heat flow in the x direction at any point (x, y, z) within the material and at any instant of time θ. Thus, Eq. (2-1) is applicable for transient as well as steady-state processes involving temperature-dependent thermal conductivities $k(T)$ and variable areas $A(x)$.

To describe the temperature distribution in an isotropic material rather than the instantaneous rate of heat flow, Eq. (2-1) is generally replaced by the one-dimensional Fourier equation

$$\partial T/\partial \theta = \partial^2 T/\partial x^2 \tag{2-2}$$

where $\partial T/\partial \theta$ is equal to zero for steady-state processes.

2.2.1 Thermal Conductivity of Solids

There is no doubt that thermal conductivity is the single most important physical property affecting conductive heat transfer. Its numerical value depends upon the composition of a substance and its pressure and temperature. For a given substance, temperature is the most significant single variable governing the thermal conductivity. Since theoretical considerations for calculating values of thermal conductivity for materials at low temperatures are still quite inadequate, it is frequently more satisfactory to obtain numerical values of this property from a table of physical properties or to undertake a direct experimental determination of the value itself for a specific application.

Experimental thermal conductivity values for most materials in low-temperature design are not only quite extensive but readily available in the literature. Some of the more useful compilations include the work by Powell and Blanpied,[2] the NBS *Compendium of the Properties of Materials at Low Temperature*,[3] the *American Institute of Physics Handbook*,[4] the *Thermophysical Properties Research Center Data Book*,[5] the National Standard Reference Data System report by Powell, Ho, and Liley,[6] and the recent literature survey by Dillard and Timmerhaus.[7] In addition, an excellent review of the theoretical aspects of thermal conductivity is given by Klemens,[8] while the measurement of thermal conductivity of solid conductors at low temperatures is capably covered by White.[9]

2.2.2 One-Dimensional Conductive Heat Transfer

An important case of one-dimensional heat flow is the conductive heat transfer through geometric shapes having parallel, isothermal surfaces which are perpendicular to the direction of heat flow. For steady-state conduction in one dimension, the first law of thermodynamics requires that the heat flow be constant and independent of the coordinate-system dimensions. Thus, Eq. (2-1) can be rewritten as

$$Q \int_{x_1}^{x_2} \frac{dx}{A(x)} = -\int_{T_1}^{T_2} k(T)\, dT \tag{2-3}$$

The right-hand side of this equation can be integrated if we define a mean thermal conductivity as

$$k_m(T_2 - T_1) = \int_{T_1}^{T_2} k(T)\,dT \tag{2-4}$$

Upon substitution and solving for Q, Eq. (2-3) becomes

$$Q = -k_m(T_2 - T_1) \Big/ \int_{x_1}^{x_2} \frac{dx}{A(x)} \tag{2-5}$$

Once the boundary temperatures are specified and the geometry known, Eq. (2-5) permits the computation of the steady-state conductive heat transfer rate in materials having both a temperature-dependent thermal conductivity and a complex shape.

Geometric shapes that are of importance in one-dimensional heat transfer at low temperatures include long, hollow cylinders (pipes) and hollow spheres (dewars). If the inner surface temperature of the hollow cylinder is constant at T_i and the outer surface is maintained uniformly at T_o, the rate of heat conduction is, from Eq. (2-5),

$$Q = -2\pi L k_m(T_o - T_i)/\ln(r_o/r_i) \tag{2-6}$$

where L is the length of the cylinder and r_i and r_o are the inside and outside radii of the hollow cylinder, respectively. The temperature distribution in the curved wall of the hollow cylinder is obtained by integrating Eq. (2-3), after appropriate geometric substitutions have been made, from the inner radius r_i and the corresponding temperature T_i to an arbitrary radius r and the corresponding temperature T, or

$$T(r) = (T_i - k_m)\,(T_i - T_o)[\ln(r/r_i)]/[k_m' \ln(r_o/r_i)] \tag{2-7}$$

where k_m' is defined as

$$k_m'[T(r) - T_i] = \int_{T_1}^{T(r)} k(T)\,dT \tag{2-8}$$

Since a hollow sphere has the largest volume per outside surface area of any geometric configuration, it has received considerable use in the cryogenic field where heat losses need to be kept to a minimum. Conduction through a spherical shell is a one-dimensional steady-state problem if the interior and exterior surface temperatures are uniform and constant. The rate of heat conduction for this case is

$$Q = -4\pi r_i r_o k_m(T_o - T_i)/(r_o - r_i) \tag{2-9}$$

if the material between the two surfaces is homogeneous.

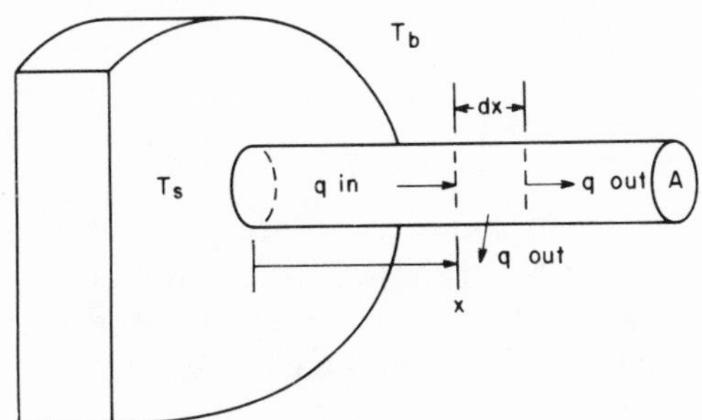

FIG. 2-1. Conductive heat transfer through an extended surface.

One of the more important applications of one-dimensional conductive heat transfer at low temperatures occurs in the use of extended surfaces or fins in the design of heat exchangers to maximize the heat transfer while minimizing the overall heat loss. The use of one-dimensional heat conduction is justified because the transverse temperature gradients are generally so small that the temperature at any cross section of the extended surface is uniform.

To determine the temperature distribution in a fin and the heat transferred by this extended surface, a heat balance is made for a small element of the fin. For illustrative purposes consider the combined conductive and convective heat transfer process associated with the pin fin shown in Fig. 2-1. Under steady-state conditions, the rate of heat flow by conduction into an element of the fin at x must be equal to the rate of heat flow by conduction from an element of the fin at $x + dx$ plus the rate of heat flow by convection from the surface between x and $x + dx$. The mathematical expression for this heat balance can be written as

$$-k_m A\, dT/dx = [-k_m A\, dT/dx + (d/dx)(-k_m A\, dT/dx)dx] + hP\, dx\,(T - T_b) \tag{2-10}$$

where h is the convective heat transfer coefficient, P is the perimeter of the pin, and Pdx represents the surface area between x and $x + dx$ in contact with the surrounding fluid. Equation (2-10) can be simplified to

$$d^2T/dx^2 = m^2(T - T_b) \tag{2-11}$$

where $m^2 = hP/k_m A$. Equation (2-11) results in a temperature distribution along the fin of

$$T - T_b = (T_s - T_b)\frac{\cosh[m(L - x)] + (h_L/mk_m)\sinh[m(L - x)]}{\cosh mL + (h_L/mk_m)\sinh mL} \tag{2-12}$$

and a heat flow rate from the pin of

$$Q = (PhAk_m)^{\frac{1}{2}}(T_s - T_b)\frac{\sinh mL + (h_L/mk_m)\cosh mL}{\cosh mL + (h_L/mk_m)\sinh mL} \tag{2-13}$$

A variety of common extended surface geometrics are used in low-temperature heat conduction applications. Some of these are detailed in the texts by Scott[10] and Barron[11] and in papers presented at past Cryogenic Engineering Conferences and published under the title *Advances in Cryogenic Engineering*.[12]

2.2.3 Conductive Heat Transfer Through Composite Structures

Equation (2-5) suggests that one-dimensional conductive heat transfer through a solid can be expressed in terms of a temperature potential $T_2 - T_1$ and a resistance to the flow of heat $R_k = (1/k_m)\int_{x_1}^{x_2} dx/A(x)$. Thus

$$Q = -(T_2 - T_1)/R_k \tag{2-14}$$

For the steady-state heat conduction through a composite structure composed of several materials having different thicknesses and thermal conductivities and arranged in successive, adjacent layers, the heat flow can be computed in a manner similar to that of an electrical current flow in a series of electrical resistances. Hence, for $T_{x_{n+1}} > T_{x_1}$,

$$Q = -\frac{T_{x_{n+1}} - T_{x_1}}{k_{m_1}^{-1}\int_{x_1}^{x_2}[dx/A(x)] + k_{m_2}^{-1}\int_{x_2}^{x_3}[dx/A(x)] + \cdots + k_{m_n}^{-1}\int_{x_n}^{x_{n+1}}[dx/A(x)]} \tag{2-15}$$

$$Q = -(T_{x_{n+1}} - Tx_1)/\sum_{i=1}^{n+1} R_{k_i} \tag{2-16}$$

If the boundary temperatures for the various materials in the composite structure are not available, a trial-and-error procedure will be required to determine the proper value for each k_m from

$$Q = \frac{T_{x_2} - T_{x_1}}{k_{m_1}^{-1}\int_{x_1}^{x_2}[dx/A(x)]} = \frac{T_{x_3} - T_{x_2}}{k_{m_1}^{-1}\int_{x_2}^{x_3}[dx/A(x)]}$$

$$= \cdots = \frac{T_{x_{n+1}} - T_{x_n}}{k_{m_1}^{-1}\int_{n}^{x_{n+1}}[dx/A(x)]} \tag{2-17}$$

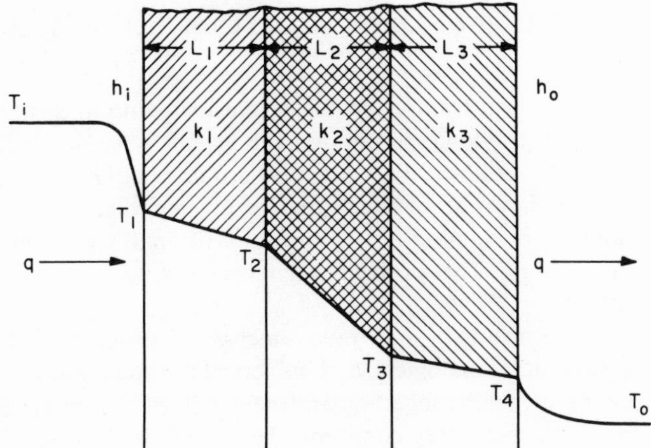

FIG. 2-2. Temperature distribution for one-dimensional heat flow through a composite system with both exterior walls exposed to fluid.

Symbols in Eqs. (2-15), (2-16), and (2-17) can be identified by inspection of Fig. 2-2.

In much of the conductive heat transfer at low temperatures the external surfaces of a composite structure (metal walls plus insulation) are in contact with fluids having different temperatures. Under this situation, heat will be transferred convectively between the external surfaces of the composite material and the fluids. This heat exchange across one surface can be expressed in terms of a convective heat transfer coefficient h and the contact surface area A as

$$Q = hA(T_{fluid} - T_{surface}) \tag{2-18}$$

Equation (2-18) also suggests that the convective heat transfer between the fluid and the solid can be written in terms of a temperature potential and a thermal resistance, $R_h = 1/hA$. The addition of the thermal resistances for the convective processes to Eq. (2-16) results in an equation applicable to the general case of heat flow across a composite structure separating a warm and a cold fluid, namely,

$$Q = \frac{T_{warm\ fluid} - T_{cold\ fluid}}{(R_h)_{warm\ fluid} + \sum_{i=1}^{n+1} R_{k_1} + (R_h)_{cold\ fluid}} \tag{2-19}$$

Similar relationships can be written for cylindrical and spherical geometries. Methods for evaluating numerical values of the convective heat transfer coefficient are presented in a later chapter.

2.2.4 Two- and Three-Dimensional Conductive Heat Transfer

The conductive heat transfer problems considered up to this point have been ones in which the temperature and the heat flow could be treated as functions of a single variable. Many practical problems fall into this category, but when the boundaries of a system are irregular or when the temperature along a boundary is nonuniform, a one-dimensional treatment may no longer be satisfactory. The problem then becomes one of two or three spatial dimensions.

Heat conduction in two- and three-dimensional systems can be treated by analytical, graphical, analogical, and numerical methods. For the common geometry of slabs, cylinders, spheres, parallelepipeds, etc., exact mathematical solutions for various boundary conditions have been worked out. However, a complete treatment of analytical solutions requires a prior knowledge of Fourier series, Bessel functions, Legendre polynomials, Laplace transform methods, and complex variable theory. A number of excellent texts dealing with mathematical solutions of conductive heat transfer problems are available.[13–20]

The derivation of the conductive heat transfer equation for a three-dimensional differential element is fairly straightforward and results in the following relationship if the heat capacity and the density of the solid are independent of temperature and a heat source $\dot{q}$ is present:

$$(\partial/\partial x)(k\,\partial T/\partial x) + (\partial/\partial y)(k\,\partial T/\partial y) + (\partial/\partial z)(k\,\partial T/\partial z) + \dot{q}(x, y, z) = C_p\rho\,\partial T/\partial\theta \tag{2-20}$$

If the thermal conductivity is assumed to be uniform, Eq. (2-20) can be written as

$$\partial^2 T/\partial x^2 + \partial^2 T/\partial y^2 + \partial^2 T/\partial z^2 + \dot{q}/k = (1/\alpha)\,\partial T/\partial\theta \tag{2-21}$$

where α is the thermal diffusivity and is equal to $k/C_p\rho$. If the conduction region contains no heat sources, this equation reduces to the Fourier equation:

$$\partial^2 T/\partial x^2 + \partial^2 T/\partial y^2 + \partial^2 T/\partial z^2 = (1/\alpha)\,\partial T/\partial\theta \tag{2-22}$$

Many problems in heat conduction are handled more conveniently in the cylindrical or spherical coordinate system. The Fourier equation for these two coordinate systems is given by the following equations:

$$\frac{\partial^2 T}{\partial r^2} + \frac{1}{r}\frac{\partial T}{\partial r} + \frac{1}{r^2}\frac{\partial^2 T}{\partial\theta^2} + \frac{\partial^2 T}{\partial z^2} = \frac{1}{\alpha}\frac{\partial T}{\partial\theta} \text{ (cylindrical)} \tag{2-23}$$

$$\frac{1}{r}\frac{\partial^2(rT)}{\partial r^2} + \frac{1}{r^2\sin\phi}\frac{\partial}{\partial\phi}\left(\sin\phi\frac{\partial T}{\partial\phi}\right) + \frac{1}{r^2\sin^2\phi}\frac{\partial^2 T}{\partial\psi^2} = \frac{1}{\alpha}\frac{\partial T}{\partial\theta} \text{ (spherical)} \tag{2-24}$$

Under steady-state conditions the temperature distribution in a body without any heat sources must satisfy the Laplace equation

$$\partial^2 T/\partial x^2 + \partial^2 T/\partial y^2 + \partial^2 T/\partial z^2 = 0 \tag{2-25}$$

An analytical solution to a conductive heat transfer problem must not only satisfy the appropriate heat conduction equation above, but must meet the boundary conditions specified by the physical conditions of the problem to be solved. The classical approach to an exact solution of the Fourier equation is the separation of variables technique.

Since many of the two- and three-dimensional heat conduction problems involve an odd geometry, variable thermal conductivity, and spatially dependent boundary conditions, their solution is best approximated by less rigorous methods. Graphical and analogical methods are well described by Kreith,[17] Rohsenow and Choi,[18] Jakob and Hawkins,[19] and Schneider,[15] while numerical methods are presented in detail by Dusinberre.[20]

An approximate solution of the steady-state conductive heat transfer equation for a two-dimensional system can be obtained graphically by plotting the temperature field freehand. The graphical method is reasonably simple for systems with isothermal boundaries, but it can also be used in cases where heat flows across a boundary of unknown temperature by convection or radiation from a source, or to a sink, of known temperature. The object of the graphical solution is to construct a network consisting of isotherms and lines of constant-heat flow. The heat-flow lines are analogous to streamlines in a fluid-flow field, i.e., they are tangent to the direction of heat flow at any point. Consequently, no heat can flow across the heat-flow lines and a constant amount of heat flows between any two heat-flow lines.

The rate of heat flow between isothermal boundaries can be obtained directly from the network by summing up the heat flow associated with each heat-flow tube or channel of the network. Thus

$$Q = (n/m)k(T_h - T_c) \tag{2-26}$$

where n is the number of heat-flow tubes in the network and m is the number of curvilinear squares in each heat-flow tube from T_h to T_c, the temperatures of the warm and cold isothermal boundaries, respectively.

There are many heat-flow problems for which solutions cannot be obtained analytically and for which experimental solutions in a thermal system would be too expensive or time-consuming. It is often possible, however, to obtain experimental solutions of such problems quite simply in an analogous system and reinterpret the solution in terms of the thermal problem. The application of an experimental solution obtained in any one system to an analogous system is the basis for the experimental–analogic method.

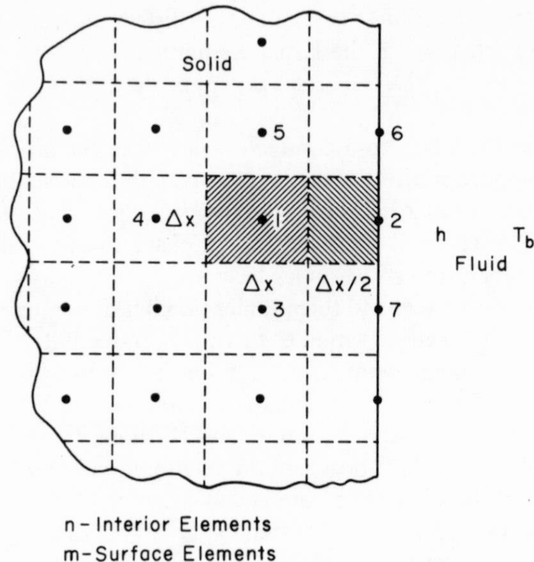

FIG. 2-3. Finite difference network for two-dimensional solid exposed to a fluid.

Heat-flow problems can be solved by several experimental–analogic methods, including the fluid-flow analogy using Moore's fluid mapper,[21] the membrane analogy,[15] which is also applicable to fields with sources,[22] and various electrical analogies like the analog field plotter.[23] Extensive reviews of these methods are given by Jakob[14] and Schneider.[15]

Numerical methods used in solving conductive heat transfer problems are often referred to as the method of finite differences. The latter term is used because in this method a continuous region is subdivided into a finite number of discrete elements and difference equations are developed from the energy laws describing the relationship between the temperature of each element and that of its adjoining neighbors.

An example of such a network of discrete elements for a two-dimensional solid including a surface exposed to a fluid is shown in Fig. 2-3. The interior of the solid is divided into n Δx square elements with the temperature at the center or nodal point of each element representing the temperature of the element. The dimensions of the m surface elements are Δx by $\Delta x/2$ to provide surface points that are uniformly spaced. Application of the first law of thermodynamics in difference form to an interior point, say 1, and assuming steady-state conductive heat transfer across the solid and constant properties, gives

$$T_2 + T_3 + T_4 + T_5 - 4T_1 = 0 \qquad (2\text{-}27)$$

The same procedure applied to a surface point, say 2, wetted by a fluid with a bulk temperature of T_b and a convective heat transfer coefficient h results in

$$T_1 - \tfrac{1}{2}T_6 + \tfrac{1}{2}T_7 + (h\Delta x/k)T_b - (2 + h\Delta x/k)T_2 = 0 \qquad (2\text{-}28)$$

Similar equations can be written for each interior and surface point, resulting in $m + n$ simultaneous, linear algebraic equations with $m + n$ unknowns. There are several methods for solving these equations, including elimination of unknowns, determinants, relaxation, iteration, and programming the problem on a high-speed digital computer. The latter is probably the most satisfactory because of its speed, convenience, and accuracy. Subroutines are generally now available for most large computers to solve systems involving a large number of simultaneous equations. To reduce the time involved in solving these simultaneous equations, the right-hand sides of these equations are usually not reduced to zero but to a residual which is sufficiently close to zero to provide a satisfactory solution.

Experience by Clark[24] in the solution of these difference equations has indicated that serious errors can result from the propagation of the effects of numerical roundoff through successive calculations and from the fact that the difference equations themselves are only approximate representations of the physical situation. Unfortunately, as the error from the discretization is reduced by selecting smaller finite elements, the number of calculations increases, thereby increasing the roundoff error. This suggests that as the size of the finite element is decreased, it will also be necessary to simultaneously reduce the magnitude of the residual if uniform convergence of the solution is to be achieved.

2.2.5 Conductive Heat Transfer with Variable Thermal Conductivity

The thermal conductivities of many materials at low temperatures vary significantly with temperature. Thus, for some cryogenic applications, the constant-property assumptions and the resulting solutions developed up to this point will not be satisfactory. The differential equation governing the steady-state conduction of heat in isotropic materials with temperature-dependent thermal properties is given by

$$(\partial/\partial x)(k\ \partial T/\partial x) + (\partial/\partial y)(k\ \partial T/\partial y) + (\partial/\partial z)(k\ \partial T/\partial z) = 0 \qquad (2\text{-}29)$$

This equation, in the form given, is difficult to solve. A simple transformation proposed by McMordie[25] and involving a new variable can be used to aid in the solution. This variable is defined as

$$E = \int_{T_R}^{T} k(T)\, dT \tag{2-30}$$

where T_R is an arbitrary reference temperature. Substitution of this variable in Eq. (2-30) results in the classical Laplace equation, or

$$\partial^2 E/\partial x^2 + \partial^2 E/\partial y^2 + \partial^2 E/\partial z^2 = 0 \tag{2-31}$$

Analytical solutions to Eq. (2-31) are available in the standard literature[16] for a wide variety of problems. Such solutions can also be found by analogical and numerical methods. The main difference in these formulations from those currently available for constant-property solutions rests in the treatment of the function E [Eq. (2-30)] to determine the temperature distribution rather than the temperature itself. For example, rather than determining the temperature at the center of each finite element in Fig. 2-3, as was done previously for the case of constant thermal conductivity, the first-law application now gives a temperature distribution for an interior point as

$$E_2 + E_3 + E_4 + E_5 - 4E_1 = 0 \tag{2-32}$$

and for a surface point as

$$E_1 + \tfrac{1}{2}E_6 + \tfrac{1}{2}E_7 + (h\Delta x/k_b)E_b - (2 + h\Delta x/k_2)E_2 = 0 \tag{2-33}$$

where

$$E_b = \int_{T_R}^{T_b} k(T)\, dT, \qquad k_b = E_b/(T_b - T_R), \qquad \text{and} \qquad k_2 = E_2/(T_2 - T_R)$$

Unfortunately, the introduction of variable k_2 in Eq. (2-33) complicates the solution of this set of equations. This complication, however, can be minimized in those situations where the convective heat transfer coefficient is large since T_2 will be nearly equal to T_b, and k_b could be substituted for k_2 as a first approximation. Thus, the unknowns in these $m + n$ set of equations are the variable E's defined by Eq. (2-30). Obviously, if the relationship between the thermal conductivity and the temperature is describable by a mathematical function, the variables can be replaced by the temperatures. In most cases, however, it will be necessary to integrate the thermal conductivity–temperature relationship numerically to obtain E as a function of the temperature.

The heat flow between any two interior points of the heat-flux network is simply the product of the thickness of the two-dimensional region and the difference between the E values of these two interior points. Similarly, the heat flow between any two surface points is one-half the product of the thickness of the two-dimensional region and the difference between the E values of these two surface points.

2.2.6 Heat Conduction in Anisotropic Solids

Solids that have transport properties exhibiting directional characteristics are said to be anisotropic. In the study of heat conduction in solids, anisotropy is associated with the thermal conductivity of the solid. For isotropic solids the heat flux at a particular point is directly proportional to the temperature gradient at that point and the heat flux vector is normal to the isothermal surface passing through that point. However, in the case of an anisotropic solid, the heat flux vector is not necessarily parallel to the temperature gradient and therefore the direction of the heat flux vector is not necessarily normal to the isothermal surface. The heat flux vector for anisotropic solids is not simply proportional to the temperature gradient. Each of the components of the temperature gradient is weighted according to the anisotropy of the particular material.[26] Thus, even though the temperature gradient is normal to a given surface, the heat flux vector is not normal to the surface and is not parallel to the temperature gradient.

There are several low-temperature engineering materials, including various crystals, woods (used as insulation material), and multilayer insulations, in which the thermal conductivity varies with direction.[27] For example, the thermal conductivity value of the multilayer insulation measured parallel to the direction of heat flow can be orders of magnitude different from the thermal conductivity value which is measured perpendicular to the direction of heat flow.[28]

The heat conduction laws for such anisotropic materials must be modified, particularly if the various directional thermal conductivity values are to be used. This will involve the insertion of the nine thermal conductivity coefficients into the basic equations. Fortunately, because of symmetry, this can usually be reduced to six conductivity coefficients. For crystals, this can be reduced even further since many crystals possess various forms of structural symmetry.

The treatment of conductive heat transfer in anisotropic materials is given in several readily available texts[13,29–31] and will not be detailed here.

2.3 UNSTEADY-STATE CONDUCTIVE HEAT TRANSFER

The solutions of unsteady-state conductive heat transfer problems are in general more difficult than the solutions of steady-state problems because of the additional independent variable of time. The temperature is a function of position in the conduction region, but, in addition, this distribution changes with time. If this change in temperature is periodic, the process is termed periodic. On the other hand, if the change is nonperiodic, the process is called transient.

The solution to such problems generally involves outlining the boundary conditions (which are often time dependent) and choosing the simplest differential equation that is applicable. In addition, in a transient process the initial temperature distribution throughout the conduction region must be given. The problem then is one of determining how this temperature distribution changes with time.

The applicable differential equation depends upon the specific conductive heat transfer situation. If the conduction region is isotropic and has a uniform thermal conductivity, Eq. (2-21) applies. If the conduction region contains no heat sources, the Fourier equation applies. This was given earlier as Eq. (2-22) for Cartesian coordinates and as Eqs. (2-23) and (2-24) for cylindrical and spherical coordinates, respectively. These equations have been solved analytically for many unsteady-state conduction processes at normal temperatures and involve numerous standard configurations.[32–42] The relations which have been developed are useful not only for engineering design of normal temperature conduction processes but also of low-temperature processes. An important example of such a process at low temperatures is the cooldown and warmup of large storage dewars for cryogens.

The analytical solution of the Fourier equation generally involves several dimensionless quantities. Two important dimensionless parameters which are commonly used to describe the transient temperature distribution are the Fourier number $(k/\rho C_p)\theta L^2$ and the Biot number hL/k. In these quantities, L is an arbitrary but characteristic dimension of the system being considered, θ is the time, and C_p is the heat capacity. One such solution incorporating these two dimensionless quantities is plotted in Fig. 2-4 and is useful in determining the relationship between the surface and centerline temperatures of a sphere.

The situation, however, that is encountered in many instances of unsteady-state heat conduction at low temperatures involves geometries that are not simple and thermal properties that are dependent upon temperature. In such cases, Eq. (2-21) can be modified with the insertion of the function E [defined by Eq. (2-30)] to give

$$\alpha(T)(\partial^2 E/\partial x^2 + \partial^2 E/\partial y^2 + \partial^2 E/\partial z^2) + \alpha(T)Q'''(x, y, z, \theta) = \partial E/\partial\theta \tag{2-34}$$

where $\alpha(T)$ is the variable thermal diffusivity, defined as $k(T)/\rho(T)C_p(T)$ and is a function of the temperature. Equation (2-34) is still nonlinear since both $\alpha(T)$ and E are functions of temperature, but it is now in a more convenient and useful form for numerical formulations.

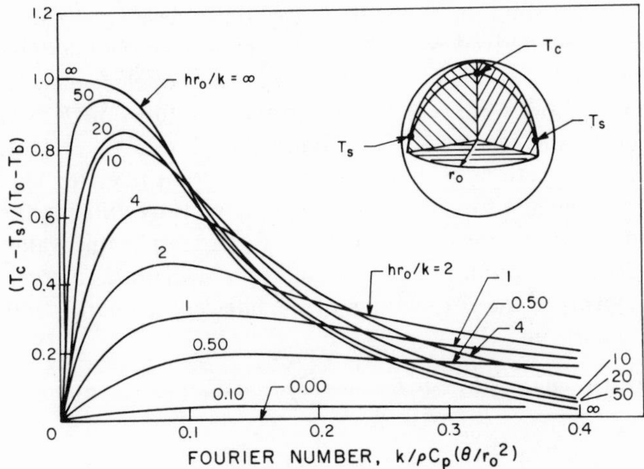

FIG. 2-4. Transient temperature distribution in a sphere.

There are certain situations at low temperatures where $\alpha(T)$ is much less variable with temperature than either $k(T)$ or $C_p(T)$. For these conditions $\alpha(T)$ can be approximated by a constant thermal diffusivity α^* and if no internal heat generation is present, Eq. (2-34) simplifies to the classical linear diffusion equation

$$\alpha^*(\partial^2 E/\partial x^2 + \partial^2 E/\partial y^2 + \partial^2 E/\partial z^2) = \partial E/\partial \theta \tag{2-35}$$

Again, a large number of analytical solutions have been generated for this equation covering various boundary conditions.[13,16]

The usefulness of Eq. (2-35) was expanded still further by Clark[43] when he showed that the various heat conduction charts formulated on the basis of uniform properties, such as the extensive charts of Schneider,[44] could be employed for the solution of a variable-property problem in terms of the function E. Multidimensional problems can also be solved by the product method for those geometries for which analytical solutions have been generated.

For those cases of unsteady-state conductive heat transfer in which the temperature variation of $\alpha(T)$ is quite large and the constant thermal diffusivity approximation cannot be made, it will be necessary to solve Eq. (2-34) by numerical methods. The most convenient form for a numerical calculation on a digital computer is to arrange the equations in an explicit formulation. This sets up a marching-type of solution and avoids the time-consuming iterative computer procedures of implicit-type formulations. The development presented here will be for the same two-dimensional region shown in Fig. 2-3 but for variable-property conditions of the solid.

The numerical solution[43] for this case is also developed from a set of algebraic equations derived by writing the first law of thermodynamics in difference form for each one of the interior and surface elements of Fig. 2-3. The resulting equations, however, are such that the function E at each one of the network points at the end of an interval of time $\Delta\theta$ is evaluated from the preexistent E functions at the beginning of the time interval. Thus, the E functions are calculated for subsequent time intervals by utilizing the results obtained from the previous time interval. For example, the value of the function E for the interior point 1 at the $(n + 1)$th time interval can be evaluated in terms of the E functions of the adjacent network points at the nth time interval from

$$[E_2{}^n + E_3{}^n + E_4{}^n + E_5{}^n + (M_1 - 4)E_1{}^n]/M_1 = E_1^{n+1} \tag{2-36}$$

where

$$M_1 = (\Delta x)^2/\alpha(T_1)\Delta\theta \tag{2-37}$$

and the superscripts n and $n + 1$ refer to the nth and $(n + 1)$th time intervals, respectively. Since $\alpha(T_1)$ will vary with changes in T_1 and this in turn will vary with time, it will be necessary to modify the value of M_1 at the start of each new series of calculations. The value of the function E_2 associated with the surface point 2 at the $(n + 1)$th time interval is obtained similarly from

$$\{E_1{}^n + \tfrac{1}{2}E_6{}^n + \tfrac{1}{2}E_7{}^n + N_2E_b{}^n + [M_2/2 - (2 + N_2)]E_2{}^n\}/(M_2/2) = E_2^{n+1} \tag{2-38}$$

where

$$M_2 = (\Delta x)^2/\alpha(T_2)\Delta\theta \tag{2-39}$$

$$N_2 = h\Delta x/k_2{}^* \tag{2-40}$$

$$k_2{}^* = [1/(T_b - T_2{}^n)]\int_{T_2{}^n}^{T_b} k(T)\,dT \tag{2-41}$$

$$E_b{}^n = \int_{T_R}^{T_b} k(T)\,dT \tag{2-42}$$

Again, $\alpha(T_2)$ will vary with changes in T_2 and this in turn will vary with time. Thus, it will also be necessary to modify the value of M_2 at the start of each new series of calculations. Clark[43] has, however, pointed out that the stability of the numerical calculations for a constant grid size requires that $M_2 \geqslant (2N_2 + 4)$ and $M_1 \geqslant M_2$. Consequently, each modification of M_1, M_2, M_3, etc., must satisfy the stability criterion. If changes in this parameter become necessary, Clark suggests that it is probably best done by modifying the length of the time interval rather than the grid size. Because of the restrictions on the length of the time interval, a considerable amount of computer time may be required to complete the calculations. Recently, Barakat and Clark[45] developed another type of explicit formulation, suitable for this kind of problem, that is unconditionally stable without restriction as to the length of the time interval used.

It should not come as any surprise that discretization and roundoff errors are encountered in unsteady-state numerical solutions in a manner similar to those experienced in steady-state numerical solutions. Because of the many tedious calculations that can be involved, digital computation is the preferred method of solution particularly if the computer is flexible with regard to input variables and data. This capability provides valuable feedback when varying the parameters to test for stability, convergence, and roundoff effects.

Finally, it is important to recognize the magnitude of the computation problem when the parameters and variables are changed. For instance, in the two-dimensional conductive heat transfer problem considered in Fig. 2-3, the total number of arithmetic calculations for a given size of network and total elapsed time of solution is proportional to $M/(\Delta x)^4$. Thus the effect of halving the network spacing is to increase the number of arithmetic calculations by a factor of sixteen. This can be quite significant even with the high-speed digital computers presently available.

2.4 CONDUCTIVE HEAT TRANSFER PROBLEMS

Having briefly reviewed the principles of conductive heat transfer pertinent to low temperatures, it is now appropriate to see how some of these principles are used in various cryogenic applications and to note the unique problems associated with these problems because of the extreme temperature differences. The most obvious and also the most important application of heat conduction principles is concerned with the area of insulation technology. Cryogenic insulations have generally been divided into five broad categories, namely; high-vacuum; multilayer, powder, foam, and special insulations. The last category includes such materials as corkboard, balsa wood, and honeycomb insulation.

Even though it is treated separately in this section, providing mechanical support among various parts of cryogenic equipment maintained at different temperatures is really an integral part of the insulation problem since the thermal efficiency of a piece of equipment must be judged on the basis of the total heat influx to the stored liquid. This problem resolves itself into a search for materials that combine the desired structural characteristics with a low thermal conductivity. This requirement can sometimes be met in compression members with the use of laminated rather than solid structural members. The conductive heat transfer in the laminated structure is a function of the contact resistance between the interface of the laminations. Finally, since heat leak into cryogenic systems can also occur from improper selection of electrical leads for heaters, thermocouples, etc., consideration of this aspect must not be overlooked.

Unsteady-state heat transfer is a common occurrence whenever a cryogenic system is cooled down or warmed up. Often, particularly in large systems, it is not only desirable to know the length of time that will be required to reach equilibrium upon cooling, but also the amount of liquid that will be required in the cooldown process. The desirability of time–temperature distributions throughout a cryogenic system during cooldown or warmup becomes clear when one considers the large thermal contractions that could occur during these transient periods. Since the investment in these systems generally is considerable, it is absolutely essential that cooldown and warmup be accomplished in such a manner as to avoid damage to the system and create no hazardous situations for operating personnel.

2.4.1 Insulation Systems

The type of insulation chosen for a given cryogenic use depends upon the specific application. Generally, the selection is aided by a knowledge of the properties of the particular insulation, such as thermal conductivity, emissivity, moisture content, evacuability, porosity, and flammability. The discussion given here will, however, be concerned mainly with the heat transfer characteristics of the various insulations and thus will place the most emphasis on the thermal conductivity property.

Heat can flow through an insulation by the simultaneous action of several different mechanisms, namely (1) solid conduction through the materials making up the insulation and conduction between individual components of the insulation across areas of contact, (2) gas conduction in void spaces contained within the insulation material, and (3) radiation across these void spaces and through the components of the insulation. Because these heat transfer mechanisms operate simultaneously and interact with each other, it is not possible to superimpose the separate mechanisms to obtain an overall thermal conductivity. Rather, it is useful to refer to an apparent thermal conductivity, which is measured experimentally during steady-state heat transfer and evaluated from the basic one-dimensional Fourier equation.

An excellent survey of insulation systems for cryogenic applications has been made by Glaser *et al.*[46] and forms the basis for much of the following discussion.

2.4.1.1 *Vacuum Insulation:* Vacuum insulation consists of an evacuated space (10^{-6} Torr or better) between two highly reflecting surfaces, one of which is at a warm temperature and the other at a cold temperature. The heat transport across this space is predominantly by radiation. Since radiative heat transfer will be emphasized elsewhere, it will receive no further discussion here. However, if the vacuum between the two surfaces is comparatively poorer than that specified above, there may be a significant contribution due to gas conduction.

For normal gaseous conduction with constant thermal conductivity there is a linear temperature gradient within the medium between the warm and cold surfaces. However, when the mean free path of the gas molecules is large compared to the distance between the two surfaces, the gas molecules become involved in collisions with the two surfaces many more times than with each other and we have the situation commonly referred to as free molecular conduction. The gaseous heat conduction under such situations for concentric spheres, coaxial cylinders, and parallel plates is given by[47]

$$Q_{gc} = [(\gamma + 1)/(\gamma - 1)](R/8\pi MT)^{1/2}\alpha p A_1(T_2 - T_1) \tag{2-43}$$

where α, the overall accommodation coefficient, is defined by

$$\alpha = \alpha_1\alpha_2/[\alpha_2 + \alpha_1(1 - \alpha_2)(A_1/A_2)] \tag{2-44}$$

and γ is the ratio of the heat capacities C_p/C_v (assumed to be constant), R is the molar gas constant, M is the molecular weight of the gas, and T, without a subscript, is the temperature of the gas at the point where the pressure p is measured. The subscripted A_1 and A_2, T_1 and T_2, and α_1 and α_2 are the areas, temperatures, and accommodation coefficients of the cold and warm surfaces, respectively.

In order for free molecular conduction to occur, the mean free path of the gas molecules must be large compared to the distance between the two surfaces. To check this condition, the mean free path λ can be determined from

$$\lambda = (3\mu/p)(\pi RT/8M)^{1/2} \tag{2-45}$$

where μ is the gas viscosity at temperature T. Equation (2-43) can be reduced to

$$Q_{gc}/A_1 = C_{gc}\alpha p(T_2 - T_1) \tag{2-46}$$

where C_{gc} is a constant given by

$$C_{gc} = [(\gamma + 1)/(\gamma - 1)](R/8\pi MT)^{1/2} \tag{2-47}$$

Values for this constant are available in the literature[48] for various cryogenic applications.

Because of the great variations in accommodation coefficients,[49] it is difficult to make accurate estimates of conductive heat transfer through a residual gas in a vacuum-insulated system. Fortunately, this is usually not a serious problem because, as a rule, the objective is to obtain a vacuum of such quality that the heat transfer by a residual gas does not contribute significantly to the overall heat transfer.

2.4.1.2 *Multilayer Insulations:* Multilayer insulations consist of alternating layers of a highly reflecting material, such as aluminum foil or aluminized Mylar, and a low-conductivity spacer material or insulator, such as fiberglass mat or paper, glass fabric, or nylon net, all under high vacuum. When properly applied at the optimum density, this type of insulation can have an apparent thermal conductivity of as low as 0.1–0.5μW/cm-K between 20 and 300 K.[46]

The very low thermal conductivity of multilayer insulations can be attributed to the fact that all modes of heat transfer—conductive, convective, and radiative—are reduced to a bare minimum. Since radiant heat transfer is inversely proportional to the number of intermediate reflecting shields and directly proportional to the emissivity of the shields, radiation is minimized by using many shields or layers of a low-emissivity material. Convection is eliminated by lowering the pressure so that the mean free path of the gas molecules is much larger than the spacing between the insulation layers. Since free molecular conduction is proportional to the residual gas pressure [see Eq. (2-46)], the multilayer insulation is highly evacuated to minimize heat transfer by this mechanism. Heat transfer through the spacer material is proportional to the thermal conductivity of the material used and inversely proportional to the resistance to heat flow at the points of contact between the spacer and the shield. The low conductivity, size, geometry, and discontinuous nature of the materials generally selected for spacers are contributing factors in reducing the solid conduction to a minimum.

For a highly evacuated multilayer insulation, heat is transferred primarily by radiation and solid conduction through the spacer material. The apparent thermal conductivity of the insulation under these conditions can be determined from

$$k_a = (N/\Delta x)^{-1}\{h_s + [\sigma e T_h{}^3/(2 - e)][1 + (T_c/T_h)^2](1 + T_c/T_h)\} \tag{2-48}$$

where $N/\Delta x$ is the number of complete layers (reflecting shield plus spacer) of insulation per unit thickness, h_s is the solid conductance for the spacer material, σ is the Stefan–Boltzmann constant, e is the effective emissivity of the reflecting shield, and T_h and T_c are the temperatures of the warm and cold sides of the insulation. From Eq. (2-48) it is evident that the apparent thermal conductivity can be reduced by increasing the layer density up to a certain point. The variation of the apparent thermal conductivity with layer density for a typical multilayer insulation is shown in Fig. 2-5.[50] It is not obvious from Eq. (2-48) that a compressive load affects the apparent thermal

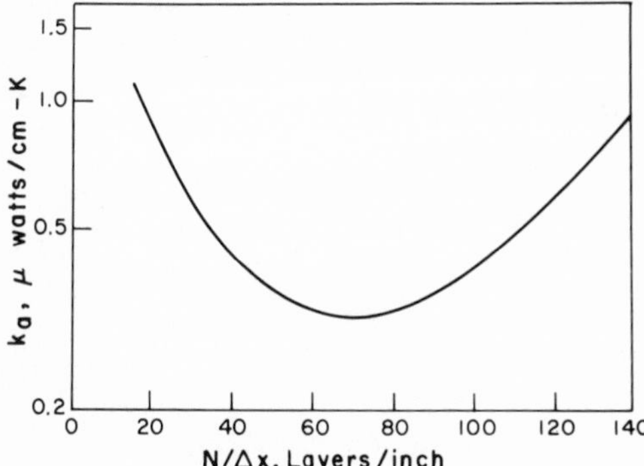

FIG. 2-5. Apparent thermal conductivity of a multilayer insulation consisting of alternate layers of glass fiber and aluminum foil between 77 and 300 K as a function of the number of alternate layers per inch.

conductivity and thus the performance of a multilayer insulation. However, under a compressive load the solid conductance increases much more rapidly than $N/\Delta x$, resulting in an overall increase in k_a. Tests have shown that when a compressive load of 2 lb/in.^{2} (0.14 kg/cm^2) is applied, the heat flux for many multilayer insulations is about 100 times greater than at the no-load condition.[51] Plots of heat flux *vs.* compressive load on a logarithmic scale result in straight lines with slopes between 0.5 and 0.67.[51]

The effects of a gas and its pressure on the performance of insulations have been investigated rather extensively by many investigators. The dependence of the apparent thermal conductivity on the residual gas pressure for a typical multilayer insulation is shown by the standard S-shaped curve of Fig. 2-6. The latter clearly indicates that multilayer insulations must be evacuated below 10^{-4} Torr if they are to be effective. As expected, gases with high thermal conductivity (e.g., helium or hydrogen) cause more rapid performance deterioration than gases with low thermal conductivity.

Finally, Eq. (2-48) indicates a possible effect of boundary temperatures on the apparent thermal conductivity. Experimental results (Fig. 2-7) with a crinkled, aluminized polyester multilayer insulation verify that the apparent thermal conductivity is approximately proportional to the third power of the warm-boundary temperature. Also, when the warm-boundary temperature is held constant, a higher apparent thermal conductivity results from increasing the cold-boundary temperature.

As has been previously indicated, multilayer insulations typically have thermal conductivities parallel to the radiation shields that are much higher

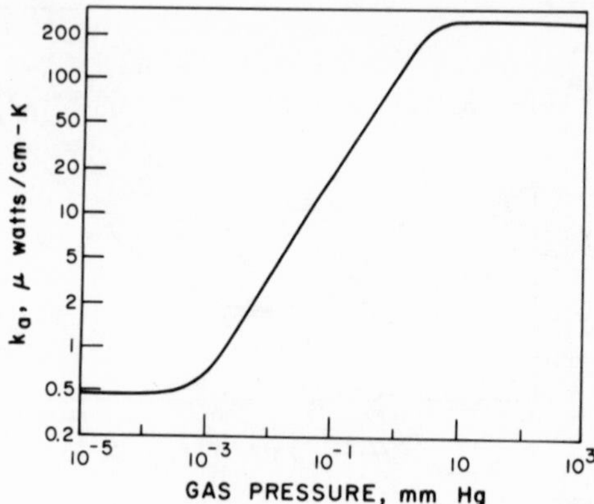

FIG. 2-6. Variation in apparent thermal conductivity with residual gas pressure for a typical (60 layer/in.) multilayer insulation between 90 and 300 K.

than those perpendicular to the shields. A parallel-to-perpendicular conductivity ratio of 10^5 is typical for a multilayer composite employing a $\frac{1}{4}$-mil aluminum foil radiation shield, whereas a conductivity ratio of 10^3 is typical for a system using double aluminized Mylar.[52] Obviously, use of these insulations on actual systems requires careful attention to the multidimensional heat transfer characteristics associated with the insulations.

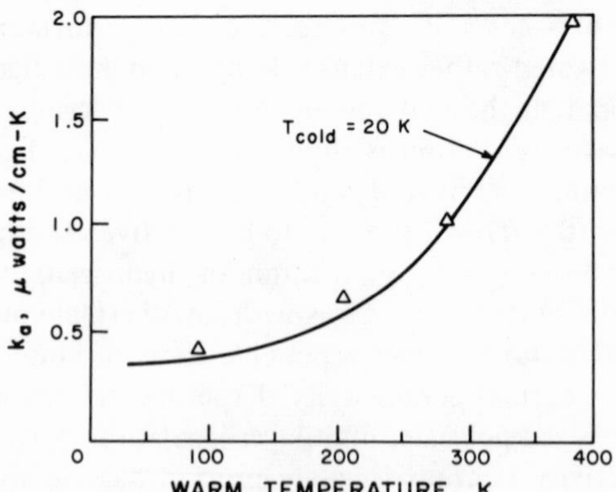

FIG. 2-7. Effect of warm boundary temperature on the apparent thermal conductivity value of crinkled, aluminized polyester multilayer insulation. The cold-boundary temperature is constant at 20 K.

In the application of a multilayer insulation system to a cryogenic container or storage vessel, the excellent one-dimensional thermal conductivity values tabulated by Glaser *et al.*[46] generally are not achieved. Usually, the effective thermal conductivity values obtained in practice are at least a factor of two greater than the one-dimensional thermal conductivity values measured in the laboratory with carefully controlled techniques. This degradation in insulation thermal performance is caused by the combined presence of edge exposure to isothermal boundaries, gaps, joints, or penetrations in the insulation blanket required for structural supports, fill and vent lines, and the high lateral thermal conductivity of these insulation systems. Introduction of reflecting shield materials having lower lateral thermal conductivity, such as metallized organic films, and the proper use of an efficient isotropic intermediary insulation system around the insulation penetrations have helped to minimize these multidimensional heat transfer effects. In the case of liquid hydrogen and helium storage, substantial improvements in the insulation system effectiveness can be achieved by utilizing the large cooling capacities of the boiloff gases.[53] Nevertheless, the effect of penetrations through the insulation remains the significant thermal problem.

2.4.1.3 *Powder Insulations:* A method of realizing some of the benefits of multiple floating shields without incurring the difficulties of awkward structural complexities is to use evacuated powder insulation. The penalty incurred in the use of this type of insulation, however, is a tenfold reduction in the overall thermal effectiveness of the insulation system over that obtained for the multilayer insulation. In applications where this is not a serious factor, such as LNG storage facilities, and investment cost is of major concern, even unevacuated powder insulation systems have found useful applications.

A powder insulation system consists of a finely divided particulate material such as perlite, expanded SiO_2, calcium silicate, diatomaceous earth, or carbon black, packed between the surfaces to be insulated. When used at 1 atm gas pressure (generally with an inert gas), the powder reduces both convection and radiation and, if the particle size is sufficiently small, can also reduce the mean free path of the gas molecules. When the powders are evacuated to pressures of 10^{-2} or 10^{-3} Torr gas conduction becomes very small and heat transfer is chiefly by radiation and solid conduction. The variation in apparent mean thermal conductivity of several powders as a function of interstitial gas pressure is shown in the familiar S-shaped curves of Fig. 2-8.[54] As the pressure of the gas within the insulation is reduced from atmospheric pressure down to about 10 Torr there is little change in the thermal conductivity. As the pressure is lowered further, a second region is reached where the thermal conductivity of the insulation varies almost linearly with pressure. This is the region of free molecular conduction for the gas in the insulation and where the apparent thermal conductivity of the

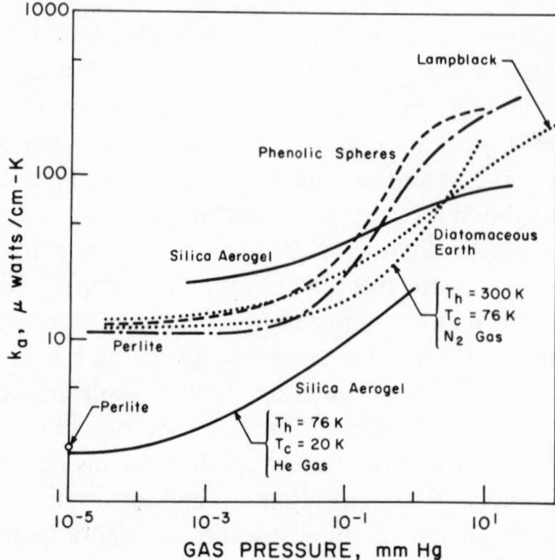

FIG. 2-8. Apparent mean thermal conductivities of several powder insulations as a function of interstitial gas pressure.

gas is linearly proportional to gas pressure. Upon further reduction in gas pressure, the effect of gaseous heat transfer becomes smaller than the effects of radiation and solid conduction. At this point, the insulation thermal conductivity begins to level off with decrease in pressure, and the apparent thermal conductivity approaches the value for radiation and solid conduction alone.

For highly evacuated powders near room temperature, the radiant contribution is larger than the solid conduction contribution to the total heat transfer rate. On the other hand, the radiant contribution is smaller than the solid conduction contribution for temperatures between 77 and 20 or 4 K. Thus, evacuated powders can be superior to vacuum alone (for insulation thicknesses greater than about 3 or 4 in.) for heat transfer between ambient and liquid nitrogen temperatures. Conversely, since solid conduction becomes predominant at lower temperatures, it is usually more advantageous to use vacuum alone for heat transfer between two cryogenic temperatures.[55]

As with multilayer insulations, there is an optimum density for evacuated powders. Figure 2-9 shows the effect of density on the apparent thermal conductivity of evacuated perlite between boundary temperatures of 300 and 76 K.[56] The apparent thermal conductivity of a variety of unevacuated perlite powder samples between the same two boundary temperatures is given for comparison in Fig. 2-10.[57] Not only are the apparent thermal conductivity values much higher in the latter figure, but the effect of the density change between 0.05 and 0.25 g/cm^3 is exactly reversed. The effect

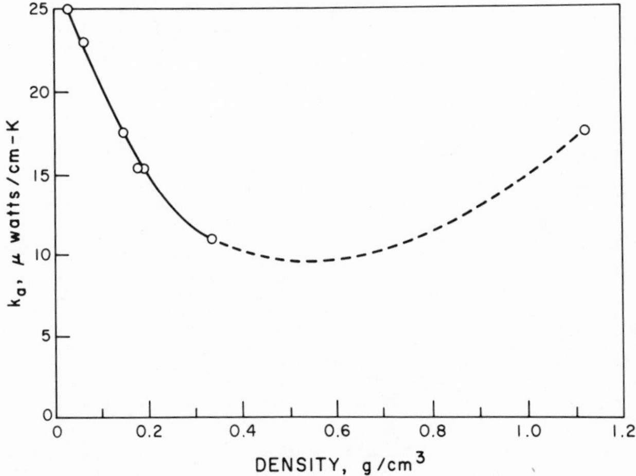

FIG. 2-9. Effect of density for evacuated perlite on the apparent mean thermal conductivity. (Over 80% of particles were 450 ± 150, boundary temperatures of 76 and 300 K, pressure less than 10^{-4} Torr.)

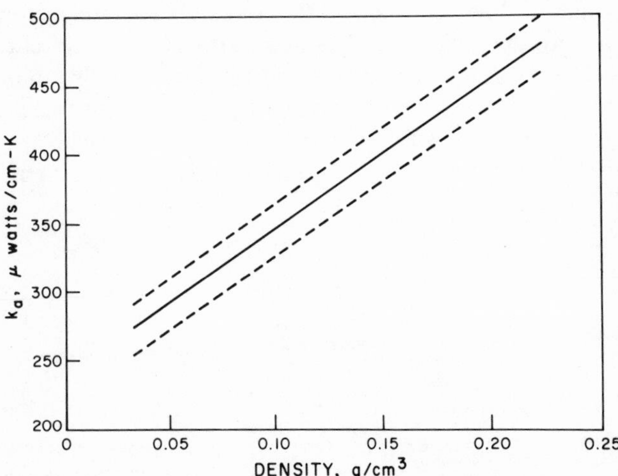

FIG. 2-10. Effect of density for nonevacuated perlite on the apparent mean thermal conductivity. (Boundary temperatures of 76 and 300 K; dashed curves indicate maximum deviation for 95% of the samples.)

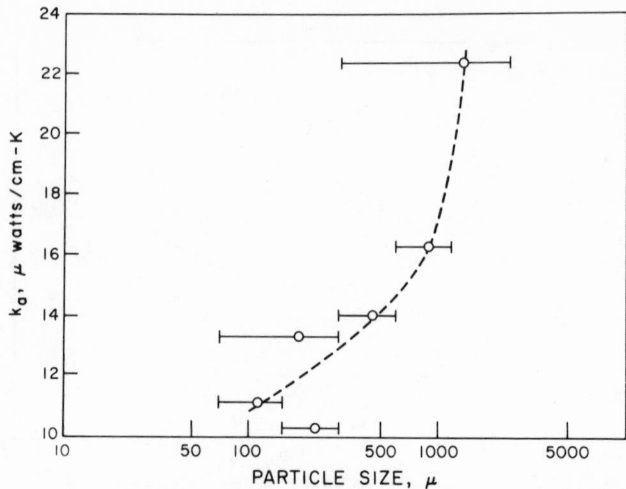

FIG. 2-11. Apparent mean thermal conductivity of evacuated perlite as a function of particle size (12 lb/ft^3 density).

of particle size of the apparent thermal conductivity between the same boundary temperatures of 300 and 76 K for evacuated perlite down to 100 μ is given in Fig. 2-11.[58]

The amount of heat transport due to radiation through the powders can be reduced by the addition of metallic powders. The apparent thermal conductivity of three opacified powders as a function of the weight fraction of the opacifier is shown in Fig. 2-12.[59] A mixture containing approximately 50% metal powder gives the optimum performance. The effect is more pronounced for Cab-O-Sil than for perlite. From a safety point of view, metallic copper as an opacifier is preferable to aluminum because the latter has a large heat of combustion in combination with oxygen.

The vacuum required with powder insulations is several orders of

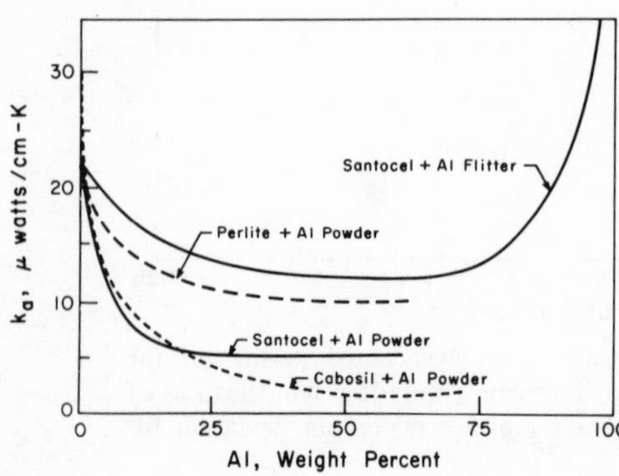

FIG. 2-12. Apparent mean thermal conductivity of several opacified powder insulations. (Boundary temperatures of 76 and 300 K.)

magnitude less than with vacuum or multilayer insulations. However, most powders easily adsorb moisture and therefore must be thoroughly dried before the required vacuum can be attained. The choice of particle size to obtain the desired density and to prevent contraction under a load, particularly when evacuated powders may be subjected to atmospheric-pressure loading, must be carefully considered. In double-walled storage vessels, powders must be compacted to a density sufficiently high to prevent formation of voids in the insulation during operation. Fine mesh filters are required to protect vacuum pumps from the highly abrasive materials.

2.4.1.4 *Foam Insulations:* Expanded-foam insulations have a cellular structure formed by evolving gas during the manufacture of the foam. Since foams are not homogeneous materials, the apparent thermal conductivity is dependent upon the bulk density of the insulation, the gas used to foam the insulation, and the mean temperature of the insulation. Heat conduction through a foam is determined by convection and radiation within the cells and by conduction in the solid structure. Evacuation of a foam is effective in reducing its thermal conductivity, indicating a partially open cellular structure; but the resulting values are still considerably higher than either multilayer or evacuated powder insulations. The opposite effect, diffusion of atmospheric gases into the cells, can cause an increase in the apparent thermal conductivity. This is even more significant with the diffusion of hydrogen and helium into the cells. Data on the thermal conductivity for a variety of foams used at cryogenic temperatures have been presented by Kropschot.[60] Of all the foams, polyurethane and polystyrene have received the widest use at low temperatures.

The major disadvantage of foams has not been their relatively high thermal conductivity compared to the other insulations but rather their poor thermal behavior. When applied to cryogenic systems, they tend to crack upon repeated thermal cycling and lose their insulation value.

2.4.1.5 *Special Insulations:* An optimum insulation system should possess not only maximum insulation effectiveness, but also minimum weight, ease of fabrication and handling, adequate service life, and reasonable cost. Since no single insulation has all the desirable thermal and strength characteristics required in many applications, composite insulations have been developed. For cryogenic applications, composite insulations consist of a polyurethane foam, reinforcement of the foam to provide adequate compressive strength, adhesives for sealing and securing the foam to the container, enclosures to prevent damage to the foam from external sources, and vapor barriers to maintain a separation between the foam and atmospheric gases.

Several external insulation systems for space applications use honeycomb structures. Phenolic resin-reinforced fiberglass-cloth honeycomb is

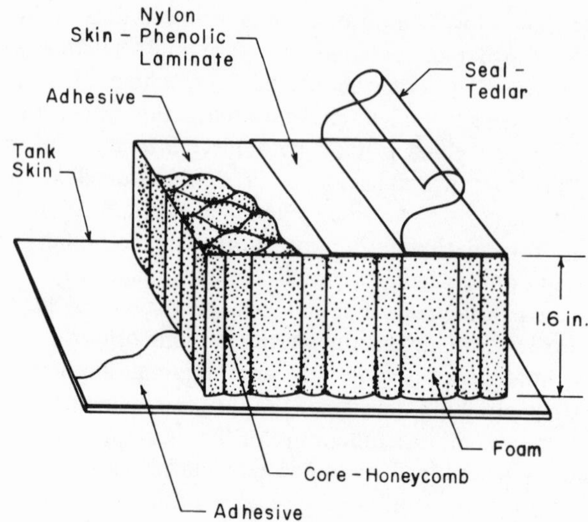

FIG. 2-13. Saturn S–II foam-filled honeycomb insulation.

most commonly used. Filling the cells with a low-density polyurethane foam further improves the thermal effectiveness of the insulation. Figure 2-13 is a schematic of a helium-purged, externally sealed honeycomb system which was developed for the Saturn S-II insulation system. Another composite lightweight external insulation developed for the liquid hydrogen stages of the Saturn V is shown in Fig. 2-14. This is a double-seal insulation consisting of an inner portion of individually sealed Mylar honeycomb cells and an outer helium purge channel of fiberglass-reinforced phenolic honeycomb. The outer helium purge channel is separated from the inner portion by a low-permeability aluminum film. In the event that both the outer and inner seals are punctured, helium under pressure flows through the puncture in the outer seal, preventing air from entering the insulation and further degrading the thermal conductivity of the insulation.

Balsa wood and corkboard are two other materials that have been considered for low-temperature insulations, with balsa wood being used to insulate liquid methane tanks of ocean-going vessels. The thermal conductivity of these materials is quite high when compared with vacuum-type insulations. Thus they can only be used when their other attributes outweigh their relatively high thermal conductivity. In the decision to use balsa wood for the insulation of large methane storage and transport vessels the selection was made on the basis of cost and dependability rather than minimum heat leak since the boiloff gas was to be either recovered or used as a fuel. More recent methane tankers have used a foam insulation that will withstand the thermal and mechanical stresses that are encountered.

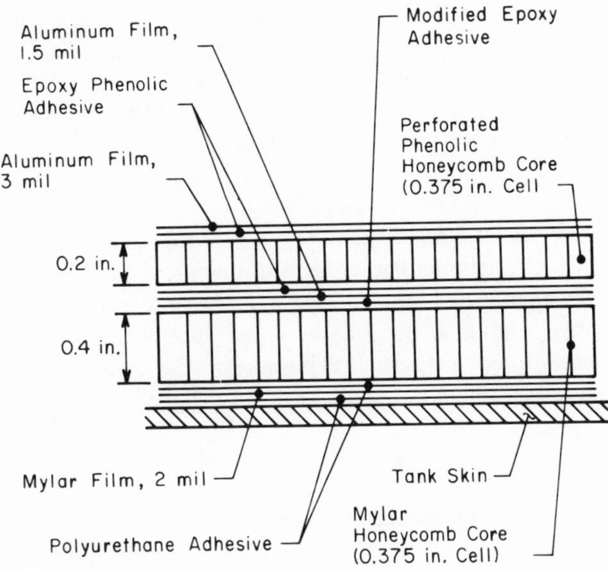

FIG. 2-14. Double-seal foam-filled honeycomb insulation for Saturn V cryogenic tankage.

2.4.2 Heat Leaks

Low heat leak is of the utmost importance in the design of most cryogenic vessels and cryostats. Consequently, considerable effort has been expended in the past to develop low-thermal-conductivity insulations. However, as noted earlier, the high performance of these insulations is often degraded by heat leaks contributed by various sources. One of the more serious contributors to this degradation is that of penetrations through the insulation in the form of structural supports, fill and vent lines, and electrical leads for heaters, liquid-level sensors, and temperature-measuring devices. Determining the magnitude of the heat leaks from these various sources and effecting the proper design to minimize these heat leaks again involves the principles of conductive heat transfer.

2.4.2.1 *Heat Leak from Structural Supports:* The support requirements for an insulated cryogenic system depend on its projected application: Laboratory equipment will be handled more gently than over-the-road or field equipment, which will be subject to accelerations and vibrations around all three axes. The supports for the latter conditions must have greater strength than the supports for laboratory equipment. Generally, the more rugged the support, the greater the heat leak. Therefore the design of the support system requires a careful balance between low heat leak and adequate strength.

The support members should be constructed of a material with a low thermal conductivity and should be as long as possible. This is evident by inspecting the integrated form of Eq. (2-5),

$$Q = -k_m A_s(T_h - T_c)/L \tag{2-49}$$

where k_m is the mean thermal conductivity of the structural support [as defined by Eq. (2-4)] between the temperature limits of T_h and T_c, the warm- and cold-end temperatures, respectively, A_s is the cross-sectional area of the support member, and L is the length of the support member. Since the support member must support the weight of the cryogenic system and the imposed acceleration loads, the required cross-sectional area for a tension member is given by

$$A_s = Ff_s/S_y \tag{2-50}$$

where F is the design load on the member, f_s is the factor of safety desired, and S_y is the yield strength of the support member. Substitution of Eq. (2-50) into Eq. (2-49) results in

$$Q = -Ff_s(T_h - T_c)/L(S_y/k_m) \tag{2-51}$$

The only factor in Eq. (2-51) affecting the choice of material for the support member is the ratio S_y/k_m, called the strength–conductivity ratio. To minimize the heat leak down the support member, this ratio should be as large as possible consistent with such other factors as ease of fabrication and cost. Values of strength–conductivity ratio for several materials suitable as support members between 300 and 90 K are given in Table 2-1. Of the metals, the austenitic stainless steels are the optimum choice for support members.

TABLE 2-1

Strength–Conductivity Comparison of Materials for Support Members[61], [a]

Material	S_y/k_m	Material	S_y/k_m
Teflon	3,350	347 stainless (cold drawn)	17,500
Nylon	11,930	1100-H16 aluminum	140
Mylar	70,700	2024-0 aluminum	210
Dacron fibers	155,000	5056-0 aluminum	300
Glass fibers	45,300	K Monel (45% cold drawn)	8,500
304 stainless (cold drawn)	12,500	Hastelloy C (annealed)	7,500
316 stainless (cold drawn)	16,300	Inconel (cold drawn)	5,700

[a] S_y in lb/in.^{2} k_m in Btu/hr-ft-°R; k_m is mean thermal conductivity between 300 and 90 K.

Certain nonmetallic materials, because of their relatively high strength–conductivity ratio, have also been used as support members for small cryogenic containers to take advantage of their excellent heat-insulating properties.

In situations where long support members cannot be used because of space limitations, heat leak can still be kept to a minimum by the use of laminated supports which consist of many layers of a poorly conducting material. These supports can only be used for compressive loads, but for such use their strength is close to that of the solid material, while their heat leak may be only a small fraction of that for the solid support. For example, a stack of stainless steel disks, each 0.0008 in. thick, compressed to 1000 lb/in.^{2} in a vacuum will conduct approximately the same amount of heat as a solid rod 50 times as long.[62] Further increase in the thermal resistance of these multiple-contact supports is achieved by lightly dusting the surfaces with manganese dioxide.

2.4.2.2 *Heat Leak from Piping:* The heat leak contribution from fill lines is a function of the cross-sectional area of the metal piping. Unless the piping is so poorly designed as to cause liquid percolation within the piping, the heat-leak contribution of the vapor in the interconnecting piping is generally considered negligible.[63] The heat transfer to the cryogenic vessel caused by conduction down the fill line is again obtained from Eq. (2-64), where A_s is now the cross-sectional area of the piping.

The heat leak contribution from separate vent lines is not as straightforward as it is for the fill lines since the cold boiloff gas generally can intercept a large amount of the heat that would otherwise have been transferred to the cryogenic vessel. The lower limit of this heat leak is determined by assuming that the boiloff gas and the vent line have the same temperature at each level of the piping. This assumes that the heat transfer between the cold gas and the piping is perfect. Calculations of this nature have been made by Scott[10] and presented in a set of useful curves which give the minimum heat leak as a function of different vent rates for helium, hydrogen, and nitrogen storage vessels.

From the above it is clear that for minimum heat leak, all interconnecting piping should have the minimum wall thickness that corresponds to the pressure rating of the vessel and that the number of interconnecting lines should be held to a minimum. In addition, the length of the piping should usually be extended by the use of loops or coils in the insulation space.

2.4.2.3 *Heat Leak from Leads:* The heat leak conducted into a cryogenic system by either a temperature-measuring device, liquid-level sensor, or small heater lead is easily calculated using Eq. (2-49) along with the appropriate thermal data for the lead. Generally, the heat leak from such a source can be minimized by careful selection of the lead material, use of the smallest lead wire diameter that is practicable, and proper tempering of the lead.

The latter is of utmost importance in the case of temperature sensors in cryogenic systems since the error most often encountered in such systems is the failure to have the temperature-sensing element at the same temperature as the object being measured because of the heat conduction along the leads. As superconducting magnets and cables get larger, the requirements for current leads into helium cryostats become more stringent. The simple relationship given by Eq. (2-49) is no longer adequate since Joule heating must also be considered in the overall lead optimization. In this case the selected lead must have a cross-sectional area which is sufficiently large to limit the Joule heating but small enough to limit the heat conduction. A general solution to this problem of a current-carrying conductor in good thermal exchange with an evaporating cryogenic fluid has recently been presented by Lock.[64] In this study he shows that the choice of the conducting lead material is less significant than the optimization of the geometric parameters of the lead. Theoretical results of the optimization procedure show that the heat leak into a liquid helium bath for copper leads, covering a wide range of residual resistivities, is about 1.2 mW/A at the optimum current, when the warm end of the lead is at room temperature.

2.4.3 Cooldown Losses

When an insulated vessel at ambient temperature is filled with a cryogenic liquid, considerable time elapses before the rate of heat transfer to the liquid reaches a steady state. Estimates of the quantity of a cryogenic liquid required to cool down a cryogenic vessel to its operating temperature require accounting for cooling the inner shell, support system, interconnecting piping, and the insulation to equilibrium temperatures. In addition to these liquid requirements, there is also an increasing cryogenic liquid loss associated with the increasing heat leak from the atmosphere through the insulation as the cooldown proceeds.

To aid the design engineer with these transient conductive heat transfer problems, Jacobs[65] has analyzed the maximum and minimum liquid requirements that are associated with the cooldown of unit quantities of copper, aluminum, and stainless steel in a vacuum-insulated cryogenic system as a function of the initial temperature and has presented this analysis in readily usable graphical form for four cryogenic liquids, helium, hydrogen, nitrogen, and oxygen. Figure 2-15 shows the specific liquid helium requirements to cool these three metals to 4.2 K from an initial temperature of up to 300 K. The maximum liquid requirement is obtained by assuming cooldown using only the latent heat of the cryogenic fluid. The minimum liquid requirement is determined by assuming that all of the available refrigerant effect (latent and sensible heats) of the liquid is utilized in the cooldown of

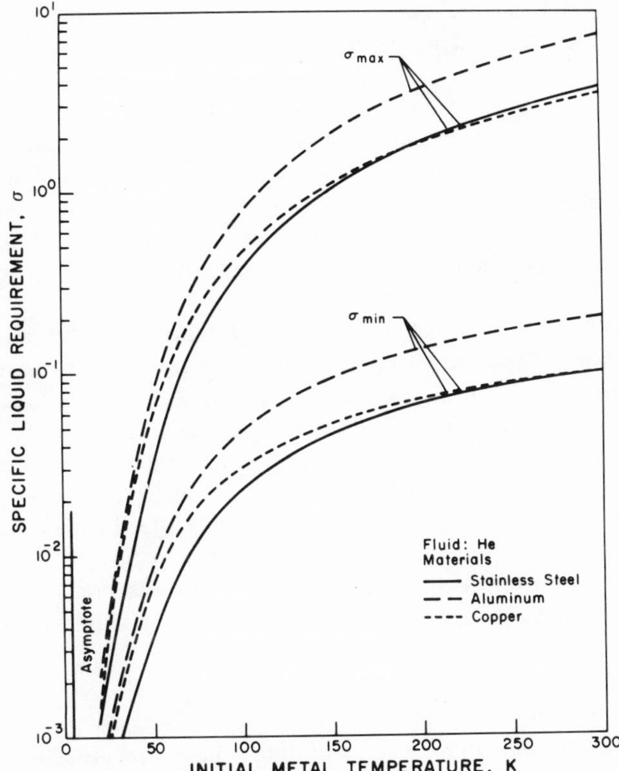

FIG. 2-15. Specific liquid helium requirement for cooldown of three structural materials as a function of initial temperature of the material.

the metal. The latter technique simply assumes that the temperature of the boiloff cryogen is equal to the temperature of the warmest part of the metal to be cooled. Again, this is the limiting case of effective heat exchange. The minimum specific liquid requirement in terms of the initial and final metal temperatures is given by

$$\sigma_{min} = -\int_{T_i}^{T_f} \{C_{p,m}[h_{fg} + \int_{T_{\text{sat}}}^{T_s} C_p \, dT_f]^{-1}\} \, dT_s \tag{2-52}$$

where σ_{min} is the minimum specific liquid requirement, T_i is the initial metal temperature, T_f is the final metal temperature, T_{sat} is the saturation temperature of the cryogenic fluid, T_s is the metal temperature, h_{fg} is the latent heat of the fluid, $C_{p,m}$ is the heat capacity of the metal, and C_p is the heat capacity of the fluid. The double integration of Eq. (2-52) requires knowledge of the heat capacity of the metal and the cryogenic fluid as a function of temperature. Since the heat capacities are not constant and are not available as analytic functions of temperature, numerical or graphical integration is necessary.

While the minimum liquid requirements represent an ideal situation, this can be approached by careful design and cooldown techniques. The former include the use of thin-wall construction and use of materials having low specific heat for those parts in contact with the cryogenic fluid. The cooldown techniques include precooling with a less expensive cryogen, and utilization of the sensible heat of the cryogenic fluid with a slower cooldown period.

The cooldown requirements of a cryogenic vessel insulated with either an evacuated multilayer or powder insulation provide an additional degree of complexity over that of the vacuum-insulated system. Consequently, analytical solutions presently available involve various simplifying assumptions. For example, the solutions presented for cooling the evacuated perlite insulation of a spherical vessel have assumed a linear temperature gradient in the insulation with constant mean specific heats and thermal conductivities.[66–68] One of the solutions[68] considers the cooldown problem by assuming the insulation to be a slab originally with a uniform temperature and suddenly being cooled on one side with a cryogenic fluid. For infinitely large dewars, a simplified relationship results:

$$\partial T/\partial \theta = (k_{\mathrm{av}}/C_{p_{\mathrm{av}}}\rho)(\partial^2 T/\partial r^2) \tag{2-53}$$

When the inner and outer surfaces of the annulus are held at constant temperature and the entire inner surface is assumed to have the same temperature, the solution to Eq. (2-53) is given as

$$q(\theta)/q_{ss} = 1 + 2\sum_{n=1}^{\infty} \exp(-\, k_{\mathrm{av}} n^2 \pi^2 \theta / C_{p_{\mathrm{av}}}\rho L^2) \tag{2-54}$$

where $q(\theta)$ is the instantaneous heat flux, q_{ss} is the steady-state heat flux, r_i and r_o are inner and outer radii of the annular space, n is an integer, and L is the annulus thickness. The time constant implicit in Eq. (2-54) is $\theta = C_{p_{\mathrm{av}}}\rho L^2/k_{\mathrm{av}}\pi^2$. The integral of the heat flux into the cryogenic vessel added to the cooldown requirements of the inner shell, supports, and interconnecting piping will give the integrated liquid loss as

$$\sigma_a = (1/h_{fg})\left[\int_{T_i}^{T_f} wC_{p,m_{\mathrm{av}}}\, dT + \int_0^{\theta_f} Q(\theta)\, d\theta\right] \tag{2-55}$$

where σ_a is the mass of cryogenic liquid used for cooldown up to the time θ_f, h_{fg} is the latent heat of the fluid, w is the mass of the metal cooled from T_i, the initial metal temperature, to T_f, the final metal temperature, and $C_{p,m_{\mathrm{av}}}$ is the mean specific heat of the metal.

A comparison between the predicted cooldown requirements of Eq. (2-54) and those obtained experimentally for a 50,000-gal liquid hydrogen storage vessel with evacuated perlite insulation has been made by Liebenberg *et al.*[69] They report that Eq. (2-54) predicts cooldown times that are too small. This is not surprising since experimental studies[70] have shown that the thermal diffusivity of evacuated perlite powder is a nonlinear function of the temperature. Thus, the use of a constant mean thermal conductivity and mean heat capacity can introduce considerable error in the predicted cooldown requirements. Knight *et al.*[71] have recently incorporated these temperature-dependent thermal properties in the general form of the conduction equation and solved the resulting nonlinear equation by approximating the first and second spatial derivatives of temperature with central difference expressions. The resulting set of ordinary differential equations was solved using a fourth-order Runge–Kutta technique. The final result gives a complete temperature history of the vessel insulation. A comparison of Knight's theoretical predictions and the experimental study reported by Liebenberg *et al.* shows reasonable agreement that is within the experimental error of the temperature-dependent thermal diffusivity values. Further experimental work is necessary before the cooldown predictions can be refined any further.

2.5 SUMMARY

Probably the most significant point to be observed with this presentation is that low-temperature conductive heat transfer problems can generally be solved through the application of the same basic principles that have been used to solve conductive heat transfer problems at normal temperatures. Situations in which new principles arise are quite unusual. Normally, the only deviations encountered in cryogenic problems are caused by changes in emphasis or differences in order of magnitude. Most design difficulties are generally the result of insufficient knowledge of the properties of materials and the effect of temperature on these properties and not because of a new principle.

It is also rather obvious that this presentation can only be a survey and that the references can only serve as a selected guide to additional data and information related to conductive heat transfer. For example, there are at least several hundred references just in the area of space applications of various multilayer and high-performance insulations. A survey of these alone would provide material for an entire volume rather than a chapter. In a similar manner, excellent tabulations of thermal properties and insulation performance important to the solution of conductive heat transfer problems at low temperatures abound in the literature and are accordingly not repeated here.

2.6 NOMENCLATURE

A = area
$A(r)$ = area as function of r
A_s = cross-sectional area of support member
$A(x)$ = area as function of x
C_{gc} = constant defined by Eq. (2-47)
C_p = heat capacity at constant pressure
$C_{p,m}$ = heat capacity of metal
C_v = heat capacity at constant volume
E = variable defined by Eq. (2-30)
$E_b{}^n$ = defined by Eq. (2-42)
e = emissivity of reflecting shield
F = design load
f_s = safety factor
h = convective heat transfer coefficient
h_{fg} = latent heat of fluid
h_L = convective heat transfer coefficient at $x = L$
h_s = solid conductance of spacer material
k = thermal conductivity
k_b = thermal conductivity at ambient temperature
k_m = defined by Eq. (2-4)
k_m' = defined by Eq. (2-8)
k_2^* = defined by Eq. (2-41)
$k(T)$ = temperature-dependent thermal conductivity
L = length
M = molecular weight
M_1 = defined by Eq. (2-37)
M_2 = defined by Eq. (2-39)
m = number of curvilinear squares
N_2 = defined by Eq. (2-40)
n = number of heat-flow tubes
$N/\Delta x$ = number of insulation layers
p = pressure
Q = heat transfer per unit time
Q''' = heat source per unit time
$q(\theta)$ = instantaneous heat flux

q_{ss} = steady-state heat flux
Q_{gc} = gaseous heat conduction
$Q(x)$ = heat transfer per unit time as function of x
R = molar gas constant
R_h = convective resistance
R_k = conductive resistance
r = radius
r_i = inside radius
r_o = outside radius
S_y = yield strength
T = temperature
T_b = ambient temperature
T_c = cold temperature
T_h = warm temperature
T_i = initial temperature
T_f = final temperature
T_R = reference temperature
$T(r)$ = temperature as a function of r
T_s = temperature of metal
T_{sat} = saturated temperature of cryogenic liquid
$T(x, y)$ = temperature as a function of x and y
x = coordinate direction
y = coordinate direction
z = coordinate direction

Greek Letters

α = thermal diffusivity and accommodation coefficient
α^* = constant thermal diffusivity
$\alpha(T)$ = temperature-dependent thermal diffusivity
γ = ratio of heat capacities
λ = mean free path defined by Eq. (2-45)
μ = viscosity
ρ = density
θ = time
θ_{D} = Debye temperature
θ_f = final time
σ = Stefan–Boltzmann constant
σ_a = mass of cryogenic liquid used for cooldown
$\sigma_{\min}$ = minimum mass of cryogenic liquid used for cooldown
ϕ = coordinate dimension defined in Eq. (2-24)
ψ = coordinate dimension defined in Eq. (2-24)

2.7 REFERENCES

1. R. B. Hinckley, "Advanced Studies on Multilayer Systems," Final Rept., NASA CR-54929 (1966).
2. R. L. Powell and W. A. Blanpied, "Thermal Conductivity of Metals and Alloys at Low Temperatures," NBS Circular 556 (September 1954) (now out of print, but available from depository libraries).
3. V. J. Johnson, ed., "A Compendium of the Properties of Materials at Low Temperatures (Phase 1), Part II." Properties of Solids, WADD Tech. Rept. 60-56 (1960).
4. R. L. Powell, in *American Institute of Physics Handbook*, 2nd ed., McGraw-Hill Book Co., New York (1963), p. 4.
5. Y. S. Touloukian, ed., *Thermophysical Properties Research Center Data Book*, Vol. 1, *Metallic Elements and Their Alloys*, Purdue University, Lafayette, Ind. (1964), Chapt. 1.
6. R. W. Powell, C. Y. Ho, and P. E. Liley, "Thermal Conductivity of Selected Materials," NSRDS–NBS 8 (November 25, 1966), Superintendent of Documents, U.S. Government Printing Office, Washington, D.C.
7. D. S. Dillard and K. D. Timmerhaus, "Thermal Transport Properties of Selected Solids at Low Temperatures," *Chem. Eng. Progr. Symp. Series*, **64** (87), 1 (1968).
8. P. G. Klemens, in *Thermal Conductivity*, Vol. 1 (R. P. Tye, ed.), Academic Press, New York (1969), p 1.
9. G. K. White, in *Thermal Conductivity*, Vol. 1 (R. P. Tye, ed.), Academic Press, New York (1969), p. 69.
10. R. B. Scott, *Cryogenic Engineering*, D. Van Nostrand Company, Princeton, N.J. (1959).
11. R. F. Barron, *Cryogenic Systems*, McGraw-Hill Book Company, New York (1966).
12. K. D. Timmerhaus, ed., *Advances in Cryogenic Engineering*, Vols. 1–19, Plenum Press, New York (1960–1974).
13. H. S. Carslaw and J. C. Jaeger, *Conduction of Heat in Solids*, 2nd ed., Oxford University Press, London (1959).
14. M. Jakob, *Heat Transfer*, Vol. 1, John Wiley and Sons, New York (1949).
15. P. J. Schneider, *Conduction Heat Transfer*, Addison-Wesley Publishing Co., Reading, Mass. (1955).
16. V. S. Arpaci, *Conduction Heat Transfer*, Addison-Wesley Publishing Co., Reading, Mass. (1966).
17. F. Kreith, *Principles of Heat Transfer*, International Textbook Co., Scranton, Pa. (1958).
18. W. M. Rohsenow and H. Y. Choi, *Heat, Mass, and Momentum Transfer*, Prentice-Hall, Englewood Cliffs, N.J. (1961).
19. M. Jakob and G. A. Hawkins, *Elements of Heat Transfer*, 3rd ed., John Wiley and Sons, New York (1957).
20. G. M. Dusinberre, *Heat Transfer Calculations by Finite Differences*, International Textbook Company, Scranton, Pa. (1961).
21. A. D. Moore, "Fields from Fluid Flow Mappers," *J. Appl. Phys.*, **20**, 790 (1949).
22. P. J. Schneider, "The Prandtl Membrane Analogy for Temperature Fields with Permanent Heat Sources or Sinks," *J. Aeronautical Sci.*, **19**, 644 (1952).
23. "Instructions for Analog Field Plotter," Catalogues 112L152G1 and G2, General Electric Company, Schenectady, New York.
24. J. A. Clark, in *Cryogenic Technology* (R. W. Vance, ed.), John Wiley and Sons, New York (1963), p. 121.

25. R. K. McMordie, "Steady-State Conduction with Variable Thermal Conductivity," *Trans. ASME, J. Heat Transfer* **84** (1), 92 (1962).
26. M. N. Ozisik, *Boundary Value Problems of Heat Conduction*, International Textbook Co., Scranton, Pa. (1968).
27. D. S. Dillard and K. D. Timmerhaus, in *Proceedings 8th Conference on Thermal Conductivity* (C. Y. Ho and R. E. Taylor, eds.), Plenum Press, New York (1969), p. 949.
28. J. G. Androulakis and R. L. Kosson, "Effective Thermal Conductivity Parallel to the Laminations and Total Conductance for Combined Parallel Heat Flow in Multilayer Insulation," Paper No. AIAA 68–765, presented at AIAA 3rd Thermophysics Conference, Los Angeles, Calif. (June 24–26, 1968).
29. J. F. Nye, *Physical Properties of Crystals*, Clarendon Press, Oxford, England (1957), p. 195.
30. W. A. Wooster, *A Textbook on Crystal Physics*, University Press, Cambridge, England (1938), p. 63.
31. S. R. DeGroot and P. Mazur, *Non-Equilibrium Thermodynamics*, North-Holland Publishing Co., Amsterdam; Interscience Publishers, New York (1963), p. 235.
32. W. H. McAdams, *Heat Transmission*, 3rd ed., McGraw-Hill Book Co., New York (1954), p. 33.
33. H. S. Carslaw, *Mathematical Theory of Heat*, Macmillan Book Co., New York (1921).
34. H. S. Carslaw, *Fourier's Series and Integrals*, Macmillan Book Co., New York (1930).
35. L. R. Ingersoll, O. J. Zobel, and A. C. Ingersoll, *Heat Conduction with Engineering and Geological Applications*, McGraw-Hill Book Co., New York (1948).
36. A. B. Newman, *Trans. AIChE*, **24**, 44 (1930).
37. H. P. Gurney and J. Lurie, *Ind. Eng. Chem.*, **15**, 1170 (1923).
38. M. Fishenden and O. A. Saunders, *An Introduction to Heat Transfer*, Clarendon Press, Oxford, England (1950).
39. M. P. Heisler, "Temperature Charts for Induction and Constant Temperature Heating," *Trans. ASME*, **69**, 227 (1947).
40. H. Gröber, S. Erk, and V. Grigull, *Fundamentals of Heat Transfer*, 3rd ed., McGraw-Hill Book Co., New York (1961).
41. L. M. K. Boelter, V. H. Cherry, and H. A. Johnson, *Heat Transfer*, 3rd ed., University of California Press, Berkeley, Calif. (1942).
42. C. M. Fowler, "Analysis of Numerical Solutions of Transient Heat Flow Problems," *Quart. Appl. Math.*, **3** (4), 361 (1945).
43. J. A. Clark, in *Advances in Heat Transfer*, Vol. 5 (T. F. Irvine, Jr., and J. P. Hartnett, ed.), Academic Press, New York (1968), p. 325.
44. P. J. Schneider, *Temperature Response Charts*, John Wiley and Sons, New York (1963).
45. H. Z. Barakat and J. A. Clark, "On the Solutions of the Diffusion Equation by Numerical Methods," *Trans. ASME, J. Heat Transfer* **88**, 421 (1966).
46. P. E. Glaser, I. A. Black, R. S. Lindstrom, R. E. Ruccia, and A. E. Wechsler, "Thermal Insulation Systems," NASA SP-5027 (1967).
47. E. H. Kennard, *Kinetic Theory of Gases*, McGraw-Hill Book Co., New York (1938).
48. R. J. Corruccini, "Gaseous Heat Conduction at Low Pressures and Temperatures," *Vacuum*, **7** (8), 19 (1957–1958).
49. R. J. Corruccini, "Properties of Materials at Low Temperatures," *Chem. Eng. Progr.*, **53** (8), 397 (1957).

50. D. I.-J. Wang, "Multiple-layer Insulations," paper presented at 1961 Conference on Aerodynamically Heated Structures, Air Force Office of Scientific Research.
51. I. A. Black and P. E. Glaser, in *Advances in Cryogenic Engineering*, Vol. 11, Plenum Press, New York (1966), p. 26.
52. R. P. Caren and G. R. Cunningham, "Heat Transfer in Multi-layer Insulation Systems," *Chem. Eng. Progr. Symp. Series*, **64** (87), 67 (1968).
53. J. A. Paivanos, O. P. Roberts, and D. I.-J. Wang, in *Advances in Cryogenic Engineering*, Vol. 10, Plenum Press, New York (1965), p. 197.
54. M. M. Fulk, in *Progress in Cryogenics*, Vol. 1 (K. Mendelssohn, ed.), Academic Press, New York (1959), p. 63.
55. R. F. Barron, *Cryogenic Systems*, McGraw-Hill Book Co., New York (1966).
56. D. Cline and R. H. Kropschot, in *Radiative Transfer from Solid Materials*, (H. H. Blair, ed.), Macmillan Book Co., New York (1962), p. 61.
57. R. H. Kropschot, in *Cryogenic Technology* (R. W. Vance, ed.), John Wiley and Sons, New York (1963), p. 239.
58. R. H. Kropschot and R. W. Burgess, in *Advances in Cryogenic Engineering*, Vol. 8, Plenum Press, New York (1963), p. 425.
59. B. J. Hunter, R. H. Kropschot, J. E. Schrodt, and M. M. Fulk, in *Advances in Cryogenic Engineering*, Vol. 5, Plenum Press, New York (1960), p. 146.
60. R. H. Kropschot, in *Applied Cryogenic Engineering* (R. W. Vance and W. M. Duke, eds.), John Wiley and Sons, New York (1962), p. 152.
61. R. W. Arnett, K. A. Warren, and L. O. Mullen, "Optimum Design of Liquid Oxygen Containers," WADC Tech. Rept. 59–62 (1961).
62. M. McClintock, *Cryogenics*, Reinhold Publishing Corporation, New York (1964).
63. P. D. Fuller and J. N. McLagan, in *Applied Cryogenic Engineering* (R. W. Vance and W. M. Duke, eds.), John Wiley and Sons, New York (1962), p. 215.
64. J. M. Lock, "Optimization of Current Leads Into a Cryostat," *Cryogenics*, **9** (6), 438 (1969).
65. R. B. Jacobs, in *Advances in Cryogenic Engineering*, Vol. 8, Plenum Press, New York (1963), p. 529.
66. WADC Technical Report 59–386 (1959).
67. S. Stoy, in *Advances in Cryogenic Engineering*, Vol. 5, Plenum Press, New York (1960), p. 216.
68. F. Kreith, J. W. Dean, and L. Brooks, in *Advances in Cryogenic Engineering*, Vol. 8, Plenum Press, New York (1963), p. 536.
69. D. H. Liebenberg, R. W. Stokes, and F. J. Edeskuty, in *Advances in Cryogenic Engineering*, Vol. 11, Plenum Press, New York (1966), p. 554.
70. R. H. Kropschot, B. L. Knight, and K. D. Timmerhaus, in *Advances in Cryogenic Engineering*, Vol. 14, Plenum Press, New York (1969), p. 224.
71. B. L. Knight, K. D. Timmerhaus, and R. H. Kropschot, "Modeling Transient Behavior in Cryogenic Storage Vessels," presented at 3rd Joint AIChE-IMIQ Meeting, Denver, Colorado, August 31–September 2, 1970.

CONVECTIVE HEAT TRANSFER TO LOW-TEMPERATURE FLUIDS 3

R. W. GRAHAM, R. C. HENDRICKS, and R. J. SIMONEAU

3.1 INTRODUCTION

The purpose of this chapter is to review forced convection and natural convection processes in low-temperature (cryogenic) fluids. The emphasis will be on forced convection because more applications for that type of cooling are found. In most instances turbulent forced convection will be discussed. Low-temperature fluids can exist as gases, two-phase fluids, low-temperature liquids, or near-critical-state liquids. These fluid states are depicted in Fig. 3-1, which is a temperature–entropy diagram for any fluid. All of these regimes will be discussed in this presentation; however, region IV is given primary consideration.

The reported research for heat transfer to fluids in the near critical state can be divided into two broad classes: forced and natural convection experiments in heated tubes and free or natural convection from heated

R. W. GRAHAM, R. C. HENDRICKS, R. J. SIMONEAU NASA Lewis Research Center, Cleveland, Ohio.

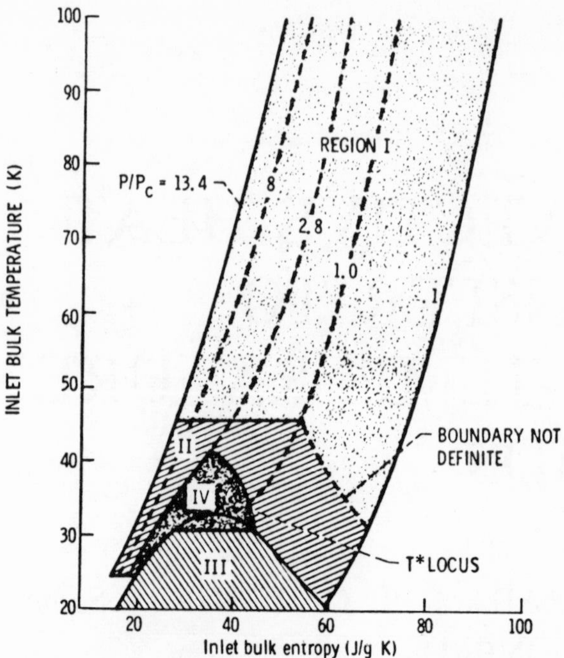

FIG. 3-1. Heat transfer regions as a function of inlet conditions. Region I, gas; region II, liquid (fluid); region III, two-phase; region IV, near critical.

wires and flat plates. The first provided turbulent experimental data for use in design and in establishing the reliability of correlations and theoretical analyses. The second group of studies focused more attention on the detailed mechanisms. The data from both these studies showed substantial contradiction. Some investigators reported that the heat transfer coefficient was enhanced, while other investigators reported that the heat transfer in this region actually approached a minimum value. These contradictions have not yet been fully resolved, but there is an ever-increasing understanding of the conditions likely to produce either very high or very low heat transfer coefficients in this region.

Theoretical analyses which show promise in unraveling the mysteries of heat transfer in the near-critical region have been proposed and involve detailed studies of the behavior in the boundary layer. No single analysis can be considered satisfactory to cover the entire region, or to explain all of the reported data at this time, however.

In addition to the difficulties associated with the contradiction regarding enhanced or attenuated heat transfer, flow oscillations have also been observed. These oscillations have been examined analytically and it has been found that the frequencies of the oscillations are reasonably well predicted by using rather conventional concepts in mechanics. The amplitudes of these oscillations have been more difficult to predict. There is, of course, an interplay between the oscillation behavior and the heat transfer behavior in this region.

Before proceeding with a discussion of the near-critical-region heat transfer, something should be said about how the boundaries of that region are defined.

It is difficult to define the boundaries which separate the near-critical region, IV, from its adjacent regions. There are several reasons for this: (1) For most fluids the data are not sufficient, (2) the transition is not abrupt and sharp demarcations are difficult to determine, and (3) the near-critical influence will persist further into the adjacent regions depending on the path (process) the fluid takes to arrive at a given state point. This last point is particularly elusive. A gas can be precooled at critical pressure down to, and possibly even below, the transposed critical temperature T^* and when run in a heated tube experiment it will behave as a precooled gas (i.e., similar to region I) to the same state point in a heated tube, the results will be quite different. Part of the near-critical heat transfer problem is how the way in which the fluid arrives at a given state point (i.e., prior history) does make a difference.

Nevertheless, classification of the heat transfer regions by state conditions is convenient, useful, and generally reasonable. For hydrogen, as shown on Fig. 3-1, the pressure boundaries are $0.8 < P/P_c < 3$. The lowest temperature boundary is T_{sat}, corresponding to $P/P_c = 0.8$, and the upper boundary is the vicinity of T^*, the transposed critical temperature. Other fluids have not been explored extensively enough to establish their boundaries or to confirm the universality of the hydrogen boundaries. In the absence of data the hydrogen boundaries can probably be taken as reasonable for other fluids.

A comment is in order here concerning boundaries. What is meant in establishing these boundaries is that outside of them, the influence of the critical point can be considered negligible and conventional variable-property correlations will prevail. It does not mean that in every case within region IV conventional methods will fail. Some combinations of parameters within region IV will be amenable to conventional approaches, but in general this region will require analyses directed specifically at the near-critical heat transfer phenomenon.

Before entering into a discussion of the heat transfer processes around the critical point, the fluid properties near the state are discussed. The connection between fluid properties and heat transfer is evident in every heat transfer correlation.

3.2 NEAR-CRITICAL FLUID PROPERTIES

3.2.1 Thermodynamics of the Critical Point

The exceptional heat transfer behavior of a near-critical fluid must ultimately be due to the influence of the unusual property behavior of a fluid near its critical point, as shown in Fig. 3-2. This is manifested both in the changes of the transport properties themselves, and in the modifications of the flow structure due to these changes. In order for any analysis to succeed, a good knowledge of the thermal properties is required.

The classical approach to the thermodynamics of the critical point is well documented by Hirschfelder *et al.*[1] and Rowlinson.[2] For a survey of some of the more recent ideas the reader is referred to Widom.[3]

The van der Waals equation of state

$$P = [\rho RT/(1 - b\rho)] - a\rho^2 \tag{3-1}$$

(where a is an attraction force constant and b is a repulsion force constant)

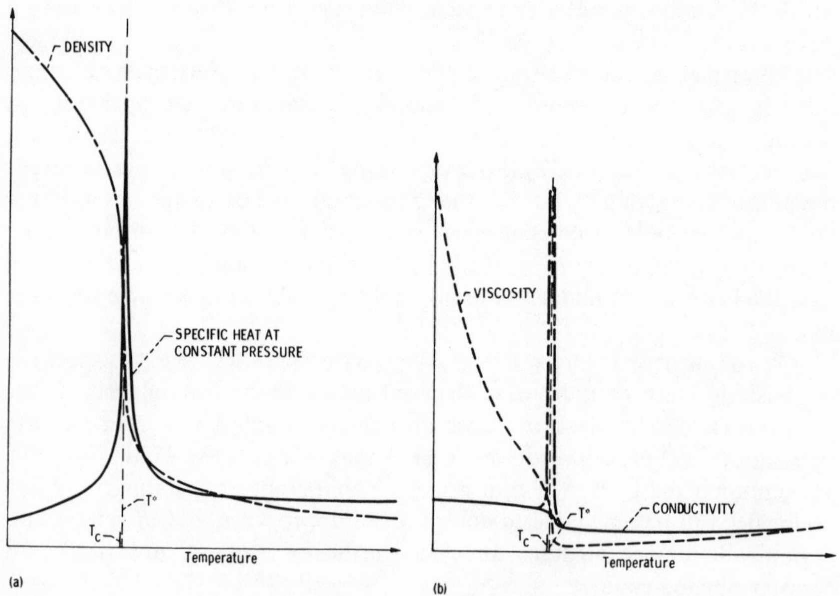

FIG. 3-2. Typical thermodynamic (a) and transport (b) properties of near-critical parahydrogen; $P/P_c = 1.05$.[6]

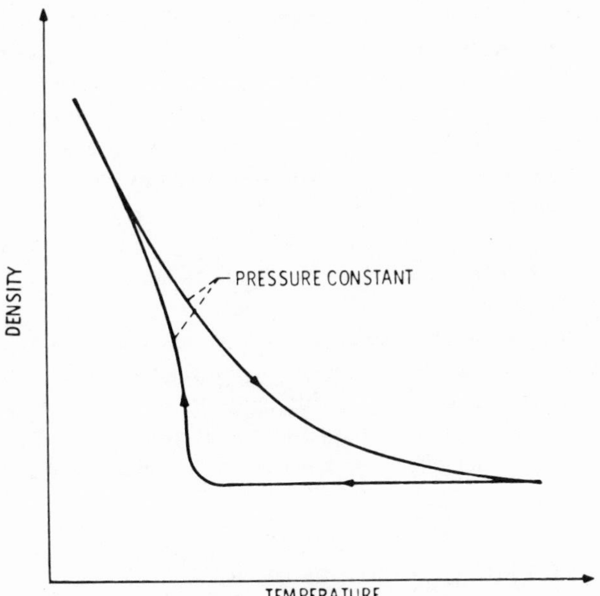

FIG. 3-3. Typical near-critical isobar exhibiting hysteresis loop according to Maass' experiment.[4]

is representative of the classical approach. Among other things, at the critical point the van der Waals equation yields an infinite specific heat at constant pressure and an infinite thermal expansion coefficient $[-\partial(\ln \rho)/\partial T]_p$, which is the reason the critical fluid is cited as an attractive heat transfer medium. The van der Waals model has serious inadequacies in the critical region. Widom's recent qualitative survey of his activities[3] proposes a three-dimensional lattice gas model to account for these difficulties.

An area of importance to heat transfer concerns the nonequilibrium phenomena associated with the critical point. Maass[4] found that a hysteresis loop in density near the critical point could be obtained by first heating and then cooling along an isobar, as shown typically in Fig. 3-3. The results were very stable and reproducible even under conditions of stirring. Another hysteresis example can be found in light scattering. Light scattering caused by severe density fluctuations near the critical point causes the fluid to become opaque as it passes through the critical point. The growth of the opaque condition, known as critical opalescence, is quite different depending on whether one is heating or cooling through the critical point. (For an excellent visual record of the critical opalescence phenomenon the reader is referred to a movie produced by Siemens Aktiengesellschaft, Postfach, West

Germany, entitled "Boiling and Evaporation Phenomena with Water.") These experiments demonstrate that, near the critical point, large relaxation times are required for a thermally disturbed system to return to equilibrium. One consequence of this is that the near-critical fluid under the dynamic conditions of heat transfer can expect to experience some degree of thermodynamic nonequilibrium. This in turn will result in some uncertainty in applying an equation of state.

Also, since the paths are different, the heat transfer process involved in cooling a near-critical fluid may be quite different from that associated with heating the same fluid.

Most of the truly severe behavior, such as the singularities in C_p, C_v, k, and β, occur precisely at the critical point. Normally one does not operate precisely at the critical point, and this tends to attenuate the influence of some of the anomalies. On the other hand, the overall large property changes, the suggested first-order effect, persist over a considerable region near the critical point.

3.2.2 P–ρ–T Data—Equations of State

The availability of actual P–ρ–T data near the critical point varies considerably from fluid to fluid. Probably the most detailed investigations near the critical point have been in carbon dioxide by Michels *et al.*[5] and in hydrogen by Goodwin *et al.*[6] For other fluids of interest, like nitrogen, there are very little, if any, actual data. For the most part the heat transfer researcher must rely on properties computed from an equation of state.

Equations of state come in all sizes and shapes. Obert[7] and Hirschfelder *et al.*[1] have good surveys. The most common are the virial type

$$P = A(T)\rho + B(T)\rho^2 + C(T)\rho^3 + D(T)\rho^4 + \cdots \tag{3-2}$$

Obert[7] and Hirschfelder *et al.*[1] list coefficients for various equations and fluids; however, their values were not determined with the critical point specifically in mind. In cryogenic fluids the Benedict–Webb–Rubin[8] equation of state, as modified by Strobridge[9] and Roder and Goodwin,[10] has been very popular:

$$\begin{aligned} P = {} & RT\rho(Rn_1T + n_2 + n_3/T + n_4/T^2 + n_5/T_4)\rho^2 \\ & + [Rn_6T + n_7 + (n_9/T^2 + n_{10}/T^3 + n_{11}/T^4)e^{-n_{16}\rho^2}]\rho^3 \\ & + n_8T\rho^4 + [(n_{12}/T^2 + n_{13}/T^3 + n_{14}/T^4)e^{-n_{16}\rho^2}]\rho^5 + n_{15}\rho^6 \end{aligned} \tag{3-3}$$

The main difficulty with these or any *curve-fit* equation is determining other properties such as specific heat C_p, since this requires derivatives which are often not very satisfactory near the critical point.

Recently there has been considerable interest in the use of the Ising model for ferromagnets to describe the lattice gas near the critical point.[11,12] In drawing the analogy with the Ising model, chemical potential appears to be a more fundamental variable than pressure. Green and his co-workers[13–15] have offered an equation in which chemical potential is expressed as a function of density and temperature. This equation holds considerable promise for the future.

3.2.3 Transport Properties

Most of the remarks made concerning P–ρ–T data can be made about actual transport-property data. Again the more popular fluids are better documented. Of particular interest are near-critical thermal conductivity data for carbon dioxide by Sengers[16] and Guildner,[17] and for hydrogen by Diller and Roder.[18]

Prior to Sengers' work it was felt that the excess of viscosity or thermal conductivity above its atmospheric value at a given temperature and pressure was a function of density alone:

$$k(P, T) - k_0(\rho \to 0, T) = f_1(\rho) \tag{3-4}$$

$$v(P, T) - v_0(\rho \to 0, T) = f_2(\rho) \tag{3-5}$$

Of course, this required a precise knowledge of ρ as a function of P and T; nevertheless, it allowed a very simple representation for viscosity and thermal conductivity. Furthermore, it could be reduced to general form by the law of corresponding states. This is still true away from the critical point, except possibly for quantum liquids such as hydrogen and helium. The functional relations of Eqs. (3-4) and (3-5) have been established for most fluids of interest by Thodos and co-workers and are summarized in a paper by Stiel and Thodos.[19] Eqations (3-4) and (3-5) remain valuable as a baseline.

The thermal conductivity data cited above exhibit an "anomalous spike" in the conductivity at the critical density along near-critical isotherms as shown in Fig. 3-4 for Senger's data.[20]

For viscosity the best evidence indicates only a weak anomaly which can be disregarded.

The computation of the anomaly is still in developmental stages. Brokaw[21] suggests treating a near-critical fluid as a dissociating polymer and that effective conductivity consists of two parts

$$k = k_f + k_r \tag{3-6}$$

Here k_r represents the contribution due to diffusion of the dissociating

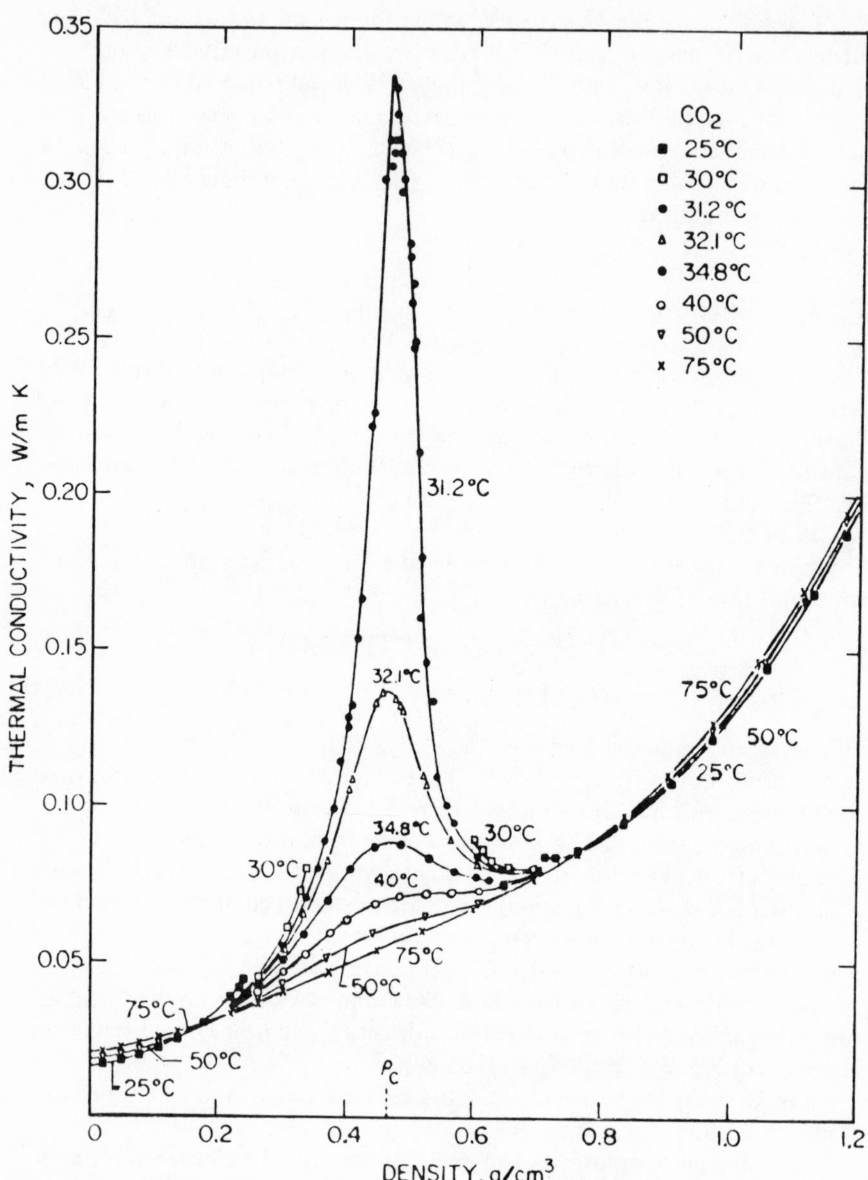

FIG. 3-4. Thermal conductivity of carbon dioxide exhibiting the anomalous spike.[20]

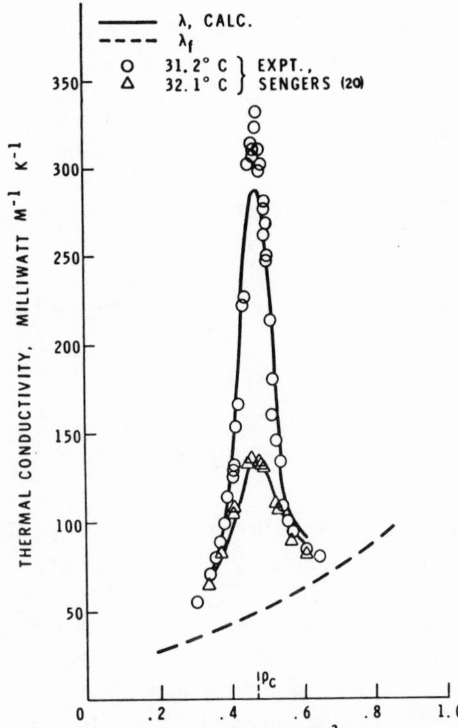

FIG. 3-5. Calculated and experimental thermal conductivities of carbon dioxide.[21]

clusters and k_f the normal conductivity is expressed by Eq. (3-4). The results of Brokaw's analysis are

$$k_r = \rho D(D_{ln}/D)C_{p_r} \tag{3-7}$$

where D is the self-diffusion coefficient, D_{ln} is the binary diffusion coefficient of the hypothetical polymer, and C_{p_r} is the specific heat in excess of the low-pressure value. Brokaw's calculations are compared to Senger's data in Fig. 3-5. Unfortunately, the theory requires a precise equation of state for computing C_{p_r} and D_{ln}/D. So far it has not been tried for other fluids.

3.2.4 Quantum States

In the discussion of "corresponding states" mention was made of the quantum liquids hydrogen and helium. The quantum properties of hydrogen do arise around its critical point, those of helium do not. In fact, the quantum

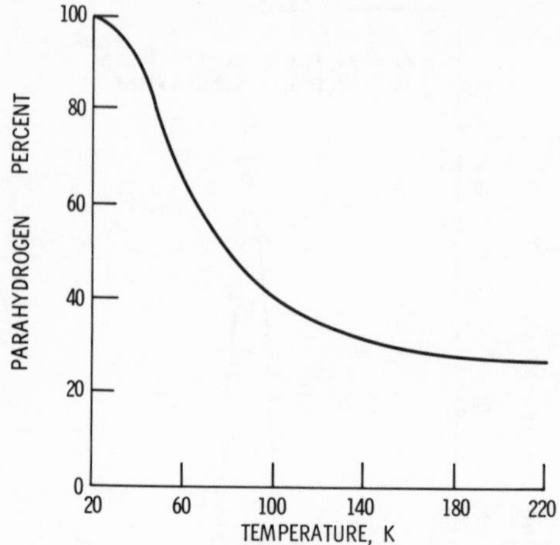

FIG. 3-6. Equilibrium percentage of parahydrogen.

properties of hydrogen are very important considerations in such applications as nuclear and chemical rockets.

In the early 1920's it was first noted that hydrogen did not appear to have a single-valued distribution for specific heat. Molecular spectrum studies were the key to understanding this apparent anomaly. It was observed that different quantum states of the hydrogen molecule resulted from the "spin" of the nuclei of the parent atoms that made up the molecule. Two conditions could exist. In one case, the spins of the two atoms could be in opposition (antiparallel) and the net spin of the molecule is zero. This is parahydrogen. In the second case, the spins are in the same direction (parallel), which results in the ortho form of hydrogen.

In the low-temperature range (from 20 to 250 K), the percentages of para and orthohydrogen vary appreciably in making up equilibrium hydrogen. The percent para over a temperature range is shown in Fig. 3-6.

Above temperatures of 260 K (470°R) the percentage of para molecules in the mixture remains fixed at 25%. This equilibrium mixture (25% para, 75% ortho) is referred to as normal hydrogen and is the equilibrium mixture encountered at room temperature and at higher conditions in the gaseous state.

The important difference among the quantum states in terms of engineering calculations lies in the specific heats.

Figure 3-7 is a comparison of the specific heats of para and equilibrium hydrogen at 1 psia. The appreciable differences are quite obvious. In an engineering application, one should be certain of the quantum state mixture. Generally, cryogenic liquid hydrogen is processed to be para to ensure long-term storage without temperature change (ortho–para conversion is exothermic). However, under some conditions the para can convert to ortho

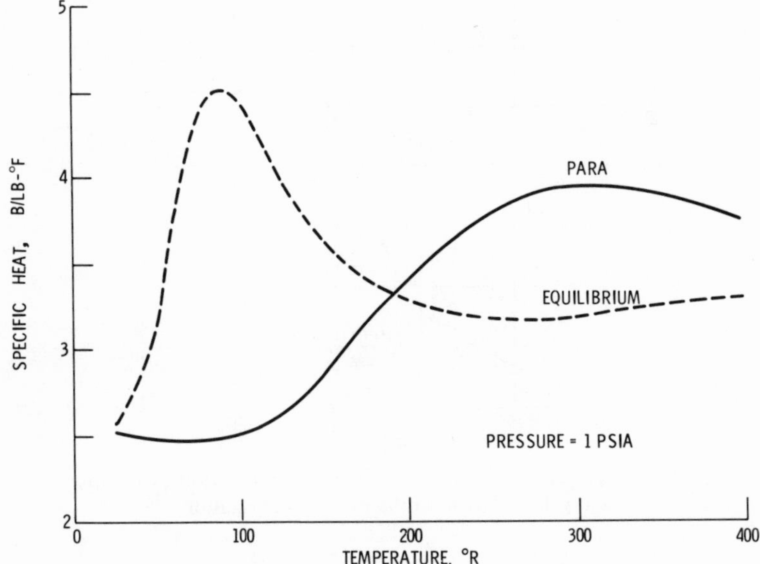

FIG. 3-7. Comparison of para and equilibrium hydrogen.

with endothermic heating. Nuclear radiation, catalysis by paramagnetic oxides, and chemisorption by metals (tungsten, nickel, and others) can promote rapid para-to-ortho conversion. Under normal conditions the conversion rate is extremely slow in the low-temperature liquid.

3.2.5 Pseudoproperties

The unusual property behavior in the near-critical region shown in Fig. 3-2 can be examined as an extension of saturation properties.

Frequent reference is made in near-critical work to the transposed critical temperature T^*. This is normally defined as the temperature where the specific heat C_p attains a maximum for a given supercritical pressure (see Fig. 3-2). It can be thought of as an extended saturation temperature.

In order to use pseudoboiling models, certain two-phase quantities in addition to the vapor pressure curve must be defined. Typical of these is the latent heat of vaporization. Figure 3-8 shows a way of defining a pseudoheat of vaporization from the enthalpy curves, as used by Thurston.[22] Other pseudo-two-phase quantities such as saturated liquid and vapor densities can be approximated in the same way.[23]

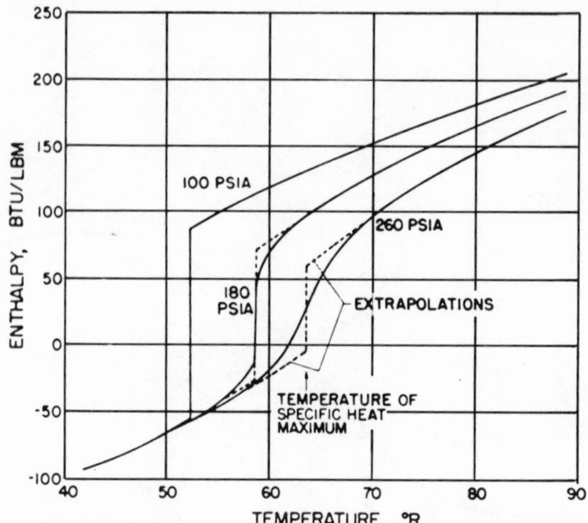

FIG. 3-8. Pseudo-two-phase properties; extrapolation procedure for equivalent latent enthalpy of vaporization.[22]

3.3 THE NEAR-CRITICAL HEAT TRANSFER REGION

3.3.1 Peculiarities of the Near-Critical Region

In region IV of Fig. 3-1 standard techniques of correlating data break down. The ordinary Dittus–Boelter equation does not correlate near-critical forced convection phenomenon; ordinary Rayleigh relationships do not correlate pool data; standard boiling equations exhibit discontinuities; and oscillations are commonplace. Standard theoretical techniques are no better than the empirical techniques. The problem, at the risk of oversimplification, is that the heat transfer coefficient has a strong and complex temperature dependence unlike an ordinary gas (Fig. 3-9).

The early experiments of Schmidt *et al.*[24] found free and natural convection in the vicinity of the critical point to exhibit a sharp increase in heat transfer coefficient. On the other hand, Powell[25] reported a sharp minimum in heat transfer coefficient in the T^* region for forced convection flow of liquid oxygen and nitrogen. Since then, many other investigators have found similar "peaks" in the axial wall temperature profile.[26–29] In direct opposition, several researchers[30–33] have reported a maximum heat transfer coefficient.

There are two associated results which may shed some light and bear further investigation. First, both Shitsman[28] and Yamagata *et al.*[29]

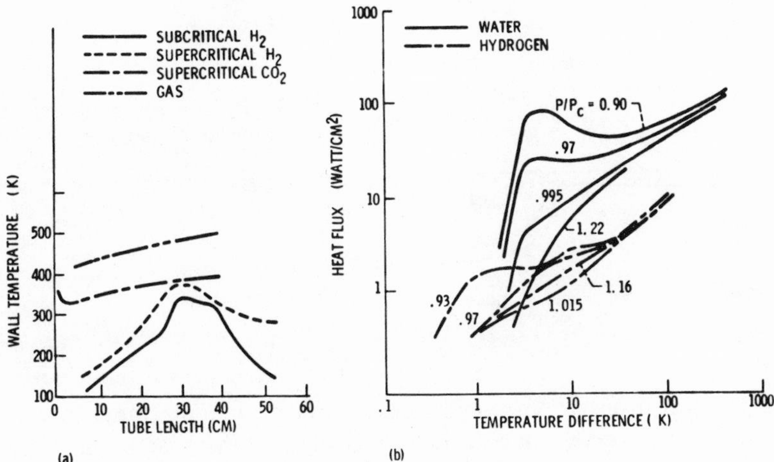

FIG. 3-9. Comparison between subcritical and supercritical heat transfer behavior in (a) forced and (b) free convection.

showed heat transfer coefficients when plotted against bulk enthalpy to be a minimum near the critical enthalpy and also reported pressure oscillations in the same region. Second, all of the reports of a maximum in heat transfer coefficient came from experiments in which the temperature difference between the wall and the bulk fluid was small when compared to the reports in which a minimum occurred.

Hsu[34] has suggested that the two results can be thought of qualitatively in boiling terms. When the temperature difference is small it can be likened to nucleate boiling, a region of very good heat transfer, thus the maximum. When the temperature difference is large it can be compared to film boiling, a region of poor heat transfer, thus the minimum. In support of this line of thinking one can examine Hauptmann's[35] data shown in Fig. 3-10. The small temperature-difference data (i.e., small heat flux) exhibit a clear maximum while the large-temperature-difference data ($q = 8.4$ W/cm^2) show, if anything, a minimum. A similar trend was found by Styrikovich *et al.*[36,37] Wood[38] and Kahn[39] showed h to be a maximum near the critical point with wall temperature varying. Shiralkar and Griffith[40] have found the heat transfer coefficient to be strongly dependent on heat flux when wall temperatures are above T^* and bulk temperatures below T^*.

Pressure oscillations are a natural phenomenon of this regime and at times can be quite large, $0.3P_{\text{test}}$ or 400 psi, in N_2O_4 as reported by McCarthy *et al.*[41] Hines and Wolf[42] reported pressure oscillations of sufficient amplitude to damage their heated tubes.

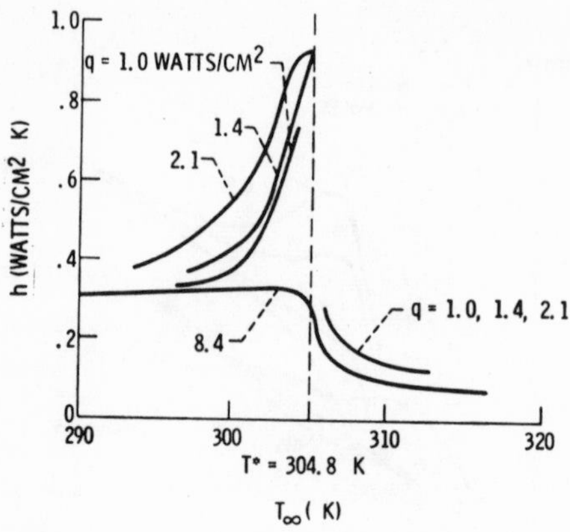

FIG. 3-10. Effect of free-stream temperature and wall heat flux on near-critical heat transfer coefficient.[35] CO_2 data; $P/P_c = 1.025$; $U_\infty = 0.45$ m/sec.

Pressure drops are large and are primarily due to momentum changes.[43] Friction losses become increasingly important as the heat transfer process moves away from the near-critical regime.

The similarities in heat transfer, pressure drop, wall temperature profiles, and pressure oscillations between subcritical and the near-critical regimes are remarkable. In the early phases of the hydrogen work, we recorded the sounds of two-phase flow and supercritical flow. The sounds were similar but supercritical sounds were not as "noisy." Goldmann[44] has noted the same phenomenon. Researchers in this area often think of the fluid as pseudo-two-phase. Some of the strongest evidence in support of a pseudo-two-phase fluid comes from thermodynamic state figures, from visual studies[46–49] and from heat flux *vs.* wall temperature plots. Griffith and Sabersky[46] and Knapp and Sabersky[47] found the heated globules to be easily taken for bubbles. The movie supplement to the work of Graham *et al.*[45] gives the viewer a statistical feel for the nature of these similarities as opposed to the instantaneous picture provided by the published photographs.[47] Cumo *et al.*[49] have documented the change in fluid structure for freon from atmospheric conditions to the near-critical region. Nishikawa and Miyabe[48] published a set of pictures for the nucleate–film boiling cycle and a comparative set at supercritical pressure ratio of $P/P_c = 1.065$. The similarities at the film boiling heat fluxes are particularly striking. The heat flux *vs.* temperature difference plots cited above, particularly those of Holt and Grosh[50] show heat transfer coefficient to be a function of heat flux—a phenomenon not found in normal gas heat transfer but a trademark of boiling. One possible explanation of the unusual wall temperature profiles discussed earlier is a "boiling" model.

The strongest argument against such a pseudo-two-phase fluid comes from thermodynamics; a supercritical fluid in equilibrium is clearly single phase. Pseudo-two-phase fluid advocates recognize this and most claim that it is a nonequilibrium situation. Experimentally it is very interesting that the single-phase proponents can turn to the same visual evidence that the two-phase advocates presented. Hauptmann[35] uses his own very dramatic pictures and reexamines other visual experiments and concludes that all of the unusual results can be explained in single-phase terms. A careful examination of all the available pictures[35,45–47] reveals as many nonsimilarities as there are similarities to boiling. The analyst really has his choice because a strong case can be made for either model. Any analytic model (pseudo-two-phase or single-phase) should work throughout region IV (Fig. 3-1) and provide smooth transition from liquid to gas if properly formulated.

Finally it should be emphasized that geometric effects (size and shape of the channel) introduce additional uncertainties that complicate the heat transfer behavior. Some of the variation in the near-critical heat transfer data is undoubtedly related to geometric effects.

3.3.2 Heat Transfer in Free and Natural Convection Systems

3.3.2.1 *Pools—Free Convection:* We divide the free and natural convection phenomena in the near-critical region into two parts. The free convection pool studies discussed in this section primarily deal with heat transfer from small test sections such as wires and filaments. In the subsequent section the heat transfer from natural convection loops will be considered.

The effect of the wide variations of properties appears to have a more direct influence in the pool heat transfer case. This is probably true because in this situation temperature differences are usually small. Pool results almost universally show an enhancement in heat transfer near the critical point. Some typical experimental data where enhancement occurred are those of Skripov and Potashev.[51]

Unlike a forced convection system, the free convection system has no constraining boundaries. Consequently, it is able to respond to favorable property variations with enhanced heat transfer.

An interesting set of near-critical free convection experiments is described in Refs. 47, 52, and 53 and used a small-diameter horizontal wire. Knapp and Sabersky[47] were the first to photograph an oscillation between laminar and turbulent flow while heating near-critical CO_2. Their data indicated a sharp increase in heat transfer in going from the all-laminar to all-turbulent region; however, the transition through the oscillating region appeared smooth. The authors indicate that this is probably because the data were average data, and that they suspect an oscillation in wire temperature to go with the laminar–turbulent oscillation. Goldstein and

Aung[52] reported similar oscillations but no sharp increase in heat transfer. The phenomenon remains unexplained at present.

Most of the free convection analyses in the literature employ rather conventional dimensionless groups. In some cases there have been modifications, primarily to account for the variable properties in the boundary region.[54–56] There appears to be general agreement that somewhat away from the critical and the transposed critical points the conventional correlations will hold for all fluids and for all systems. Thus away from the critical point McAdams' basic equation should hold:

$$\mathrm{Nu}_f = C\,\mathrm{Ra}^n \tag{3-8}$$

Closer to the critical point, modifications of the following basic form are employed:

$$\mathrm{Nu}_x = C\,\mathrm{Gr}_x{}^a \mathrm{Pr}_x{}^b [T\alpha/(T_w - T_\infty)]^c \tag{3-9}$$

For example, Larson and Schoenhals[57] conducted an experiment with a vertical ribbon in near-critical water. The constants they found appropriate were

$$a = \tfrac{1}{3}, \qquad b = 0.247, \qquad c = 0.137, \qquad x = 0.5$$

Very close to the critical point Simon and Eckert[58] recommend

$$\frac{q_l[(T - T_c)/T_c]^{1/2}}{\rho_c k_w(\partial T/\partial \rho)_p} = 3.25 \times 10^{-9} \frac{\mathrm{Ra}_w}{(\mathrm{Pr}_w)^{1/2}} \tag{3.10}$$

3.3.2.2 *Loops—Natural Convection:* Historically, there has been considerable interest in the use of a natural convection loop or column. Here a fluid in the supercritical region can operate at high heat flux.

Experiments and analyses in natural convection loops begin with the early work of Schmidt in Germany. As in pools, the results of loop experiments are universally an enhancement of heat transfer near the critical point when compared to noncritical fluids under similar heat transfer conditions.

In an effort to correlate near-critical data, Holman and Boggs[59] rearranged Schmidt's basic loop equations to the form

$$\mathrm{Nu} = 16\mathrm{Re}^2\mathrm{Pr}\mathrm{Gr}^{-1}(l_T/L)(d/y) \qquad \text{(laminar)} \tag{3-11}$$

$$\mathrm{Nu} = 0.079\mathrm{Re}^{11/4}\mathrm{Pr}\mathrm{Gr}^{-1}(l_T/L)(d/y) \qquad \text{(turbulent)} \tag{3-12}$$

To test the effects of geometry, Tanger *et al.*[60] compared sulfur hexafluoride data obtained in two different loops. For a fixed l_T and d, the geometric factors of Eqs. (3-11) and (3-12) of loop 1 were approximately

1.8 times those of loop 2. They found no effect from variations in geometry. For the near-critical region they recommend

$$\mathrm{NuGr}_f/\mathrm{Pr} = 0.00982\mathrm{Re}^{11/4} \qquad (3\text{-}13)$$

3.3.3 Heat Transfer in Forced Convection Systems

The experimental work in near-critical forced convection can be broken down into two major categories. First are the conventional heated tube experiments which have been used so successfully in determining Nusselt correlations for gases. In the second category are experiments which examine one or two details of the heat transfer process in order to explain the mechanism of near-critical heat transfer.

In the past 17 years, over 30 experiments in forced convection heat transfer to supercritical fluids have been conducted which fall into the broad classification known as heated tube experiments. All of these experiments were very similar in design, operating procedure, measurements, and, to some extent, results. These experiments can be surveyed as a group, pointing out specifics where necessary. The range of conditions for these experiments is quite extensive. The most popular fluids have been water, hydrogen, and carbon dioxide.

3.3.3.1 *General Characteristics of Near-Critical Forced Convection Heat Transfer:* From our experience with the heat transfer to near-critical hydrogen flowing through electrically heated tubes, the following general observations are made.

The test sections were oriented vertically and the flow was upward and turbulent. The test sections were heated uniformly (constant heat flux). The wall temperature distribution along the tube had a humplike shape, which was similar to what was observed with two-phase hydrogen. In Fig. 3-11 the wall temperature distributions for two-phase, near-critical, and gaseous hydrogen are compared. Note how similar subcritical run 539 and supercritical run 700 are. In contrast, run 1059, which is gaseous hydrogen, exhibits a monatomically increasing wall temperature, which is characteristic of all constant-heat-flux, gaseous, turbulent heat transfer data. Figure 3-11 could have been drawn using heat transfer coefficient as the ordinate. The h curves would be mirror images of the wall temperature plots. The axial position of the peak wall temperature (minimum h) was observed to be a function of the heat flux level, weight flow rate, and inlet enthalpy conditions. However, the peak did seem to occur at the axial position near where the bulk temperature reached the transposed critical temperature.

It was observed that the measured heat transfer coefficients in the tube were higher than those computed from a forced convection correlation

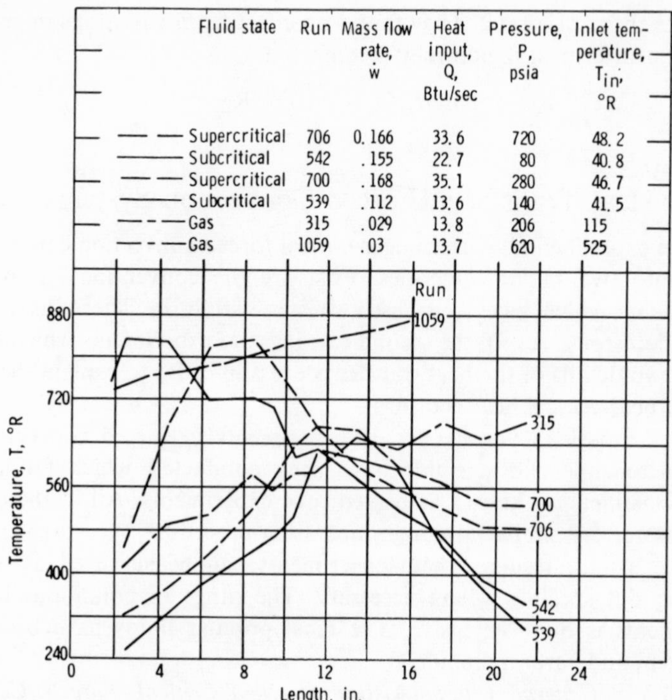

FIG. 3-11. Surface temperature distribution with axial position for near-critical and gaseous hydrogen. (Tube inside diameter, 0.335 in.)

employing film properties. (Some researchers have compared the near-critical data to correlations employing bulk properties for which the data are often smaller than the prediction.) More will be said about correlation procedures in the next section.

It should be pointed out that fluid oscillations were observed in these heated tube tests. Sometimes it was difficult to avoid them in establishing an operating condition. Changing the heat flux level, rather than manipulating the flow rate, was observed to be the most effective way of getting out of an unstable condition. A broad band of instability frequencies was observed which ranged from 0.5 Hz up to several kHz. The operating conditions where these instabilities occurred were not predictable, nor were the frequency levels. The high frequencies appeared to correspond to a radial model.

The correlation attempts can be treated in a general manner as to form. Early work centered on reference-temperature modifications to the basic Dittus–Boelter equation:

$$\mathrm{Nu}_x = C\,\mathrm{Re}_x{}^a\mathrm{Pr}_x{}^b(T_w/T_b)^p \tag{3-14}$$

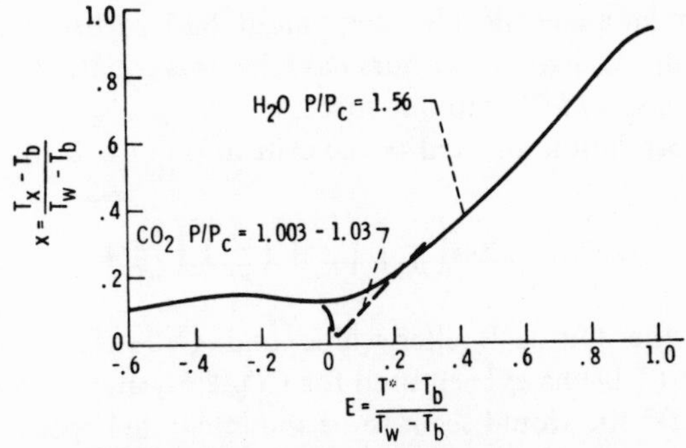

FIG. 3-12. Effect of relative proximity to the critical point on Eckert reference temperature.[31,95]

where x means that the fluid properties are evaluated at $T_x = T_b + x(T_w - T_b)$. Eckert[61] suggested that x may be a function of a dimensionless temperature frequently called the Eckert parameter:

$$x = f[(T_x - T_b)/(T_w - T_b)] \tag{3-15}$$

Bringer and Smith[62] and Schnurr[63] have plotted these functions as shown in Fig. 3-12. While this treatment handles quite a bit of data, it has never been generalized to other data and no analysis is available to predict the function.

Another early correlation directed specifically at heat transfer near the critical point was that of Miropolskii and Shitsman.[64] Using water experiments these authors proposed

$$\text{Nu}_b = 0.023\text{Re}_b^{0.8}\,\text{Pr}_{\text{min}}^{0.4} \tag{3-16}$$

where Pr_{min} is the minimum Prandtl number defined as

$$\text{Pr}_{\text{min}} = \begin{cases} \text{Pr}_b; & \text{Pr}_b < \text{Pr}_w \\ \text{Pr}_w; & \text{Pr}_w < \text{Pr}_b \end{cases}$$

This amazingly simple modification to the Dittus–Boelter equation produced some remarkable results.

Shitsman successfully applied this correlation to the water data of Dickinson and Welch,[65] the oxygen data of Powell,[25] and the carbon dioxide data of Bringer and Smith.[62] Particularly interesting was the fact that the correlation comprehended both the maximum in heat transfer coefficient reported by Dickinson and Welch and the minimum reported by Powell. Later, however, using water, Shitsman also discovered experimentally the same temperature "spikes" that Powell reported for oxygen. He called this a region of "impairment" to heat transfer and likened it to boiling crisis. He cautioned his readers that his correlation would not correlate this region despite the fact that it had earlier handled Powell's similar conditions. One of the authors has tried this equation on the hydrogen data of Hendricks *et al.*[26] without success.

Another correlation directed at the critical point is of the form

$$\mathrm{Nu} = \mathrm{Nu}_b\left(\frac{\mu_w}{\mu_b}\right)^a\left(\frac{k_w}{k_b}\right)^b\left(\frac{\rho_w}{\rho_b}\right)^c\left(\frac{\bar{C}_p}{C_{pr}}\right)^d \tag{3-17}$$

where Nu_b is the Dittus–Boelter value and $\bar{C}_p = (\mathrm{H}_w - \mathrm{H}_b)/(T_w - T_b)$. Versions of Eq. (3-17) have been used for CO_2,[32] water,[33] and propane.[66] The key characteristic would seem to be the integrated specific heat $\bar{C}_p$.

The concept of using not only integrated specific heat but all physical properties on an integrated average basis was put forth by Brokaw[67] for a reacting N_2O_4 system. The properties would be expressed as

$$\varphi = [1/(T_w - T_b)]\int_{T_b}^{T_w} \varphi(t)\, dt \tag{3-18}$$

where $\varphi(t)$ is any fluid property. This has a tendency to smooth out the sharp near-critical property changes.

Hess and Kunz[68] examined the heat transfer results for near-critical hydrogen on the basis of a differential boundary layer analysis. The eddy diffusivity for the differential description was based upon the mixing-length model. Van Driest had suggested that the mixing length was sensitive to a viscous damping parameter A^+. Hess and Kunz did not execute a complete differential analysis with this parameter. However, they did relate A^+ to the kinematic viscosity ratio v_0/v_b and suggested a general correlation of the form

$$\mathrm{Nu}_f = 0.0208\mathrm{Re}_f^{0.8}\mathrm{Pr}_f^{0.4}[1 + 0.0146(v_w/v_b)] \tag{3-19}$$

This forced convection correlation has worked quite well for near-critical hydrogen provided the bulk temperature is above the transposed critical temperature.

Earlier, some of the similarities between two-phase and near-critical heat transfer results were mentioned. These similarities have encouraged the development of pseudoboiling models for the near-critical region. Correlations used for boiling heat transfer results have been adapted for use in the near-critical region. An example of such an adaptation is presented by Hendricks *et al.*,[26,43] who converted the Martinelli two-phase flow parameter to a pseudo-two-phase fluid model. Instead of considering vapor and liquid, light- and heavy-density species were conceived to make up the near-critical fluid. The heavy density was assumed to be the liquid-melt density, while the light density was the perfect-gas density. Appropriate mixtures of the heavy and light species made the bulk density variation observed around the critical point. The pseudoquality was the mass fraction of the light species present in the mixture, viz.,

$$\frac{1}{\rho_b} = \frac{x_2}{\rho_{\mathrm{pg},b}} + \frac{1 - x_2}{\rho_{\mathrm{melt}}} \tag{3-20}$$

The heat transfer correlation for the near-critical region is expressed as

$$\mathrm{Nu}_f = \mathrm{Nu}_{fm} f(\chi'_{tt}) \tag{3-21}$$

where

$$\mathrm{Nu}_{fm} = 0.023(\rho'_{fm} u d/\mu_f)^{0.8}(\mathrm{Pr}_f)^{0.4}$$

and ρ'_{fm} is defined to be

$$1/\rho'_{fm} = (x_2/\rho_{\mathrm{pg},f}) + (1 - x_2)/\rho_{\mathrm{melt}}$$

3.3.3.2 *Penetration Model for Near-Critical Fluids:* Graham[69] suggested that the enhanced heat transfer coefficient (as compared to a film correlation) might be explained on the basis of a penetration model. It was proposed that the enhancement be represented by an additive term to the conventional film-property forced convection heat transfer coefficient. In general, the average heat transfer coefficient can be written as

$$h_{\mathrm{av}} = C_1 h_f + C_2 h_{\mathrm{enh}} \tag{3-22}$$

This h_{enh} was attributed to a penetration mechanism. In brief, the penetration mechanism was assumed to be a disruption of the sublayer by the periodic migration of fluid packets from the outer edge of the boundary layer to the wall where they pick up thermal energy in a transient fashion. Figure 3-13 is a schematic of that process. After a short contact with the wall they move out to the edge of the boundary layer and dissipate their energy. This cyclic thermal pumping is somewhat analogous to the movement of the elements in a fluidized bed where the original penetration model has been employed in describing that mass and heat transfer mechanism.

Equation (3-22) was modified to be

$$h_{\mathrm{av}} = (A_p/A_T)h_p + [1 - (A_p/A_T)]h_f$$

where A_p is the average surface area where penetration is in effect and A_T is the total surface area. The penetration heat transfer coefficient is

$$h_p = (2/\pi)[(\rho C_p k)_b/\tau_{\mathrm{contact}}]^{1/2} \tag{3-23}$$

This expression for h_p is derived from the transient conduction equation with the appropriate boundary conditions. The film coefficient h_f is computed from the conventional Nu, Re, Pr correlation.

Time and space limitations prevent a full elaboration of the method.

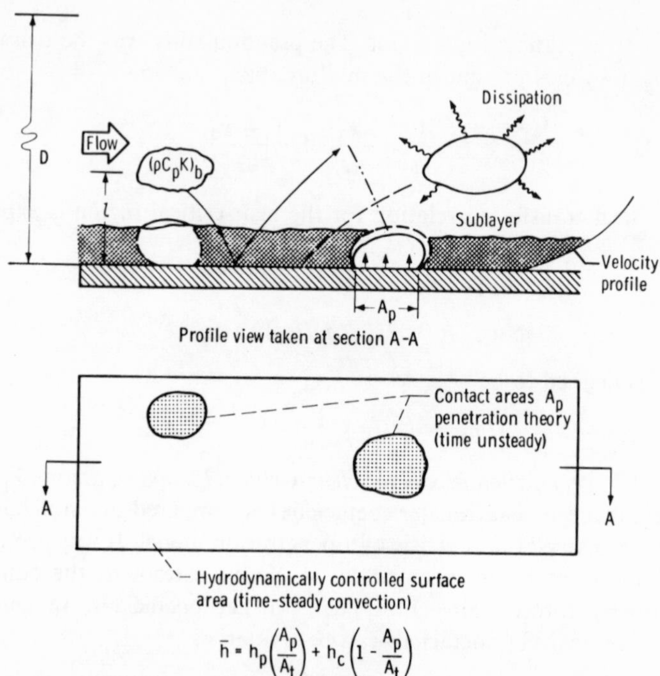

FIG. 3-13. Concept of simultaneous hydrodynamic and penetration mechanisms.

However, it has been employed in correlation near-critical hydrogen, carbon dioxide, and water heat transfer data. The method and all the associated assumptions are presented by Graham.[69]

3.3.3.3 *Detailed Investigations into Mechanisms:* There have been several experiments directed to more specific and detailed information than available from the heated tube experiments. Probably the most useful details concerning the flow are velocity and temperature profiles. The only experiments in which profiles have been measured were performed by Wood[38] and Wilson.[70] In both cases they surveyed radially across a vertically oriented heated tube near the exit. Wood used CO_2 at an L/D of 30.7 and Wilson used H_2 at an L/D of 128.5. Wilson's data, in addition to temperature and dynamic head profiles, included hot-wire measurements. In both cases, the major result was the appearance of the so-called M-shaped velocity profiles when the bulk temperature was near the transposed critical temperature as shown typically in Fig. 3-14 and predicted analytically by Hsu and Smith.[81] These are similar to the profiles presented in Ref. 71. Recent experiments of Bourke *et al.*[72] indicate the hot wire to be a useful tool in this regime.

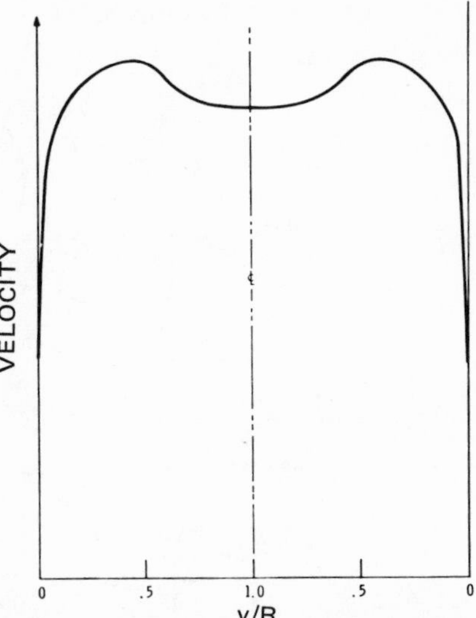

FIG. 3–14. Typical M-shaped profiles as found by Wood[38] and Wilson.[70]

Kahn,[39] using CO_2, conducted experiments designed to isolate some of the many interacting variables and to highlight the influence of a single parameter. His method was to flow CO_2 at its critical point between parallel plates of large aspect ratio with one plate heated and the other cooled at exactly the same rate so that no increase in the heat content of the fluid occurred. The heat transfer coefficient peaked sharply when the hot wall (which was always the upper wall) approached the transposed critical temperature. The peaks became sharper as the temperature difference between the walls was decreased.

A different result was observed by Hauptmann,[35] who flowed near-critical CO_2 over a horizontal heated flat plate. Among his results is an indication of a 30% increase in heat transfer for the same conditions with the heater oriented upward. Hauptmann's major contribution was a set of excellent color high-speed movies coupled with good heat transfer data. His data have already been discussed with respect to Fig. 3-10.

3.3.4 Near-Critical Heat Transfer in Relation to Conventional Geometric Effects

3.3.4.1 *Curved Tubes:* Systematic studies of the effects of curvature on a near-critical fluid are lacking. Miller (see Ref. 73) has recently completed such a study but the data are not yet available. Several studies have been made at higher pressures and for gases. The basic effect of curvature is to instigate a secondary flow such that the boundary layer is thinned at the

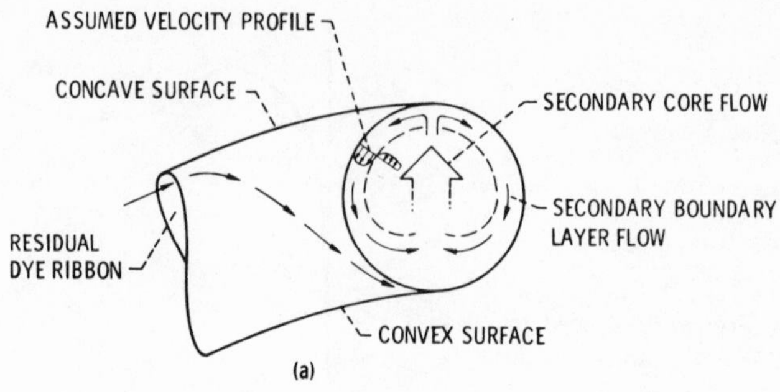

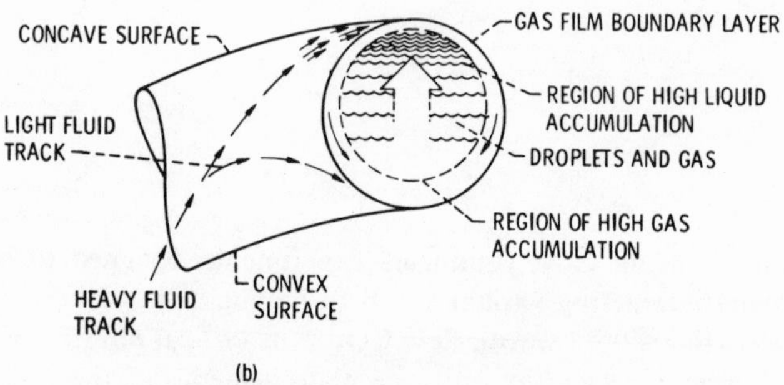

FIG. 3-15. Models of fluid flow in curved tubes. (a) Single-phase flow; (b) two-phase or near-critical flow.

concave surface (see Fig. 3-15) and thickened at the convex surface. The flow patterns indicate that the core moves toward the concave surface and then flows back along the periphery discharging into the region of the convex surface. While the classical works of Itō,[73] on turbulent flows and Dean,[74] on laminar flows deal with single-phase, fully developed fluids, the results of their analyses have been applied with varying degrees of success to the near-critical fluids.

Basically, Itō found that, for high-Reynolds-number fluids, the average friction factor for a curved tube was increased over that for a straight tube according to the relation

$$f_c/f_{st} = [\mathrm{Re}(r/R_c)^2]^{0.05} = I^{0.05} \tag{3-24}$$

where r is the pipe radius and R_c is the bend radius.

Hendricks and Simon[75] have pointed out some effects of curvature on near-critical hydrogen, which can be summarized as follows:

1. Rather conventionally, at high Reynolds numbers the concave surface enhances and the convex surface degrades heat transfer. The

magnitude of these effects depends on: fluid conditions, curvature-to-tube radius ratio, and angular position along the bend.

2. Visual studies with small-bend-angle tubes, using liquid nitrogen, indicate that this fluid is centrifuged to the concave wall rather than swirled in the normal secondary flow patterns.
3. Entrance conditions, profile similarity, and fluid history greatly influence the heat transfer coefficients. Furthermore, the effects of curvature persist downstream of the bend and appear to be propagated upstream as well.

In subsequent tests, it was found that h could vary as though the fluid was oscillating around the tube in some harmonic manner. Such oscillations were apparently instigated by a nonuniform heat flux pattern at the upstream heating flange. They could also be induced by a nonuniform inlet velocity profile.

The effects of the ratio of the radius of curvature to the tube radius for hydrogen are shown in Fig. 3-16. At large R_c/r the effect is small and approaches that of the straight tube. At smaller R_c/r, the effects are quite large (2:1) but data seem to indicate two possible paths at lower R_c/r in Fig. 3-16. Which one is correct awaits experimental verification.

McCarthy *et al.*[77] investigated near-critical N_2O_4 and noted a nonuniform surface temperature phenomenon along the convex surface and a persistence of the curvature effects in the downstream region. Their experiment was complicated by the dissociation of N_2O_4; however, they still found up to a 2:1 increase in h_c due to curvature. In a report[76] investigating liquid hydrogen at $P/P_c \approx 5$, a 2:1 increase was also found in heat transfer

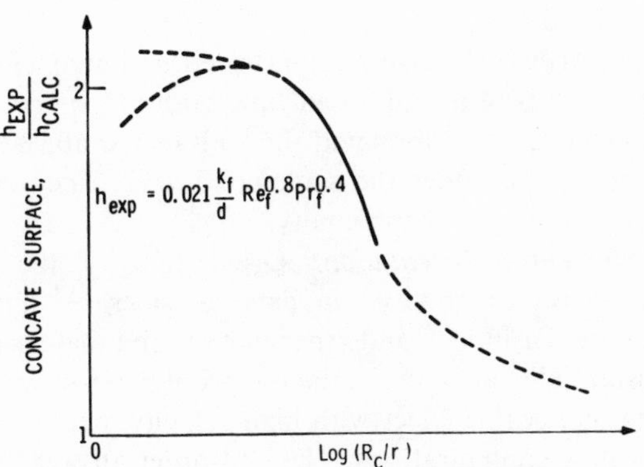

FIG. 3-16. Concave surface heat transfer coefficient as a function of R_c/r.

due to curvature; however, no effect was found due to asymmetric heating on noncircular flow passages.

Curved-tube investigations at Rocketdyne using hydrogen, showed the heat transfer coefficient to increase as

$$h_c \propto h_{st}(1 + 0.1L/D), \qquad L/D \leqslant 10 \tag{3-25}$$

up to $L/D = 10$ and becomes fully developed thereafter. For L/D values greater than 10, the following correlations were suggested:

$$\mathrm{Nu}_{cv} = \mathrm{Nu}_{st}[1 + 1.87(r/R_c)^{0.519}](0.0822\mathrm{Re}_b{}^{0.155}) \tag{3-26}$$

$$\mathrm{Nu}_{cx} = \mathrm{Nu}_{st}[1 + 2.21(r/R_c)^{0.894}](0.0822\mathrm{Re}_b{}^{0.155}) \tag{3-27}$$

where Nu_{st} is computed from Eq. (3-14) with the following values:

$$C = 0.0204, \qquad x = 0.4, \qquad a = 0.0098$$

From the hydrogen data in the literature, Taylor[78] proposed the following correlations using the Itō parameter:

$$\mathrm{Nu}_{cv}/\mathrm{Nu}_{st} = I_b{}^{0.05} \qquad \text{concave} \tag{3-28}$$

$$\mathrm{Nu}_{cx}/\mathrm{Nu}_{st} = I_b{}^{-0.05} \qquad \text{convex} \tag{3-29}$$

where Nu_{st} is for a straight pipe. While the above equations were developed for hydrogen, similar correlations would be anticipated for other fluids.

3.3.4.2 *Twisted Tapes and Rifle Boring:* If one can get the bulk fluid near the wall, an augmentation in the heat transfer coefficient will occur. Bartlit and Williamson[79] induced swirl flow in LH_2 using a twisted tape in a long tube and found the heat transfer coefficients to be approximately those predicted by the standard Dittus–Boelter equation. This is in contrast to the same pipe without a swirl inducer, where the experimental h was much less than the predicted h (Fig. 3-17).

Michaud and Welch[80] have demonstrated a similar effect with near-critical water. The lands of a rifle bored tube both induce swirl and trip the flow; these geometry effects eliminated the wall temperature "spike" noted in a similar smooth tube under the same conditions. These enhancements, of course, require a pressure drop penalty.

3.3.4.3 *Body Force Orientation:* Orientation of the heat transfer apparatus with respect to gravity can have considerable influence on the results. Shiralkar and Griffith[40] and others have found that parallel buoyancy effects can substantially alter the velocity profiles. However, the present authors found no noticeable effect with high-velocity, near-critical hydrogen in up and down flow configurations. The first-order answer lies in the ratio of buoyancy to inertia forces, which for our case was quite small, $\mathrm{Gr}_f/\mathrm{Re}_f{}^2 = 4 \times 10^{-4}$. Shiralkar and Griffith[40] found that the effect began to be notice-

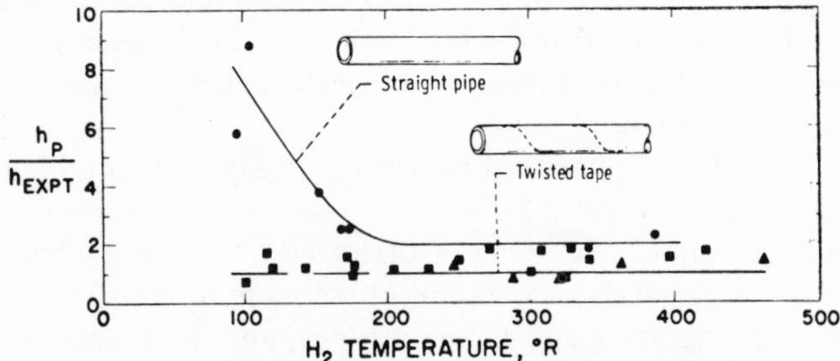

FIG. 3-17. Comparison of twisted tape insert to straight pipe heat transfer in near-critical hydrogen.[80] ▲ Swirl inducer for the first 14 feet ; ● no swirl inducer; ■ swirl inducer for 46 ft.

able when $Gr/Re^2 = 4 \times 10^{-3}$ for water and 10^{-2} for CO_2. These values and conclusions are in agreement with the computations of Hsu and Smith[81]

3.3.4.4 *Entrance Effects:* Papell and Brown[82] investigated entrance effects in near-critical hydrogen. The experimental h values at high heat fluxes were 1.5 times higher for an undeveloped entrance than for a hydrodynamically developed entrance, as illustrated in Ref. 82. This effect persisted an abnormally long distance down the tube. Little effect was found for the low-heat-flux case.

3.3.5 Theoretical Considerations in Free Convection (Laminar Flow)

It appears to be physically unlikely that forced convection laminar flow will be generated with near-critical fluids. This comment is also pertinent to two-phase forced convection systems. Density changes and attendant instabilities inherent to two-phase and near-critical fluids produce turbulence.

Continuing our discussion with regard to near-critical fluids, it is apparent that laminar flow will be most likely associated with free or natural convection. Koppel and Smith[31] have analytically considered the possibility of laminar flow in pipes. No laminar experimental data have been generated to date to compare with this analysis (nor are such data likely). Nevertheless the analysis is a very important piece of work. Interestingly, the analysis did predict the occurrence of minimum and maximum heat transfer coefficients at intermediate axial positions along the tubular geometry. The paper made a strong case for the importance of variable-property analyses as compared to constant-property solutions.

Simoneau and Williams[83] did an analysis of Couette flow of near-critical hydrogen between two flat plates. The objective was to see what extreme variations in fluid transport properties would do to heat transfer predictions. It turned out that the heat transfer around the critical point did reduce sharply but this change was not nearly as large as the variations in the transport properties.

Quite a number of analytical papers have been written on laminar free convection along a vertical plate in which the fluid considered is near the critical thermodynamic state. One of the earliest of these papers is that by Sparrow and Gregg.[84] In fact, several of the more recent papers have been extensions of the Sparrow and Gregg analysis, and have made use of their analytical procedures. The vertical plate analyses have been applied to several near-critical fluids, including Freon, CO_2 and water.

Fritsch and Grosh[85] applied the vertical plate, laminar flow analysis to water. Their analytical results included pressures and temperatures above and below the critical point. In a subsequent paper Fritsch and Grosh[86] also presented experimental free convection data for water. They used a constant-temperature vertical flat plate and ran the experiment at critical pressure (3208 psia) and two additional pressures above the critical value. The fluid bulk temperature was varied above and below the critical value.

In comparing the experimental results to the analysis, there was general agreement, although the experimental values seemed to be consistently higher than the analytical ones by about 20%.

Parker and Mullin[55] extended the Sparrow and Gregg analysis to apply to near-critical Freon 114. They computed the thermal boundary layer profiles for four combinations of wall-to-stream temperatures.

Freon 12 was analyzed by Brodowicz and Bealokoz.[87] The experimental heat transfer coefficient was observed to be greater than the predicted value when the wall temperature was at the critical value. They ascribed this difference to the onset of turbulent agitation in the boundary layer. The temperature difference between the wall and the bulk was 20°C, which is a fairly high ΔT. Consequently, turbulent agitation is to be expected for such thermal conditions. This observation is added evidence to support the earlier comment that it is difficult to obtain laminar flow conditions with near-critical fluids.

Another paper in which some experimental information accompanied the analysis of a vertical flat-plate configuration is that by Hasegawa and Yoshioka.[88] The analytical method was somewhat different from the other in the literature in that a perturbation technique was employed in solving the nonlinear differential equations. The properties of water and CO_2 were used in the analytical solution. They compared the analytical predictions to the water data of Ref. 86 and to CO_2 data generated at their research laboratory in Kyushu University. The authors examined the possibility of using a reference enthalpy in computing the heat transfer. They concluded that a reference enthalpy could be assigned for a narrow range of temperature conditions around the critical point.

3.3.6 Theoretical Considerations in Forced Convection

In the free and natural convection areas reasonable success has been realized using conventional variable-property approaches. In forced convection, complex flow interactions caused by the rapid expansion of the near-critical fluid are superimposed on the normal forced convection patterns and conventional forced convention approaches seem inadequate. A fair amount of analytic attention has been turned toward explaining these phenomena. This will be briefly recapped here. As pointed out in the previous section, our considerations will pertain to turbulent flow only.

3.3.6.1 *Mixing-Length Analyses:* By far the most widely used approach is the Prandtl mixing-length concept. The basic idea is that the Reynolds stress terms of the Navier–Stokes equations can be made to look like laminar shear terms by the introduction of a turbulent viscosity or eddy diffusivity. The shear and heat flux expressions assume the general form

$$\tau = \mu \, du/dy + \rho\epsilon \, du/dy \tag{3-30}$$

$$q = -k \, dT/dy - \rho C_p \epsilon_h \, dT/dy \tag{3-31}$$

Subject to the following nondimensionalization, Deissler[89,90] proposed that these equations apply throughout a supercritical flow field:

$$u^+ = u/u^*, \qquad y^+ = u^* y/v_0, \qquad T^* = (1/\beta)[1 - (T/T_0)]$$
$$\beta = q_0 u^*/C_p T_0, \qquad \text{and} \qquad \epsilon^+ = \epsilon/v_0$$

where u^* is the shear velocity

$$u^* = (\tau_0/\rho_0)^{1/2} \tag{3-32}$$

The dimensionless equations become

$$\tau/\tau_0 = [(u/u_w) + (\rho/\rho_w)\epsilon^+] \, du^+/dy^+ \tag{3-33}$$

$$q/q_0 = [(k/k_w \mathrm{Pr}_w) + (\rho/\rho_w)(C_p/C_{pw})\epsilon] \, dT^+/dy^+ \tag{3-34}$$

Deissler[89] numerically integrated these equations for water at $P/P_c =$ 1.56 with moderate success. He suggested using a reference temperature T_x (discussed in Section 3.3.3 on heat transfer in forced convection systems) for evaluating properties, and Eckert consolidated the data spread by suggesting the following dimensionless temperature as discussed earlier:

$$E = (T_x - T_b)/(T_w - T_b) \tag{3-35}$$

In that same year, Goldman[91] introduced a technique to transform the variable-density problem to the constant-density form. This was done by using a variable density in the shear velocity:

$$u^* = (\tau_0/\rho)^{1/2} \tag{3-36}$$

The forms of Eqs. (3-33) and (3-34) remain unchanged but all the dimensionless variables are changed. They are usually designated u^{++}, y^{++}, and T^{++} to indicate the change. Goldman also suggested a different form for the heat transfer coefficient which evolves from the Dittus–Boelter[92] equation:

$$h' = q_0 D^{0.2}/G^{0.8} \tag{3-37}$$

These modifications have been popular but universal success has not been achieved.

Most of the research in mixing-length theories for near-critical fluids has centered on adapting the eddy diffusivity to take account of the near-critical phenomenon.

Generally, eddy diffusivities ϵ can be divided into two categories: (1) continuous, and (2) multiple part. In the continuous case, Van Driest[93] argues that one form of ϵ should apply over the entire regime, while for the multiple-part case, Deissler[89] indicates a near-the-wall region and an away-from-the-wall region.

The Van Driest continuous approach involves

$$\epsilon = K^2y^2\{1 - \exp[-(y/A)^2]\}\, du/dy \tag{3-38}$$

$$K = \text{Von Karman constant} \approx 0.4$$

and the Deissler multiple-part approach proposes that

$$\epsilon = n^2uy[1 - \exp(-n^2uy/v)], \qquad \epsilon/v < \gamma$$

$$\epsilon = K^2\frac{(du/dy)^3}{(d^2u/dy^2)^2}\left\{1 - \exp\left[-\frac{(K^2/v)(du/dy)^3}{(d^2u/dy^2)^2}\right]\right\}, \qquad \epsilon/\nu \geqslant \gamma \tag{3-39}$$

$$n = \text{empirical constant} = 0.109$$

Equations (3-38) and (3-39) were developed primarily for constant-density analyses. They have been used successfully for variable-property gases. The near-critical research has required modification of these basic forms. A few are discussed here.

Hess and Kunz[68] found that for near-critical hydrogen, the damping factor of Eq. (3-38) had to include a term dependent on kinematic viscosity. Their work in this area led to the correlating equation most frequently used for near-critical hydrogen, Eq. (3-19).

Hsu and Smith[81] investigated a technique for transforming the variable-density eddy diffusivities to the constant-density forms of Eqs. (3-38) and (3-39). The eddy diffusivity expressions can be written as

$$\epsilon_m = (1 + F_m)\epsilon, \qquad \epsilon_H = (1 + F_H)\epsilon \tag{3-40}$$

where ϵ is given by Eq. (3-39), and

$$\begin{aligned} F_m &= \frac{d(\ln \rho)/dy^+}{d(\ln u^+)/dy^+} \\ F_H &= \frac{d(\ln \rho)/dy^+}{d(\ln C_pT^+)/dy^+} \end{aligned} \tag{3-41}$$

Note that Eq. (3-40) reduces properly to the constant-density form, and that either form of ϵ, Eq. (3-38) or (3-37), may be used. Hsu and Smith[81] performed a first-order force balance on a differential volume to derive an equation relating shear, buoyancy, and pressure drop. The success of their equation in correlating the data of Bringer and Smith[62] indicates that such a modification is warranted, and is another step in the proper direction. The main problem here seems to be that while each researcher has success correlating his own data, subsequent researchers have trouble with different data.

Each technique has a great deal of merit and correlates much of the data presented in the respective papers. Using near-critical hydrogen, NASA and also Szetela of United Aircraft have tried the techniques of Deissler[89,90] and Goldman[91] with very limited success. The technique of Hess and Kunz[68] was tried and success was limited, in the main, to those data for bulk temperatures above the T^* regime. The integral average property approach of Goldman[91] gave similar limited results. The modifications of Hsu and Smith[81] were also tried, again with only limited success. Modifications of the modifications such as using different dampening factors, including upstream effects, etc. were also attempted; however, the results were very disappointing.

3.3.6.2 *Acceleration—Straining rates:* Deissler[94] found the straining rate (acceleration or deceleration) to be important in turbulent heat transfer. For example, the strain effects of longitudinal acceleration on homogeneous turbulence intensities are shown in Fig. 3-18. The ordinate is the square of the turbulence intensity of the transverse velocity component normalized with respect to an initial intensity condition, $[(\bar{v}')^2/u^2]/[(\bar{v}')^2_{u0}/u_0{}^2]$. It can be shown that this dimensionless parameter is related to the square of a Stanton number ratio $\mathrm{St}^2/\mathrm{St}_0{}^2$, where the denominator is the initial Stanton number before acceleration:

$$\left(\frac{\mathrm{St}}{\mathrm{St}_0}\right)^2 = \frac{(\bar{v}')^2/u^2}{(\bar{v}')^2_{u0}/u_0{}^2} = f\left[\frac{du/dx}{(du/dx)_0}\right] \tag{3-42}$$

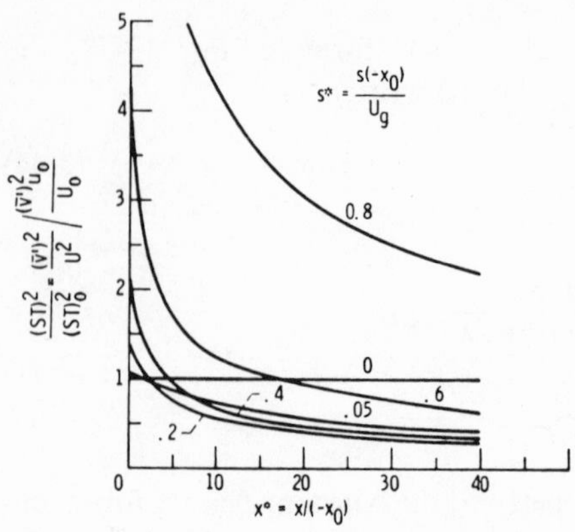

FIG. 3–18. Effect of acceleration or longitudinal strain on heat transfer[84] for positive strain rates.

The abscissa of Fig. 3-18 is axial position; the parameter S^* is a strain rate.

The figure shows that the local intensity or Stanton number progressively decreases when subject to a strain rate.

Under heating conditions, a near-critical fluid would undergo large longitudinal accelerations while expanding in a constant-area pipe. Experiments by Weiland[95] and Taylor[96] with gases at high heating rates suggest that such expansion effects occur. Deissler[94] showed that the data trends in Refs. 95 and 96 could be explained by straining or acceleration effects. Perhaps some of the unusual distributions of heat transfer coefficient noted for near-critical hydrogen relate to strain effects.

The near-critical region just discussed (region IV, Fig. 3-1) is bounded by the two-phase region, the gaseous region, and the liquid region. The heat transfer for the two-phase region is discussed in part II of this book. Recommended correlations for the gaseous and liquid regions are presented here.

The data of McCarthy and Wolf[97,98] and Taylor[99] for hydrogen and helium give extensive heat transfer and pressure drop coverage of region I. The recommended equation is that of Taylor, which includes entrance effects and reduces to the successful equation of McCarthy and Wolf[97,98] at $X/D > 25$:

$$\mathrm{Nu} = \mathrm{Nu}_b(T_w/T_b)^{-0.57-(1.59/X/D)} \tag{3-43}$$

where

$$\mathrm{Nu}_b = 0.023\mathrm{Re}_b^{0.8}\mathrm{Pr}_b^{0.4} \tag{3-44}$$

Taylor has tested his correlation with low-temperature gases and found the agreement to be quite satisfactory.

Taylor[100] and McCarthy and Wolf[98] also determined the friction factor for the data of region I. The equation of Taylor is recommended. It approaches the classic equation of Koo–Drew and McAdams at low T_w/T_b. Taylor's equation is

$$f/2 = 0.0007 + (0.0625/\mathrm{Re}_w^{0.32})(T_b/T_w)^{1/2} \tag{3-45}$$

McCarthy *et al.*[97] also investigated the influence of surface roughness of heat transfer and found up to a 20% increase in heat transfer coefficient depending on the ratio of roughness to tube diameter.

The liquid regime (region II, Fig. 3-1) has not been studied with cryogenic fluids. As is the case for all the fluid regimes, cryogenic fluid temperatures themselves do not make the heat transfer unique. Thus it is presumed that the cryogenic liquid region is quite similar to that of any fluid—water, for instance—at the same wall to bulk temperature ratio. At low and moderate heat fluxes, the Nusselt correlation for liquids is based on bulk properties. It is presumed that this approach is applicable to cryogenic liquids until proven otherwise. With cryogenic fluids it is possible to have large temperature ratios which can influence the experimental results significantly. The reader is cautioned about the use of liquid correlations for these cases.

3.4 NOMENCLATURE

A = area
A_p = average surface area
A_T = total surface area
C_p = specific heat at constant pressure
$\bar{C}_p$ = integrated specific heat
C_v = specific heat at constant volume
D = self-diffusion coefficient
D_{ln} = binary diffusion coefficient
d = diameter
f = friction factor
$f(\,)$ = function of
Gr = dimensionless Grashof number
H = enthalpy
h = heat transfer
h_c = average heat transfer coefficient in curved duct
h_{cv} = heat transfer coefficient on concave surface
h_{cx} = heat transfer coefficient on convex surface
k = thermal conductivity
k_0 = atmospheric value of thermal conductivity
L = length
L = total length of natural circulation loop
l_T = length of loop test section
Nu = dimensionless Nusselt number
P = pressure
P_c = thermodynamic critical pressure

Pr = dimensionless Prandtl number
q = heat flux
R = gas constant
Ra = dimensionless Rayleigh number
Re = dimensionless Reynolds number
St = dimensionless Stanton number
T = temperature
T_c^* = thermodynamic critical temperature
T^* = transposed critical temperature
T_{sat} = saturation temperature
u = velocity
u^* = shear velocity
u^+ = dimensionless velocity u/u^*
x = quality
y = half-length of loop test section
y^+ = dimensionless distance from wall, $y(\tau_0/\rho)^{1/2}/v$

Greek Letters

β = coefficient of thermal expansion
α = hydraulic diameter and slip parameter
ϵ = eddy diffusivity
μ = absolute viscosity
v = kinematic viscosity
v_0 = atmospheric value of kinematic viscosity
ρ = density
ρ_{melt} = density of liquid in equilibrium with solid at corresponding pressure
τ = shear stress
φ = any fluid property

Subscripts

av = average value
b = evaluated at bulk temperature
c = thermodynamic critical value
calc = calculated value
cv = concave surface
cx = convex surface
enh = enhancement value
exp = experimental value

f = fluid
f = evaluated at film temperature
g = gas
l = liquid
m = mean or average value
pg = perfect gas value
st = straight pipe
test = prescribed value during experimental test
tp = two-phase
v = vapor
w = evaluated at the wall
x = evaluated at position x
x = fluid property evaluated at reference temperature $T_x = T_b + x(T_w - T_b)$
0 = evaluated at the wall

3.5 REFERENCES

1. O. Hirschfelder, F. Curtiss, and R. Bird, *Molecular Theory of Gases and Liquids*, John Wiley and Sons, New York (1954).
2. J. S. Rowlinson, *Liquids and Liquid Mixtures*, Butterworth Scientific Publication (1959).
3. B. Widom, *Science*, **157** (3787), 375 (1967).
4. O. Maass, *Chem. Rev.* **23**(1), 17 (1938).
5. A. Michels, B. Blaisse, and C. Michels, *Proc. Roy. Soc.* (*London*), **160A**(902), 358 (1937).
6. R. D. Goodwin, D. E. Diller, H. M. Roder, and L. A. Weber, *J. Res. NBS, A. Phys. and Chem.*, **67**(2), 173 (1963).
7. E. F. Obert, *Concepts of Thermodynamics*, McGraw-Hill Book Co., New York (1960).
8. M. Benedict, G. B. Webb, and L. C. Rubin, *J. Chem. Phys.*, **8**(4), 334 (1940).
9. T. R. Strobridge, NBS Tech. Note 129 (January 1962).
10. H. M. Roder and R. D. Goodwin, NBS Tech. Note 130 (December 1961).
11. B. Griffiths, *J. Chem. Phys.* **43**(6), 1958 (1965).
12. B. Griffiths, *Phys. Rev.*, **158**(1), 176 (1967).
13. M. S. Green, M. Vicentini-Missoni, and J. M. H. Levelt Sengers, **18**(25), 1113 (1967).
14. J. M. H. Levelt Sengers and M. Vicentini-Missoni, *Proceedings 4th Symposium on Thermophysical Properties*, ASME (1968), 79.
15. M. Vicentini-Missoni, J. M. H. Levelt Sengers, and M. S. Green, *Phys. Rev. Lett.*, **22**(9), 389 (1969).
16. J. V. Sengers, Ph.D. Dissertation, Univ. of Amsterdam, The Netherlands (1962).
17. A. Guildner, *J. Res. NBS, A. Phys. and Chem.*, **66**(4) 341 (1962).
18. D. E. Diller and H. M. Roder, *Advances in Cryogenic Engineering*, Vol. 15 Plenum Press, New York (1970), p. 58.

19. I. Stiel and G. Thodos, *Progress in International Research on Thermodynamic and Transport Properties*, ASME (1962), 352.
20. J. V. Sengers, *Recent Advances in Engineering Science*, Vol. 3 (A. C. Eringen, ed.), Gordon and Breach Science Publ. (1969), p. 153.
21. S. Brokaw, paper presented at International Conference on Properties of Steam, Tokyo, Japan, Sept. 9–13, 1968.
22. R. S. Thurston, Ph.D. Dissertation, Univ. of New Mexico, Albuquerque, New Mexico (1966).
23. R. C. Hendricks, R. W. Graham, Y. Y. Hsu, and R. Friedman, NASA TN D-765 (May 1961).
24. E. Schmidt, E. R. G. Eckert, and U. Grigull, Trans. No. F-TS-527-RE, Air Material Command, Wright-Patterson AFB, Ohio (April 26, 1946).
25. B. Powell, *Jet Propulsion*, **27**(7), 776 (1957).
26. R. C. Hendricks, R. W. Graham, Y. Y. Hsu, and A. A. Medeiros, *ARS J.*, **32**(2), 244 (1962).
27. W. S. Miller, J. D. Seader, and D. M. Trebes, *Bull. Inst. Intern. Froid, Annexe No. 2*, **1965**, 173.
28. M. E. Shitsman, *High Temp.*, **1**(2), 237 (1963).
29. K. Yamagata, K. Nishikawa, S. Hasegawa, and T. Fugii, paper presented at Semi-International Symposium, Sept. 1967, Japan Society of Mechanical Engineers.
30. G. Domin, *Brennstoff-Wärme-Kraft*, **15**(11), 527 (1963).
31. L. B. Koppel and J. M. Smith, *Intern. Developments in Heat Transfer, ASME* **1963**, 585.
32. B. S. Petukhov, E. A. Krasnoschekov, and V. S. Protopopov, *Intern. Developments in Heat Transfer, ASME*, **1963**, 569.
33. H. S. Swenson, J. R. Carver, and C. R. Kakarala, *Trans. ASME, J. Heat Transfer*, **87**(4), 477 (1965).
34. Y. Y. Hsu, *Intern. Developments in Heat Transfer, ASME*, **1963**, D-188.
35. E. G. Hauptmann, Ph.D. Dissertation, California Institute of Technology, Pasadena, California (1966).
36. M. A. Styrikovich, M. E. Shitsman, and Z. L. Miropolskii, *Teploenergetika*, **3**,32 (1956).
37. M. A. Styrikovich, Z. L. Miropolskii, and M. E. Shitsman, *Mitt. Ver. Grosskessel-besitzer*, **61**, 288 (1959).
38. D. Wood, Ph.D. Dissertation, Northwestern Univ., Evanston, Illinois (1963).
39. A. Kahn, Ph.D. Dissertation, Univ. of Manchester, England (1965).
40. B. S. Shiralkar and P. Griffith, Paper 68-HT-39, ASME (August 1968).
41. J. R. McCarthy, D. M. Trebes, and J. D. Seader, Paper 67-HT-59, ASME (August 1967).
42. W. S. Hines, and H. Wolf, *ARS J.*, **32**(3), 361 (1962).
43. R. C. Hendricks, R. W. Graham, Y. Y. Hsu, and R. Friedman, NASA TN D-3095 (1966).
44. K. Goldmann, Rept. NDA-2-31, Nuclear Development Corp. of America (1956).
45. R. W. Graham, R. C. Hendricks, and R. C. Ehlers, NASA TN D-1883 (1964).
46. J. D. Griffith and R. H. Sabersky, *ARS J.*, **30**(3), 289 (1960).
47. K. Knapp and R. H. Sabersky, *Intern. J. Heat Mass Transfer*, **9**(1), 41 (1966).
48. K. Nishikawa and K. Miyabe, *Mech. Fac. Eng. Kyusa Univ.* **25**(1) 1 (1965).
49. M. Cumo, G. E. Farello, and G. Ferrari, paper presented at the 11th National Heat Transfer Conference, Minneapolis, Minnesota, August 3, 1969.
50. V. E. Holt and R. J. Grosh, *Nucleonics*, **21**(8), 122 (1963).

51. V. P. Skripov and P. I. Potashev, NASA TT F-11333 (1967).
52. R. J. Goldstein and W. Aung, paper 67-WA/HT-2, ASME (November 1967).
53. U. Grigull and E. Abadzic, presented at Symposium on Heat Transfer and Fluid Dynamics of Near Critical Fluids, Inst. Mech. Eng., Bristol, England (March 1968).
54. S. Hasegawa and K. Yoshioka, in *Proceedings of 3rd Intern. Heat Transfer Conference*, Vol. 2, AIChE (1966), p. 214.
55. J. D. Parker and T. E. Mullin, presented at Symposium on Heat Transfer and Fluid Dynamics of Near Critical Fluids, Inst. Mech. Eng., Bristol, England (March 1968).
56. E. M. Sparrow and J. L. Gregg, *Trans. ASME*, **80**(4), 879 (1958).
57. J. R. Larson and R. J. Schoenhals, *Trans. ASME J. Heat Transfer*, **88**(4), 407 (1966).
58. H. A. Simon and E. R. G. Eckert, *Intern. J. Heat Mass Transfer*, **6**(8), 681 (1963).
59. J. P. Holman and J. H. Boggs, *Trans. ASME, J. Heat Transfer*, **82**(3), 221 (1960).
60. G. E. Tanger, J. H. Lytle, and R. I. Vachon, *Trans. ASME, J. Heat Transfer*, **90**(1), 37 (1968).
61. E. R. G. Eckert, *Trans. ASME*, **76**(1), 83 (1954).
62. R. P. Bringer and J. M. Smith, *AIChE J.*, **3**(1), 49 (1957).
63. N. M. Schnurr, *Trans. ASME, J. Heat Transfer*, **91**(1), 16 (1969).
64. Z. L. Miropolskii, V. J. Picus, and M. E. Shitsman, in *Proceedings of 3rd International Heat Transfer Conference*, Vol. 2, AIChE (1966), p. 95.
65. N. L. Dickinson and C. P. Welch, *Trans. ASME*, **80**(3), 746 (1958).
66. D. Finn, Ph.D. Dissertation, Univ. of Oklahoma, Norman, Oklahoma (1964).
67. R. S. Brokaw, NACA RM E57K19a (1958).
68. H. L. Hess and H. R. Kunz, *Trans. ASME, J. Heat Transfer*, **87**(1) 41 (1965).
69. R. W. Graham, NASA TN D-5522 (1969).
70. M. Wilson, Ph.D. Dissertation Univ. of New Mexico, Albuquerque, New Mexico (1969).
71. L. E. Gill, G. F. Hewitt, and P. M. C. Lacey, Rept. AERE-R-3955, United Kingdom Atomic Energy Authority (1963).
72. P. J. Bourke, D. J. Pulling, L. E. Gill, and W. H. Denton, paper presented at Symposium on Heat Transfer and Fluid Dynamics of Near Critical Fluids, Inst. Mech. Eng., Bristol, England (March 1968)
73. H. Itō, *J. Basic Eng.*, **81**(2), 123 (1959).
74. W. R. Dean, (Ser. VII), **5**(30), 674 (1928).
75. R. C. Hendricks and F. F. Simon, in *Multi-Phase Flow Symposium* (N. J. Lipstein, ed.), ASME (1963), p. 90.
76. "Heat Transfer to Cryogenic Hydrogen Flowing Turbulently in Straight and Curved Tubes at High Heat Fluxes," NASA CR-678 (1967).
77. J. R. McCarthy, *et al.*, Rept. 6529, Rocketdyne Div., North American Aviation, NASA CR-78634 (September 15, 1966).
78. F. Taylor, *J. Spacecraft Rockets*, **5**(11), 1353 (1968).
79. J. R. Bartlit and K. D. Williamson, Jr., *Advances in Cryogenic Engineering*, Vol. 11, Plenum Press, New York (1966), p. 561.
80. E. Michaud and C. P. Welch, short communication at the Seminar on Near Critical Fluids, 1968 Cryogenic Engineering Conference, Case Western Reserve Univ., Cleveland, Ohio.
81. Y. Y. Hsu and J. M. Smith, *Trans. ASME, J. Heat Transfer*, **83**(2), 176 (1961).
82. S. S. Papell and D. D. Brown, paper presented at the 11th National Heat Transfer Conference, Minneapolis, Minnesota (August 3, 1969).

83. R. J. Simoneau and J. C. Williams, III, *Intern. J. Heat Mass Transfer*, **12**, 120 (1969).
84. E. M. Sparrow and J. L. Gregg, *Trans. ASME*, **80**, 879 (1959).
85. C. A. Fritsch and R. J. Grosh, in *Proceedings of 1961 International Heat Transfer Conference*, ASME (August 1961), Vol. 5, paper 121.
86. C. A. Fritsch and R. J. Grosh, *Trans. ASME, Heat Transfer*, **85**, 289 (1963).
87. K. Brodowicz and J. Bealokoz, *Archuvium Budowy Masyn* **X**, 4 (1963).
88. S. Hasegawa and K. Yoshioka, in *Proceedings of 1966 Intern. Heat Transfer Conference*, ASME (August 1966), Vol. III, No. 2.
89. R. G. Deissler, *Trans. ASME*, **76**(1), 73 (1954).
90. R. G. Deissler, NACA Rept. 1210 (1955).
91. K. Goldmann, *Chem. Eng. Progr. Symp. Ser.*, **50**(11), 105 (1954).
92. F. W. Dittus and L. M. K. Boelter, *Univ. of California Publications in Engineering*, Vol. 2, UCLA Press, Los Angeles, California (1930), p. 443.
93. E. R. Van Driest, Heat Transfer and Fluid Mechanics Institute, University of California, Los Angeles, (1955), paper XII.
94. R. G. Deissler, NASA TN D-2800 (1965).
95. F. Weiland, Jr., *Chem. Eng. Progr. Symp. Ser.*, **61**(60), 97 (1965).
96. M. F. Taylor, NASA TN D-2280 (1964).
97. J. R. McCarthy and H. Wolf, Rept. RR 60-12 (NP-10572), Rocketdyne Division, North American Aviation (December 1960).
98. J. R. McCarthy and H. Wolf, *ARS J.*, **30**(4), 423 (1960).
99. M. F. Taylor, NASA TN D-4332 (1968).
100. M. F. Taylor, *Intern. J. Heat Mass Transfer*, **10**(8), 1123 (1967).

PART II

Two-Phase Phenomena

TERMINOLOGY AND PHYSICAL DESCRIPTION OF TWO-PHASE FLOW 4

W. FROST and W. L. HARPER

4.1 INTRODUCTION

This chapter defines the common terminology and generally accepted physical description of two-phase flow and heat transfer phenomena. The various kinds and regimes of boiling are described first, followed by the various flow patterns and definitions associated with the flow of a saturated two-phase mixture.

4.2 KINDS OF BOILING

Boiling is the formation of vapor due to conversion of the liquid phase to the vapor phase by heat transfer or pressure changes. There are various kinds of boiling encountered in practice; the most common are nucleate boiling and film boiling. These are classified as natural convection, also called pool boiling, or forced convection boiling, and these are further subdivided depending on whether the boiling occurs under saturated or subcooled conditions.

W. FROST and W. L. HARPER The University of Tennessee Space Institute, Tullahoma, Tennessee.

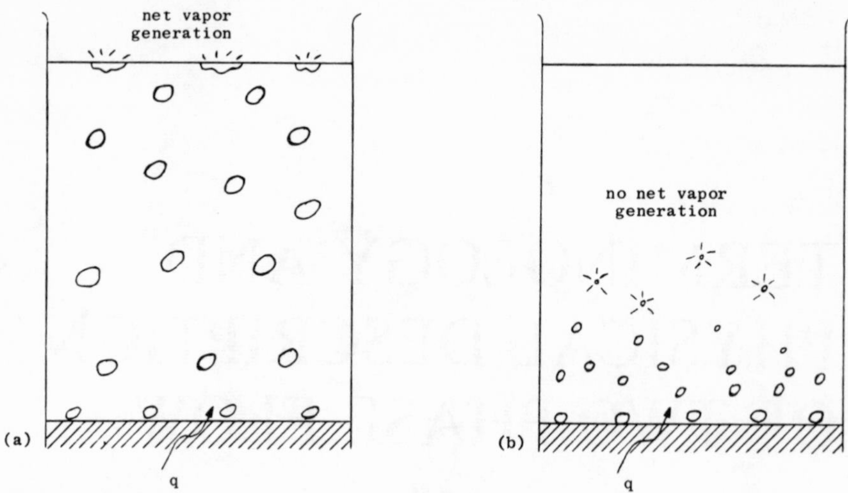

FIG. 4-1. Pool boiling: (a) Saturated (net generation of vapor); (b) subcooled, local, or surface boiling.

Nucleate boiling refers to vapor formation as bubbles at specific nuclei on a heated surface submerged in liquid. If the boiling takes place in a large pool or container where all fluid motion is induced by natural and bubble-induced convection, the process is called natural convection or pool boiling. Most often with pool boiling the bulk temperature of the liquid is equal to or slightly above the saturation temperature corresponding to the system pressure and the boiling is classified as saturated pool boiling. Bubbles formed at the heated surface grow as they rise through this slightly heated bulk fluid and escape at the free liquid surface (Fig. 4-1a). Thus there is net generation of vapor and often the term pool boiling with net vapor generation is used instead of saturated boiling.

If means for maintaining the bulk temperature of the pool below the

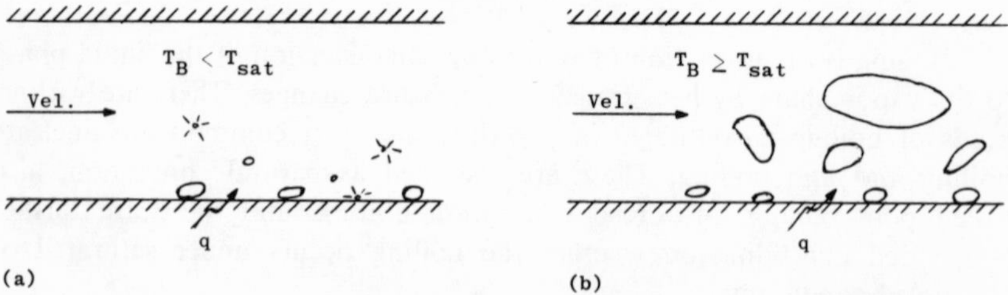

FIG. 4-2. Forced convection boiling: (a) Subcooled, local or surface boiling (no net vapor generation); (b) saturated or bulk boiling with net vapor generation and many vapor–liquid flow patterns being possible.

saturated value without introducing forced convection are devised, the system undergoes subcooled pool boiling. The degree of subcooling is defined as the difference between the saturation temperature and the bulk fluid temperature (i.e., $\Delta T_{sub} = T_{sat} - T_{bulk}$). In this form of boiling bubbles grow in a layer of superheated liquid next to the heated surface but quickly collapse near or on the surface upon expanding into the subcooled surroundings; hence subcooled boiling is also referred to as local or surface boiling (Fig. 4-1b).

Forced convection nucleate boiling occurs when fluid is forced to flow past a highly heated surface and bubbles again form at specific nuclei on the hot surface. Subcooled boiling is easily achieved under these conditions since the bulk fluid is continuously replaced by low-temperature liquid as depicted in Fig. 4-2a. The designations of subcooled nucleate forced convection, surface, or local boiling are assigned interchangeably to this type of boiling.

Saturated forced convection boiling or bulk boiling results when the bulk liquid temperature of the flowing fluid equals or slightly exceeds the saturation value (Fig. 4-2b). Many different liquid–vapor flow patterns occur with different associated mechanisms of heat transfer as described later.

A second kind of boiling encountered is film boiling which occurs both in natural flow and forced flow systems. This mechanism of boiling is

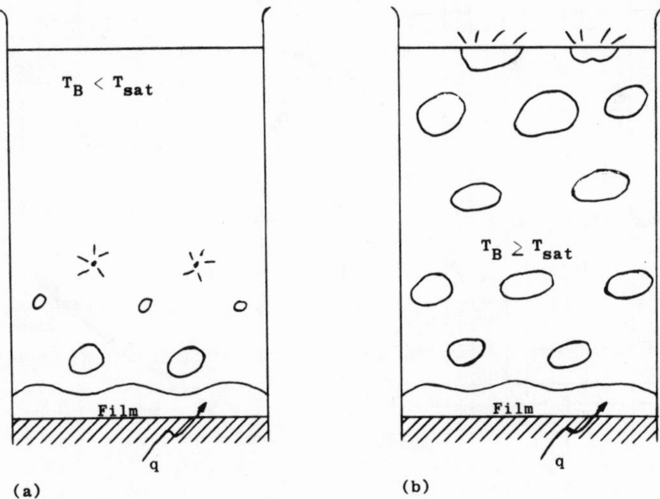

FIG. 4-3. Film boiling. A film of vapor covers the bottom surface. (a) Subcooled film boiling (no net generation of vapor); (b) saturated film boiling (net vapor generation).

characterized by a film of vapor which separates the liquid from contact with the heated surface. Energy is transferred through the highly thermal resistant vapor film to the liquid. Large masses of vapor detach from the wavy liquid–vapor interface and either rise in a saturated pool or collapse in a subcooled pool (Fig. 4-3). With forced flow in ducts the liquid is channeled through a central core separated from the walls by an annular layer of vapor.

4.3 REGIMES OF BOILING

The various kinds of boiling occur at different regimes on a plot of heat flux versus the temperature difference between the heated wall and the saturated fluid (i.e., $\Delta T_{sat} = T_w - T_{sat}$). Consider, for example, Fig. 4-4, which is a typical plot of heat flux versus ΔT_{sat} for pool boiling of nitrogen. The extreme left part of the curve represents a regime of convective heat transfer due to circulation of superheated liquid rising to the free liquid surface where evaporation takes place. Heat transfer in this regime is predicted by the methods described in Chapter 3. An increase in wall temperature results in the formation of vapor bubbles at a small number of specific sites on the surface. These bubbles condense before reaching the liquid surface, as in subcooled boiling. Concomitant with the first appearance of

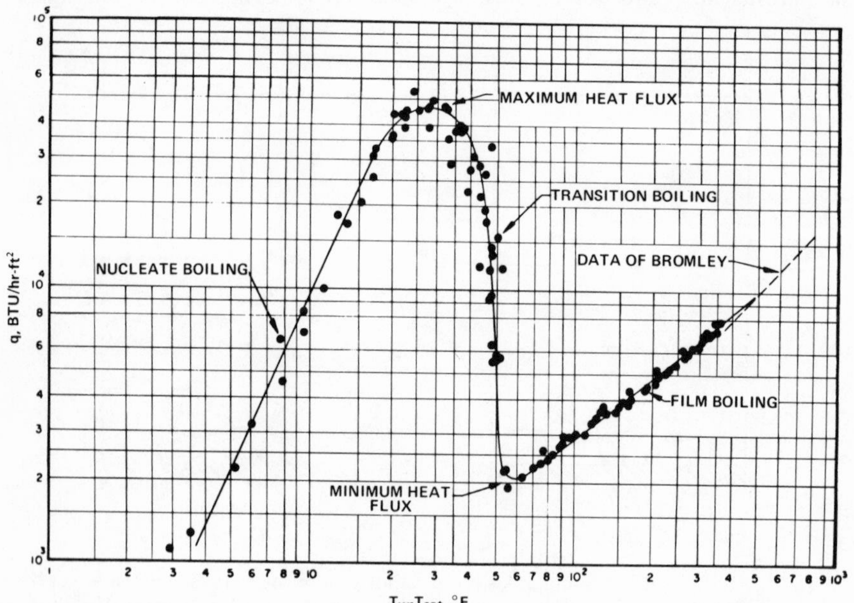

FIG. 4-4. Characteristic boiling curve for liquid nitrogen.[12]

bubbles, the curve of heat flux begins to depart from the gradual slope encountered in the natural convection regime, indicating the inception of boiling. Further increase in wall temperature results in many more bubbles which rise to the free interface, transporting vapor to the surroundings. Conditions of fully developed saturation boiling are thus quickly reached and the heat flux is observed to increase rapidly with relatively little increase in wall temperature. The marked increase in heat transfer rate is associated with bubble growth and agitation and is discussed in Chapter 5.

At a certain value of heat flux a maximum is reached which is associated with a transition from nucleate boiling to film boiling. The transition occurs in two ways, depending upon the method of heating. With a constant heat flux, such as electrical resistance heating by electronic components or nuclear fission heating by reactor cores, the transition to film boiling is almost instantaneous. The temperature difference jumps immediately to a large ΔT_{sat} which is required to maintain the constant rate of heat transfer across the suddenly imposed, high thermal resistance vapor film. Figure 4-4 shows that this temperature difference is on the order of 1000–2000°F. For cryogens with low saturation temperatures the heated surface will in general not melt from the sudden temperature rise; however, with a temperature-sensitive electronic component, internal physical destruction may occur. Thus the maximum heat flux is to be avoided in most instances.

A smooth transition from nucleate to film boiling occurs if the heat flux is generated by condensing vapors or by convection from high-temperature fluids. Since this process is self-regulating, portions of the heated surface will undergo film boiling while others undergo nucleate boiling. The areas of film and nucleate boiling are unstable and with increasing wall temperature the surface covered with a film becomes progressively larger until fully developed film boiling is reached at the minimum of the curve. In this transitional regime of partial film boiling the heat flux decreases with increasing wall temperature. Little study of the transitional regime has been made and correlations for describing it are unknown. No discussion of the regime is given in subsequent chapters and the only recommended approach to prediction of transitional boiling is recourse to a basic heat flux curve such as Fig. 4-4. In the absence of data, fairing in a curve between the maximum and minimum points of the curve which can be predicted by the methods of Chapters 6 and 8, respectively, is suggested.

Figure 4-5 depicts the mechanism by which the transition from nucleate to film boiling occurs for the different conditions of heat flux. The former is referred to as a heat-flux-controlled transition and the latter as a temperature-controlled transition.

The heat flux at the maximum point has been given a multitude of designations, including critical heat flux, peak heat flux, burnout heat flux,

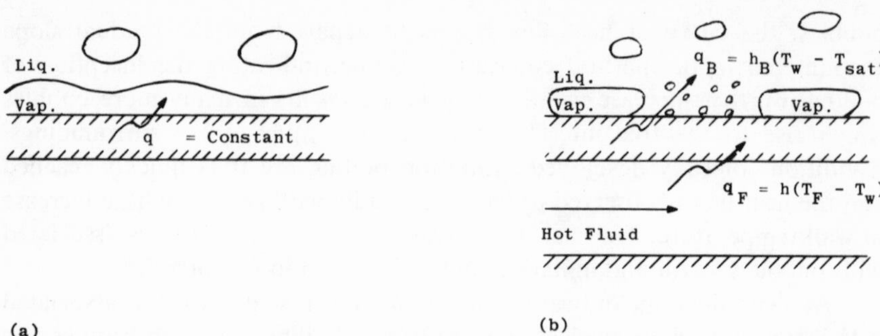

FIG. 4-5. Mechanism of burnout: (a) Heat-flux-controlled surface: If q is constant, when the excess vapor forms, h_B drops and T_w becomes extremely high in order to maintain the heat flux; $q = h_B(T_w - T_{sat})$. (b) Temperature-controlled surface: as the vapor film forms, T_w goes up, therefore q_F drops; hence the process is self-regulating.

maximum heat flux, first boiling crisis (common in Russian literature), and departure from nucleate boiling heat flux. Chapter 6 deals with the mechanism and methods of predicting the maximum heat flux where these terms are used interchangeably. The minimum of the curve is also referred to with a variety of designations, including minimum film boiling heat flux, second boiling crisis, and spheroidal heat flux. Chapter 8 describes the minimum film boiling heat flux.

The remaining regime of the boiling curve is the fully developed film boiling curve. Here the heat flux increases less rapidly with increasing ΔT_{sat}, indicating a much less efficient mechanism of heat transfer from that of nucleate boiling. For this reason, practical applications of film boiling are not frequently encountered with common fluids like water; however, film boiling can often occur with cryogenic fluids because of their low saturation temperatures. Chapter 7 provides methods of correlating film boiling heat flux.

It should be noted that the pool boiling curve exhibits a hysteresis effect for a heat-flux-controlled surface as indicated by the dashed lines in Fig. 4-7. The return from film to nucleate boiling does not follow the same path as the transition from nucleate to film boiling, but returns abruptly from the minimum point along the dashed arrow shown in the figure.

4.4 FORCED CONVECTION BOILING

The forced convection boiling curve is similar to that for pool boiling; however, the effects of velocity, subcooling, and different vapor–liquid flow patterns make the various regimes somewhat more difficult to define.

4.4.1 Subcooled

Experimentally observed trends for subcooled forced convection boiling where the exit quality of the fluid is very low are shown in Fig. 4-6. The initial portion of the curve represents pure forced convection where heat transfer can be predicted from the correlations presented in Chapter 3. The steep rise in the curve represents the onset of nucleate boiling. Inspection of the curves illustrates that nucleate boiling is relatively insensitive to the degree of subcooling or the fluid velocity, although the inception of boiling does vary with bulk temperature. When plotted against ΔT_{sat}, however, the data for forced convection boiling cluster around one curve, indicating that

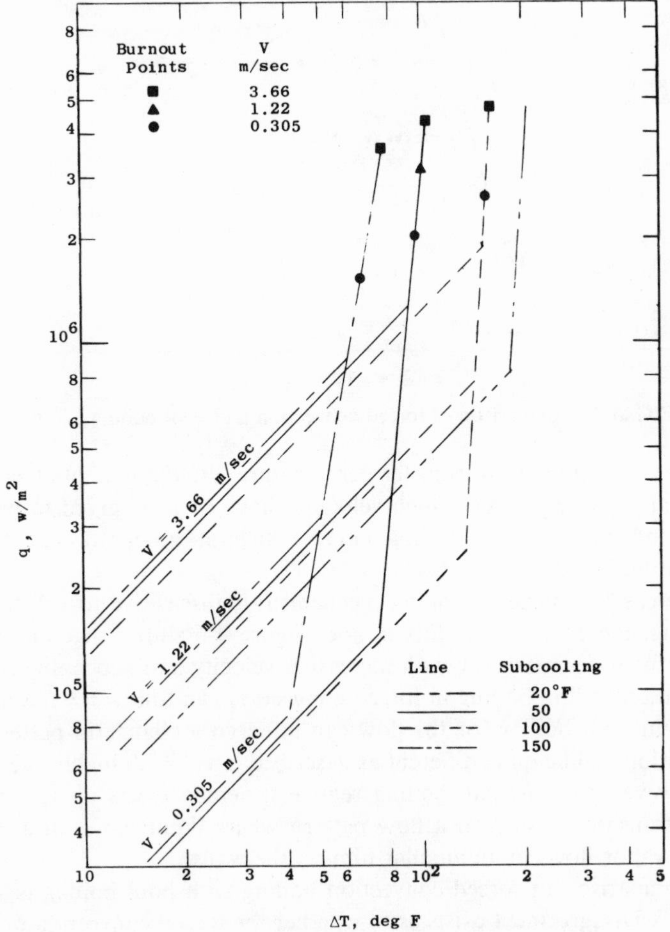

FIG. 4-6. Characteristic subcooled, forced convection boiling curves.

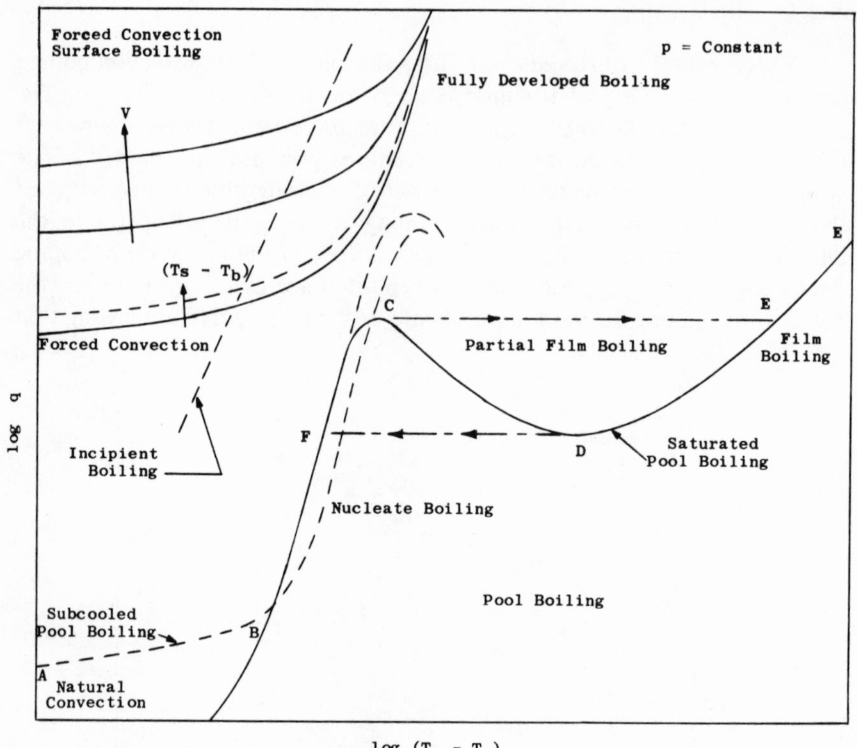

FIG. 4-7. Comparison of forced convection and pool boiling curves.

the nucleate boiling heat transfer temperature driving potential is ΔT_{sat} rather than $T_w - T_{bulk}$. Very high velocities have been observed to produce a small influence on the boiling curve which at moderate velocities is imperceptible.[1]

Whereas the nucleate boiling mechanism is insensitive to velocity and subcooling, the critical heat flux is not. Figure 4-6 shows that the critical heat flux increases markedly with increasing velocity and subcooling. Either the transition to film boiling in forced convection can follow the mechanism observed in pool boiling, or the flow can undergo a change in pattern and the transition will be quite different as described later. With highly subcooled flow, however, the nucleate boiling regime generally exists up to burnout and the transition results in a flow pattern where liquid flows in a central core and vapor flows in an annular film on the walls.

A comparison of forced convection boiling with pool boiling is shown in Fig. 4-7. Disagreement exists as to whether the forced convection nucleate boiling curve is an extension of the pool boiling curve.

4.4.2 Saturated

Figure 4-8 illustrates the typical nature of boiling heat transfer under saturated conditions. Consider subcooled fluid entering a duct with a constant heat flux on the wall. Initially the bulk temperature will be below the saturation value and single-phase forced convection will take place. The fluid is continuously heated as it flows upward and eventually subcooled boiling begins when the liquid near the wall becomes sufficiently superheated to form bubbles. Further up the duct the bulk fluid becomes saturated and vigorous boiling takes place with many bubbles leaving the wall and moving with the flow. Due to coalescence, the bubbly flow pattern soon changes to one of bullet-shaped vapor slugs surrounded with saturated liquid. Nucleate

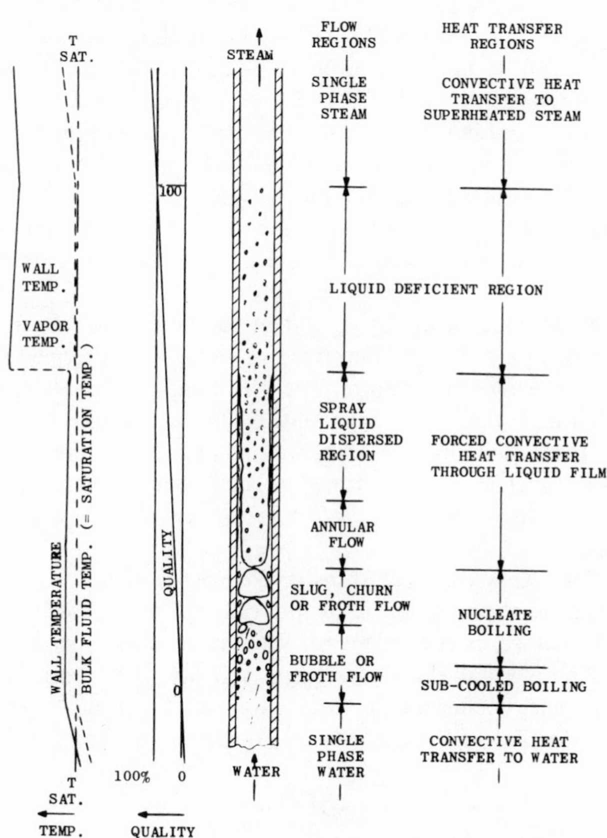

FIG. 4-8. Characteristic forced convection saturated boiling flow patterns and regimes.[7]

boiling still occurs at the wall and bubbles continue to feed the vapor slugs until the vapor flow rate becomes so high that all liquid is forced to an annular film on the wall and vapor with entrained liquid droplets flows in the central core. Bubbles continue to grow in the liquid film until the increasing vapor quality results in vapor velocities which suppress nucleation, forcing a new mechanism of heat transfer. Heat transfer is now due to convection or conduction through the thin liquid film with evaporation at the liquid–vapor interface. This mode of heat transfer also produces a high heat transfer coefficient and ΔT_{sat} remains relatively constant.

The liquid film on the wall diminishes under the effects of evaporation and entrainment until the film breaks up into rivulets with associated dry spots. The heat transfer coefficient drops drastically and the wall experiences a sharp rise in temperature. The tube may overheat and possibly be damaged extensively, causing "burnout." These conditions of heat transfer and fluid flow are referred to as dry-out, net quality burnout, or critical or peak heat flux conditions. The term dry-out is probably preferable since this mechanism of burnout has been well established by experiment.[2] It is quite different from that experienced in subcooled nucleate boiling where the mechanism of burnout is believed to be a hydrodynamic instability phenomenon (see Chapter 6). The important distinction between the two is that subcooled burnout is very rapid and often occurs with destructive temperature rise, whereas net quality burnout occurs slower and often produces tolerable wall temperature.

With the destruction or drying out of the liquid film the flow pattern becomes one of liquid droplets entrained in vapor and is termed either mist flow or liquid-deficient flow region. The wall temperature, after the sudden excursion to a high value due to dry-out, begins to decrease slightly. This is a result of the liquid droplets evaporating and producing higher vapor velocities with increasing heat transfer coefficients. Finally all liquid is evaporated and single-phase convective heat transfer in superheated vapor takes place.

The point where the heat transfer changes from nucleate boiling to forced convection vaporization is of interest. The latter mechanism, being convective in nature, is dependent on velocity or mass flow rate, whereas nucleate boiling is relatively insensitive to velocity. If the ratio of the two-phase heat transfer coefficient h_{TP} to that for only liquid flowing at the same rate in the same size duct h_{LO} is plotted *vs.* the parameter

$$1/\chi_{tt} = [x/(1 - x)]^{0.9}(\rho_l/\rho_v)^{0.5}(\mu_l/\mu_v)^{0.1} \tag{4-1}$$

the transition point is determined where the curve becomes sensitive to flow rate (Fig. 4-9).

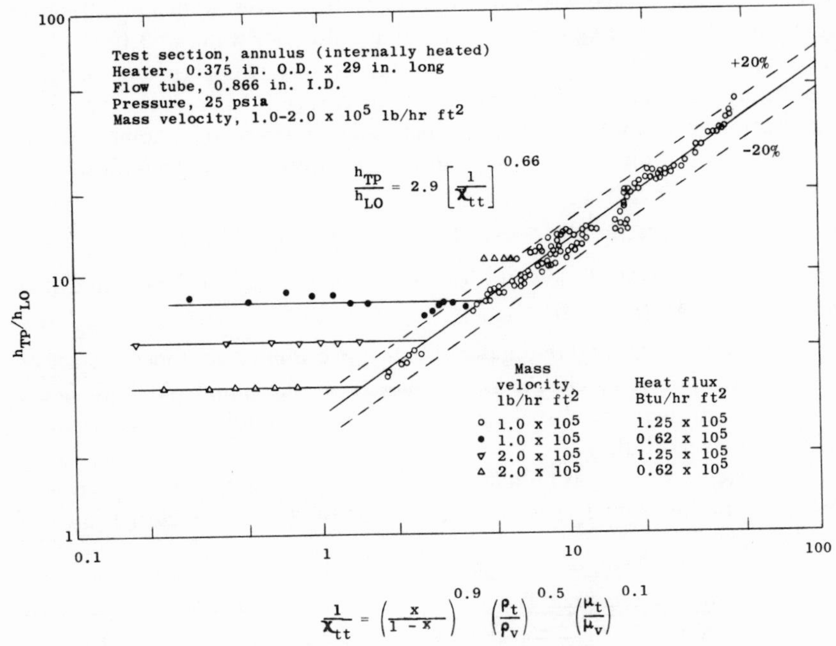

FIG. 4-9. Characteristic transition from nucleate boiling to forced convection vaporization.[11]

4.5 CHARACTERISTICS OF TWO-PHASE FLOW

The preceding discussion has illustrated that a variety of flow patterns, ranging from vapor bubbles in a continuous liquid medium to liquid droplets in a continuous vapor medium, occur with saturated boiling in forced flow. These flow patterns are distinguished either by fundamental differences in the transfer processes (phenomenological description) or by characteristic geometric distributions of the various phases (visual description). The visual description does not aiways have associated with it a change in the basic transfer mechanism of momentum, heat, or matter, and vice versa. Moreover, the transition region from one flow pattern to the next is often unstable, making precise definition of the range of the patterns difficult. Factors which can effect the beginning of a flow pattern regime are: (1) inlet conditions, (2) pipe dimensions, geometry, and inclination, (3) flow rate, (4) fluid properties, and, (5) method in which the individual phases are introduced into the channel.

Numerous visual descriptions of flow patterns are cited in the literature, practically all of which were observed for isothermal or adiabatic air–water mixtures. Some observations of nonadiabatic flows of water and Freon have been reported[3,4] but the authors are unaware of any experiments designed specifically to establish flow patterns for cryogens. However, individual observations which were secondary experimental goals[5,6] give some evidence that the classical air–water flow patterns also occur in cryogenic flows.

4.5.1 Flow Patterns in Horizontal Tubes

Scott[7] gives the following description of the generally accepted flow patterns (Fig. 4-10).

(a) *Bubble flow.* Discrete gas bubbles move along the upper surface of the pipe at approximately the same velocity as the liquid. At high liquid rates, bubbles may be dispersed throughout the liquid, a pattern often referred to as froth flow.

(b) *Plug flow.* The gas bubbles tend to coalesce as gas flow rate increases, to form gas plugs which may fill a large part of the cross-sectional area of smaller tubes.

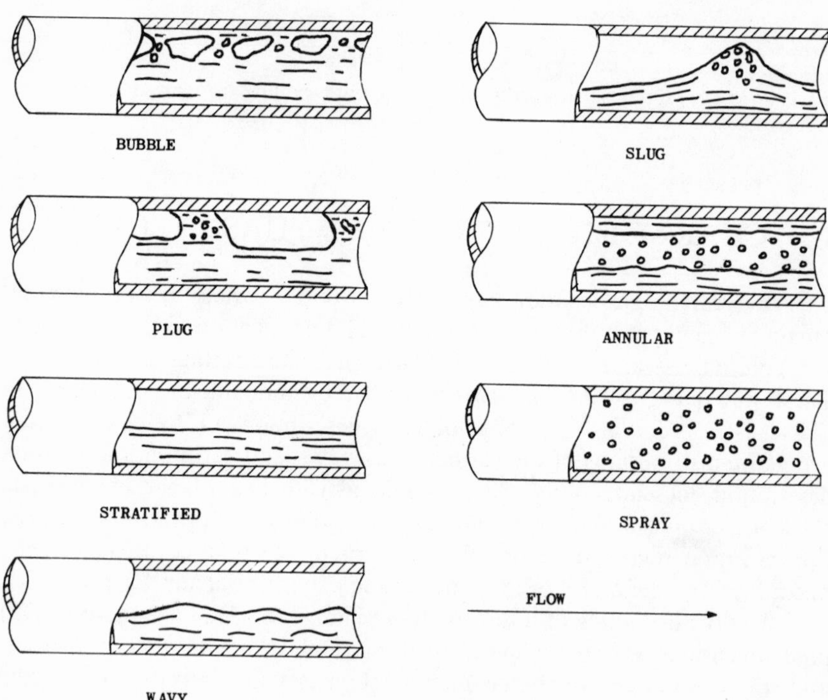

FIG. 4-10. Horizontal flow patterns.

(c) *Stratified flow*. Complete stratification of gas and liquid, with the gas occupying a constant fraction of the cross-sectional area in the upper portion of pipe, over a smooth liquid–gas interface. It occurs at lower liquid rates than bubble or plug flow, and more readily in larger tubes.

(d) *Wavy flow*. Increasing gas rate produces waves of increasing amplitude at the stratified gas–liquid interface, because of the higher gas velocity.

(e) *Slug flow*. Wave amplitudes increase to seal the tube, and the liquid wave is picked up by the rapidly moving gas to form a frothy slug which passes through the pipe at a much greater velocity than the average liquid velocity. Slug flow is also formed from plug flow as the gas flow rate is increased at constant liquid rate.

(f) *Annular flow*. Gravitational forces become less important than interphase forces, and the liquid is mainly carried as a thin film along the tube wall. The gas moves at a high velocity in the core of the tube and carries with it some of the liquid as a spray. Film flow is a term also applied to this pattern.

(g) *Mist or spray flow*. More and more liquid is carried in the gaseous core at the expense of the annular film, until nearly all of the liquid is entrained in the gas. This pattern has also been called dispersed flow or fog flow.

Predictions of the flow pattern which may exist for given values of the independent variables of the system are generally made from a graph out-

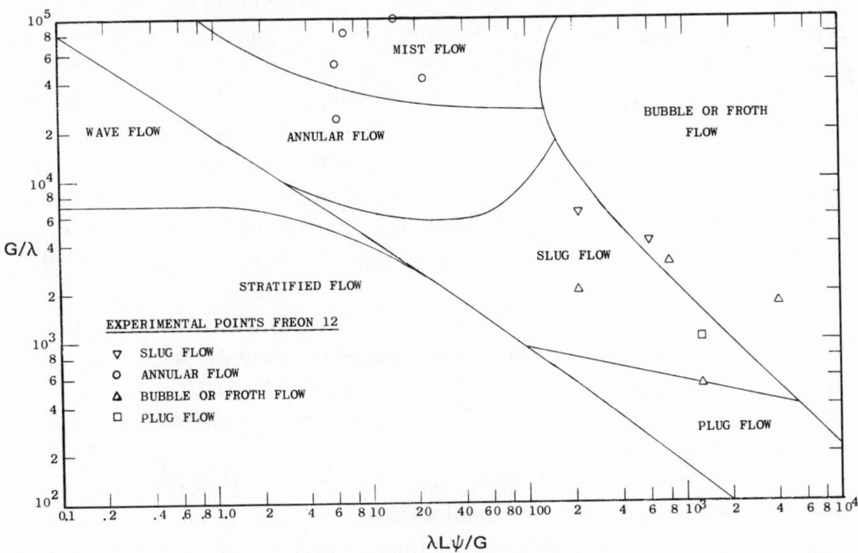

FIG. 4-11. Baker's flow regime map.

lining regions where various flow patterns might be expected to occur. The Baker chart is the best-known flow regime graph for horizontal flow and is shown in Fig. 4-11. The coordinates are expressed in terms of the mass fluxes ($1b_m$/hr-ft^2) of gas G_g and liquid G_l, and the parameters

$$\lambda = [(\rho_g/\rho_A)(\rho_l/\rho_w)]^{0.5} \tag{4-2}$$

$$\psi = (\sigma_w/\sigma)[(\mu_l/\mu_w)(\rho_w/\rho_l)^2]^{1/3} \tag{4-3}$$

are empirical correction factors which, in principle, generalize the Baker plot to fluid mixtures other than air and water. Leonhard and McMordie[4] have plotted Freon 12 data on the Baker chart (Fig. 4-11) with reasonably good results.

Although the Baker plot provides a qualitative guide to flow patterns, Wallis[8] points out that such a two-dimensional plot is quite inadequate for a general representation of a phenomenon dependent on at least a dozen variables. Notable among these are inlet conditions[9] and pipe diameters, particularly below 1 in. A more detailed discussion of flow pattern maps is given by Scott.[7]

4.5.2 Flow Patterns in Vertical Tubes

Typical vertical flow patterns are illustrated in Fig. 4-12. Again Scott[7] gives an excellent description of the various patterns.

(a) *Bubble flow*. Gas is dispersed in the upward-flowing liquid in the form of individual bubbles of various sizes. As gas flow increases, the bubbles increase steadily in number and size.

(b) *Slug (or plug) flow*. The gas bubbles coalesce to form larger, bullet-shaped slugs having a parabolic outline at the head. These slugs increase in length and diameter, and their upward velocity increases as the gas rate increases. The slugs are separated by liquid plugs which contain gas bubble inclusions. As the gas slug moves along the tube, liquid flows down through the thin liquid annulus surrounding it into the bubble-filled liquid plug beneath.

(c) *Froth flow*. When back flow of the liquid around the slugs nearly stops, the slug becomes unstable, and gas slugs seem to merge with the liquid into a patternless turbulent mixture having the general nature of a coarse emulsion. The elements of this structure are in a continual process of collapse and reformation.

(d) *Annular (or climbing film) flow*. The gas travels up the core of the tube at a high velocity and the liquid forms an annular film around the tube walls. Initially, this film may be fairly thick, and have long waves on which are superimposed a pattern of fine capillary waves. As gas flow rate increases,

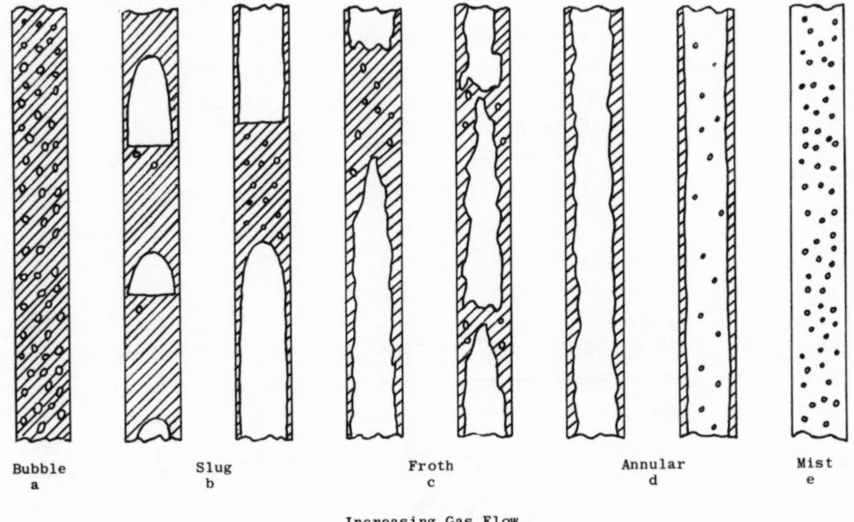

FIG. 4-12. Vertical flow patterns.

the film becomes thinner and the amount of the liquid entrained as droplets in the central gaseous core increases.

(e) *Mist* (*spray*, *fog*, *fully dispersed*) *flow*. At very high gas rates the amount of liquid entrainment increases until apparently all the liquid is carried up the tube as a mist or fog. Although a thin liquid film may exist on the wall, its presence is not obvious in this region.

The transition from froth to annular flow, in particular, seems to cover a fairly wide range of flow conditions and lacks reproducibility.

Griffith and Wallis[10] proposed the flow pattern diagram for vertical tubes shown in Fig. 4-13. The diagram was developed for air–water mixtures in $\frac{3}{4}$–1-in. tubes; however, the coordinates are expressed as dimensionless parameters which in theory make the diagram applicable to cryogens. No comparison of Fig. 4-13 with cryogenic data appears to be available in the open literature. Baker[3] has studied boiling Freon in rectangular vertical channels. He did not observe slug flow, but only annular and mist flow. Wallis[8] and Hewitt and Hall Taylor[9] give some discussion of basic transitional flow studies.

Since two-phase flow patterns specifically for cryogens are presently unavailable, the design engineer must resort to the aforementioned flow charts, bearing in mind the limitations involved.

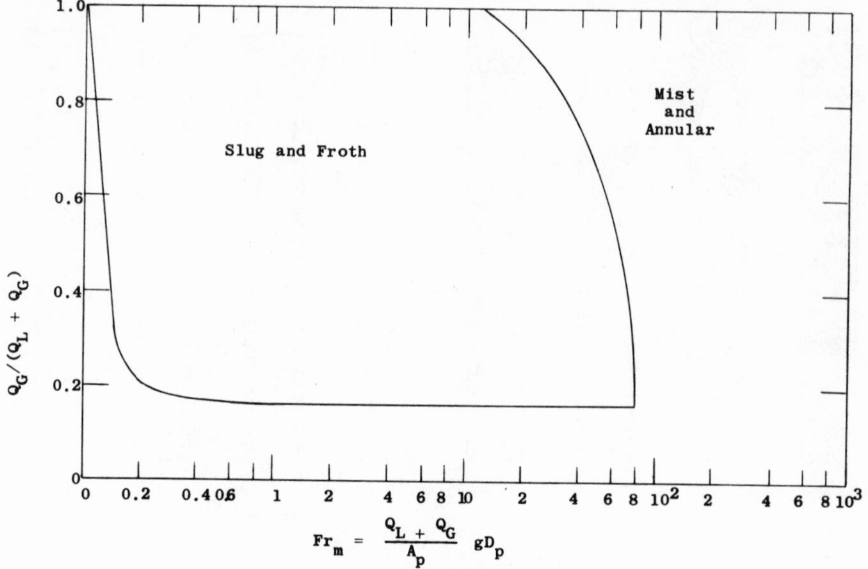

FIG. 4-13. Vertical flow regime map.[10]

4.6 DEFINITIONS

This section gives a number of definitions which are commonly used in the two-phase flow literature.

The volumetric concentration is defined as the fraction of the elemental volume of mixture occupied by the lighter phase. Since a volume element of size less than a droplet or bubble is meaningless, α is generally measured as an average over the entire cross section of the duct for an increment of length δl. Hence

$$\alpha = A_g/A_{\text{total}} \tag{4-4}$$

In gas–liquid flows α is called the void fraction.

Quality, x, is another significant variable in two-phase flow and is defined as the fraction of the total mass flow composed of vapor at a given cross section. Therefore quality is given by

$$x = W_g/W_{\text{total}} \tag{4-5}$$

Two velocities are of interest in a flowing mixture: the superficial or volumetric flux j, and the component velocity v. The volumetric flux is given by

$$j_g = Q_g/A_{\text{total}}; \qquad j_f = Q_f/A_{\text{total}} \tag{4-6}$$

and the component velocity by

$$v_g = Q_g/A_g; \qquad v_f = Q_f/A_f \tag{4-7}$$

Note that j is based on total cross-sectional area and v on the component cross-sectional area. Thus

$$j_g = (A_g/A_{\text{total}})(Q_g/A_g) \tag{4-8}$$

$$j_g = \alpha v_g \tag{4-9}$$

and

$$j_f = (1 - \alpha)v_f \tag{4-10}$$

A relationship between quality and void fraction can be derived by algebraic manipulation of the definitions,

$$(1 - x)/x = (v_f/v_g)(\rho_f/\rho_g)(1 - \alpha)/\alpha \tag{4-11}$$

The ratio v_g/v_f is called the slip velocity ratio.

Other definitions are given in the following chapters.

4.7 NOMENCLATURE

A = area
j = volumetric flux
Q = volumetric flow rate
q = heat transfer
T_{bulk} = bulk temperature
T_{sat} = saturation temperature
T_w = wall temperature
v = component velocity
w = mass flow rate
x = quality

Greek Letters

α = volumetric concentration
μ = viscosity
μ_w = viscosity of water at atmospheric temperature and pressure
ρ = density
ρ_A = density of air at atmospheric temperature and pressure
ρ_w = density of water at atmospheric pressure and temperature
σ = surface tension

σ_w = surface tension of water at atmospheric pressure and temperature
χ_{tt} = Martinelli parameter

Subscripts

f = fluid
g = gas
l = liquid
v = vapor

4.7 REFERENCES

1. W. M. Rohsenow, *Developments in Heat Transfer*, The M.I.T. Press, Cambridge, Massachusetts (1964).
2. R. V. MacBeth, in *Advances in Chemical Engineering*, Vol. 7, Academic Press, New York (1968).
3. J. L. Baker, Argonne National Laboratory, ANL-7093 (1965).
4. K. E. Leonhard and R. K. McMordie, in *Advances in Cryogenic Engineering*, Vol. 6, Plenum Press, New York (1961), p. 481.
5. M. A. Tripplett, private communications, AEDC, Tullahoma, Tennessee (1971).
6. W. G. Steward, in *Advances in Cryogenic Engineering*, Vol. 10, Plenum Press, New York (1965), p. 313.
7. D. S. Scott, in *Advances in Chemical Engineering*, Academic Press, Vol. 9 (1963), p. 199.
8. G. B. Wallis, *One-Dimensional Two-Phase Flow*, McGraw-Hill Book Co., New York (1969).
9. G. F. Hewitt and N. S. Hall Taylor, *Annular Two-Phase Flow*, Pergamon Press, Oxford, England (1970).
10. P. Griffith and G. B. Wallis, *Trans. ASME Heat Transfer*, **83C**, 307 (1961).
11. L. S. Tong, *Boiling Heat Transfer and Two-Phase Flow*, John Wiley and Sons, New York (1965).
12. H. Merte and J. A. Clark, in *Advances in Cryogenic Engineering*, Vol. 7, Plenum Press, New York (1962), p. 546.

NUCLEATE POOL BOILING 5

W. M. ROHSENOW

5.1 INTRODUCTION

The process of evaporation associated with vapor bubbles in a liquid is called nucleate boiling. Here attention will be focused on boiling at heated solid surfaces of interest in engineering applications.

When a pool of liquid at saturation temperature is heated by an electrically heated wire or flat plate, data for q vs. $(T_w - T_{\text{sat}})$ usually appear as shown in Fig. 4-7 (lower solid curve). As ΔT increases, the initial natural circulation regime AB changes to the nucleate boiling regime. The appearance of the first bubble at B requires a significant finite superheat.

As ΔT or heat flux is increased, more and more nucleation sites become active producers of bubbles. The peak nucleate boiling heat flux is reached when bubbles stream forth from so many nucleation sites that the liquid is unable to flow to the heated surface.

The dashed line shown for pool boiling of a subcooled liquid lies near the curve for saturated liquids; it may lie either to the left or right, depending on different natural convection effects resulting from different surface geometries.

Experimental data for forced convection—inside of ducts or across wires and plates—when plotted on the same coordinates, appear as shown in Fig. 4-7. If the liquid is subcooled, temperature T_b, the heat flux is represented by

$$q_{\text{before boiling}} = h[(T_w - T_{\text{sat}}) + (T_{\text{sat}} - T_b)] \tag{5-1}$$

W. M. ROHSENOW Heat Transfer Laboratory, Massachusetts Institute of Technology, Cambridge, Massachusetts.

At high heat flux, the forced convection curves for a particular pressures and at various subcoolings and velocities appear to merge on log-log plots into a single curve called the fully developed boiling curve. For some systems this fully developed region lies approximately on an extension of the pool boiling line for the same surface.

Boiling data for cryogenic fluids show the same behavior as those for noncryogenic fluids, as evidenced by Fig. 4-4 for liquid nitrogen. Consequently the theories and correlations developed on the basis of experiments with noncryogenic fluids presented in the following pertain to cryogens also. This is illustrated where possible with experimental data for cryogens.

The literature on boiling heat transfer is now quite voluminous. For references in addition to those discussed here the reader is referred to the quite complete annotated bibliography assembled by Gouse.[1]

5.2 NUCLEATION

In carefully cleaned test systems, heating of pure liquids can produce high superheat at temperatures up to a limit of stability where homogeneous nucleation, vapor formation, takes place. For the case of heat transfer associated with boiling at solid heating surfaces, observed wall temperatures are very much lower than these. Further observations of nucleate boiling show bubble streams emerging at single spots on the surface. Microscopic observation of these spots[2] has revealed a scratch or a cavity where a bubble formed. Earlier, Corty and Foust,[3] Bankoff,[4] and others developed the

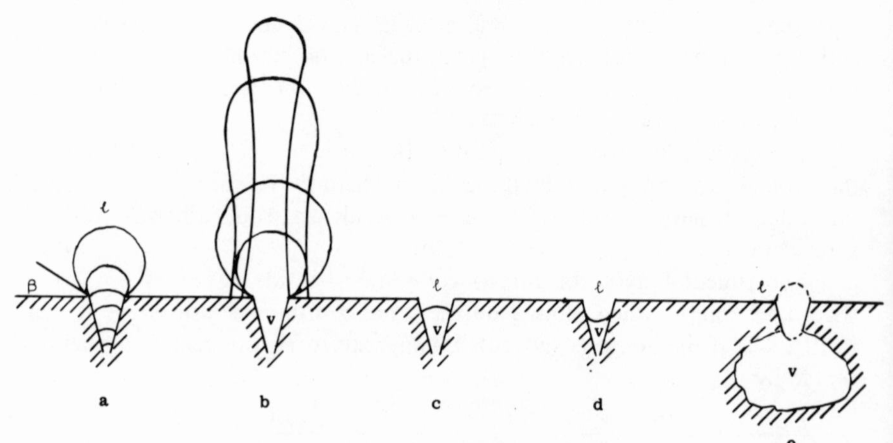

FIG. 5-1. The formation of bubbles of vapor over cavities in a heated surface.

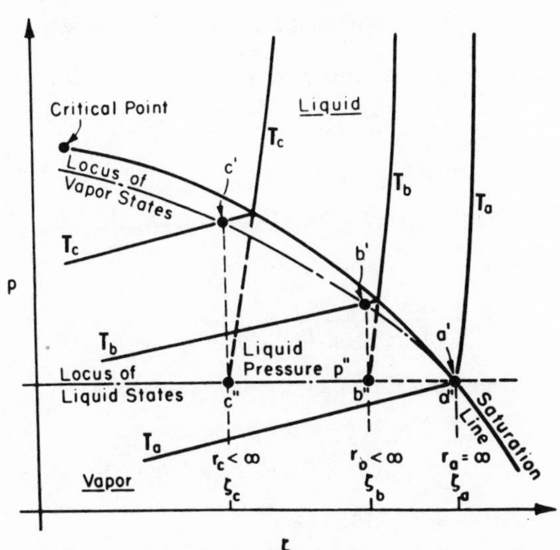

FIG. 5-2. Approximation of vapor–liquid equilibrium at a curved interface of a vapor bubble in a one-component system.

postulate that bubbles at a heating surface emerge from cavities in which a gas or vapor phase preexists (Fig. 5-1a), and a bubble emerges and departs (Fig. 5-1b), after which liquid closes in over the cavity, trapping vapor, which is the source of the next bubble. Surveys of nucleation phenomena are given in Refs. 5 and 6.

To put these notions on a quantitative basis, consider the equilibrium and growth of a spherical bubble in a large body of liquid. The following equilibrium conditions apply: equal and uniform temperatures in both phases, equal chemical potential ($\mu = H - TS$) in both phases, and

$$P_v - P_l = 2\sigma/r \tag{5-2}$$

With these the equilibrium conditions can be shown graphically as in Fig. 5-2, where states b', b'' and c', c'' are corresponding vapor and liquid states for two different radii. The production of a figure such as Fig. 5-2 is rather tedious. A simplified procedure derives from the observation that the locus of vapor states is very close to the saturation line. If it is assumed that the vapor temperature is the plane-surface saturation temperature corresponding to the pressure of the vapor P_v, a much simpler analysis follows.

Calculate the ΔT corresponding to $P_v - P_l$ given by Eq. (5-2), integrating the Clausius–Clapeyron relation

$$dT/dp = Tv_{fg}/h_{fg} \tag{5-3}$$

along the p–T saturation curve.

The following are results of such an integration for various assumptions.

(a) If $h_{fg}/v_{fg}T$ is constant and $T \approx T_{sat}$,

$$T_v = T_{sat}[1 + (2\sigma/r)(v_{fg}/h_{fg})] \tag{5-4a}$$

(b) If h_{fg}/v_{fg} is constant,

$$T_v = T_{sat} \exp [(2\sigma/r)(v_{fg}/h_{fg})] \tag{5-4b}$$

(c) $v_{fg} \approx v_g = R_v T/p$,

$$T_v = \frac{T_{sat}}{1 - (T_{sat}R_v/h_{fg})\ln[1 + (2\sigma/rp_{sat})]} \tag{5-4c}$$

(d) For the same conditions as (c) with $2\sigma/p_l r \ll 1$

$$T_v = \frac{T_{sat}}{1 - (T_{sat}R_v/h_{fg})(2\sigma/rp_{sat})} \tag{5-4d}$$

Other alternatives are as follows:

(e) Use Eq. (5-3) with an empirical fit of data for the saturation curve as $\log_{10} p$ (atm) $= A - B/T$ (K).

Using Eq. (5-3), this can be solved by trial and error, or with assumption (c):

$$\Delta T = \frac{T_{sat}^2}{B} \log_{10}\left(1 + \frac{2\sigma}{rp_{sat}}\right)\left[1 + \frac{T_{sat}}{B} \log_{10}\left(1 + \frac{2\sigma}{rp_{sat}}\right)\right] \tag{5-5}$$

for LN_2, $A = 3.879$ and $B = 297.83$; for steam, $A = 5.775$ and $B = 2156.00$ at 1 atm, and $A = 5.425$ and $B = 2004.54$ at 68 atm; for LO_2, $A = 4.038$ and $B = 368.33$.

(f) Use Eq. (5-3) with tabulated values from property tables for the saturation curve.

A comparison of results of the above calculation methods is given in the following table for $r = 0.0005$ in. and water at 1 and 68 atm:

Method	T_v, K	
	$p = 1$ atm $T_{sat} = 373.3$ K	$p = 68$ atm $T_{sat} = 558.12$ K
(a) Eq. (5-4a)	375.88	558.14
(b) Eq. (5-4b)	375.91	558.14
(c) Eq. (5-4c)	375.83	558.16
(d) Eq. (5-4d)	375.95	558.13
(e) Eq. (5-5)	375.77	558.13
(f) Tables + Eq. (5-3)	375.81	558.14

Typical values for liquid nitrogen and oxygen using Eq. (5-4a) are as follows:

	LN_2	LO_2
p, atm	1	1
T_{sat}, K	77.4	90.2
T_v, K	77.46	90.30

The above equations represent approximate expressions for the superheat required for equilibrium of a bubble of radius r. From Eq. (5-5), nuclei of radius greater than r should become bubbles and grow; those of smaller radius should collapse.

Experiments have shown that at a heated surface in water at atmospheric pressure, boiling begins around 30°F (17°C) above saturation temperature. For this condition Eq. (5-5) predicts an equilibrium bubble radius of 10^{-4} in. This is about 10,000 times larger than the maximum cavity size expected from molecular fluctuations.[7] Volmer estimated the cavity formation rate for a size of 10^{-4} in. to be approximately 1 in.$^{-3}$ hr^{-1}. From this, we readily conclude that free vapor nuclei arising from molecular fluctuations are not important as nucleation cavities. On the other hand, gas nuclei will be very significant as nucleation cavities.

If inert gas molecules are present in the liquid and hence in the bubble, Eqs. (5-2) and (5-4d) are modified as follows:

$$p_v - p = (2\sigma/r) - p_g \tag{5-6}$$

$$T - T_{sat} = (R_v T^2/p h_{fg})[(2\sigma/r) - p_g] \tag{5-7}$$

which shows that the superheat required for a bubble of a given size to grow is decreased by the presence of the gas partial pressure p_g.

The preceding calculation procedure relates to nucleation at a cavity in the heating surface in the following way. Griffith[6,8] showed that the radius of curvature of a vapor interface for positions inside and outside of the cavity (Fig. 5-1a) can be plotted as in Fig. 5-3 as $1/r$ vs. vapor volume. In Eqs. (5-4) (5-6), and (5-7) the superheat required to cause a bubble to grow is determined by $1/r$. The maximum point in the $1/r$ curve (Fig. 5-3) then determines the minimum superheat required for a bubble to grow at this cavity. The significant feature of Fig. 5-3 is that while the shape of the curve depends on cavity geometry and contact angle β, the magnitude of r at the maximum point of the curve is equal to the radius of the cavity for any $\beta \leqslant 90°$. Hence the cavity radius should determine the required superheat for nucleation for liquids which wet the surface.

Equation (5-6) has been verified experimentally by Griffith and Wallis[8] for single cavities in a uniform temperature system.

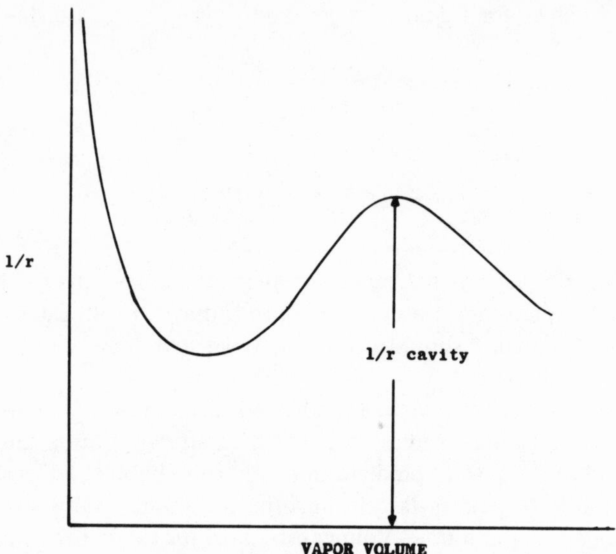

FIG. 5-3. Interface curvature vs. bubble volume for cavity shown in Fig. 5-1.

It is possible for a cavity to continue to contain vapor even when the surface is highly subcooled. If the contact angle is such that the curvature is as shown in Fig. 5-1d, the vapor pressure may be very much less than the liquid pressure as the radius decreases down into the cavity. Here Eq. (5-2) has a minus sign, and vapor will be present at temperatures very much less than the normal saturation temperature of the liquid. Hence this cavity is immediately ready to produce bubbles on its next heating. In a similar way, reentrant cavities, such as the one shown in Fig. 5-1e, may also withstand high subcooling without collapsing the cavity. The cavity in Fig. 5-1c would immediately fill with liquid upon cooling.

5.3 INCIPIENT BOILING IN FORCED CONVECTION

Consider a liquid flowing in a tube. As the heat flux or wall temperature is raised, boiling or nucleation begins at a particular value of the wall temperature. We wish to investigate this condition.

The following procedure was developed by Bergles and Rohsenow[9] based on a suggestion of Hsu and Graham.[10] For flow inside a tube the expression for heat flux is

$$q = h(T_w - T_l) = -k_l(\partial T/\partial y)_{y=0} \tag{5-8}$$

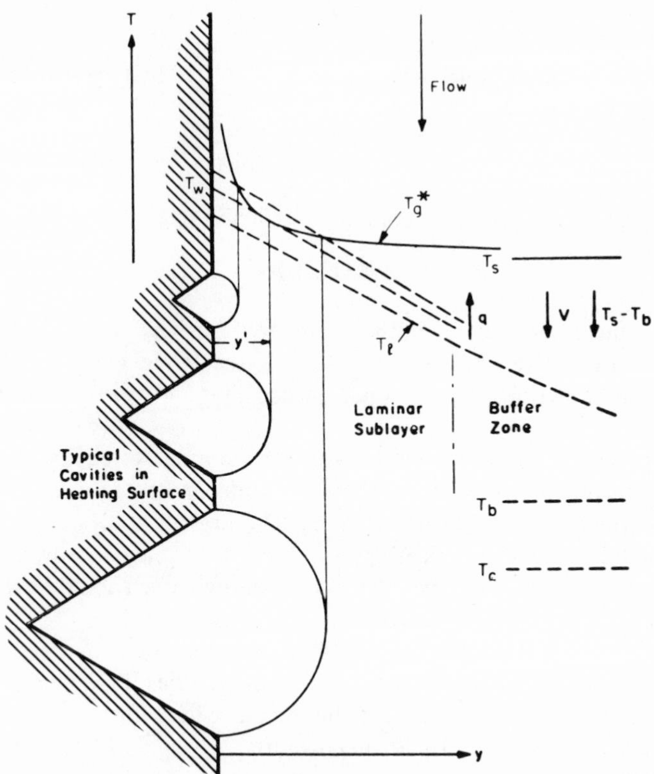

FIG. 5-4. Initiation of bubble growth in forced-convection surface boiling.[9]

Here h is the heat transfer coefficient, which is a function of the geometry, the fluid properties, and the flow rate; and k_l is the liquid thermal conductivity. For a given liquid temperature, both the temperature gradient $(\partial T/\partial y)_{y=0}$ and the wall temperature T_w increase as the heat flux increases. A series of curves representing the temperature distribution very near the heated wall is shown for increasing heat flux in Fig. 5-4. Also shown in Fig. 5-4 is a curve labeled $T_g{}^*$, which is a plot of Eq. (5-4) with the radius of the cavity taken as the distance from the heated surface. A possible theory is that nucleation takes place when the temperature curve in the liquid is tangent to the curve representing Eq. (5-4). The implication is that the surface contains cavities of various sizes and when the temperature at the outer surface of the bubble reaches the critical value given by Eq. (5-4), the bubble grows at the cavity whose radius is represented by the distance between the wall and the point of intersection.

At the point of tangency, the radius of the first cavity to nucleate [solve Eqs. (5-4a) and (5-8) simultaneously] is

$$r_{\text{nucl}} = [2\sigma T_{vfg} k / h_{fg}(q/A)]^{1/2}$$

and the heat flux at incipient boiling is predicted to be

$$q_i = (h_{fg} k / 8\sigma T_{vfg})(T_w - T_{\text{sat}})^2$$

For heat fluxes greater than that given by this incipient condition, the radii that are active are on either side of the value given by the above equation (intersections of the upper dashed line and the T_g* curve of Fig. 5-4).

On commercially prepared surfaces a wide range of cavity sizes can be expected. Incipient boiling, then, should be independent of surface condition for most commercially finished surfaces. The present analysis should therefore be valid in most practical applications. It is noted, however, that the position and slope of the remainder of the boiling curve should be dependent on the size range of active nuclei.

Equations (5-4) and (5-8) could be differentiated to determine the conditions at the point of tangency. However, this procedure does not lead to an explicit formulation for the incipient heat flux q_i. Consequently, a graphical solution was found much more convenient. The results for q_i were calculated for water over a pressure range of 15–2000 $\text{lb}_f/\text{in.}^2$ abs and can be expressed quite accurately as follows:

$$q_i = 15.60 p^{1.156} (T_w - T_{\text{sat}})^{2.30/p^{0.0234}} \tag{5-9}$$

where q is in Btu/hr-ft^2, p is in $\text{lb}_f/\text{in.}^2$ abs, and the temperatures are in °F.

It should be noted that although velocity does not appear in Eq. (5-9), the equation is valid for all fluid velocities.

As heat flux is increased above the condition for incipience (the upper dashed line in Fig. 5-4), cavities of a range of sizes become nucleated. A word of caution is in order here. The preceding calculation procedure presumes that cavities exist at sizes out to the point of tangency. At low heat fluxes, as in natural convection, the slope of the temperature distribution, Eq. (5-8), may be so small that the point of tangency occurs at cavity sizes larger than those present. Then the heat flux and wall superheat must be raised until the two curves intersect at the largest available active cavity. Hence for low heat flux cases, as in natural convection, the procedure usually predicts superheats and q_i lower than the observed magnitude.

The intersection of the two curves of Fig. 5-4, Eqs. (5-4a) and (5-8), at the maximum available cavity $r_{\max}$ gives

$$q_i = (k_l/r_{\max})(T_v - T_{\text{sat}}) - (2\sigma T_{vfg}k/h_{fg}r_{\max}^2) \tag{5-10}$$

A number of other suggestions have been made for predicting the incipient boiling condition. Most of them reduce to the scheme shown in Fig. 5-4 except that T_l is plotted vs. y/n and $T_g{}^*$ is plotted vs. r. In the above n is taken as unity. In these other proposals the suggested magnitude of n ranges from 0.67 to 2. Brown[11] tried these suggestions with experimental data for various fluids and found n to range between 1 and 3.

Frost and Dzakowic[12] found for cryogenics that the assumption that bubble growth begins when $T_l = T_g{}^*$ at $y = \text{Pr}_r{}^n$, with $n = 2$, correlated data from the literature reasonably well.

The above procedure then appears to provide an estimate of the lower ΔT and q_i values at incipience. Actual ΔT and q_i values may be greater than these predicted magnitudes. These higher wall superheats may be required to provide sufficient bubble growth acceleration to remove it from the wall. Also, thermocapillarity, as described later, may provide convection currents not accounted for in the above model.

Another study of nucleation[13] attempted to account for flow patterns around the bubble surface protruding from the cavity in forced convection. From comparison with only a small amount of data this study led to the following prediction for wall superheat in forced convection:

$$(T_w - T_{\text{sat}})_{\text{incip}} = 1.33(q/k)^{0.4}(\tau_w/\rho_l v^2)^{0.1}(2\sigma T_{\text{sat}}v_{fg}/h_{fg})^{0.6} \tag{5-11}$$

More study of this relation with comparison with a wider variety of data is needed.

Deane *et al.*[14] investigated the possibility that cavities could be deactivated to a predictable size by preconditioning by raising the pressure and temperature to specified levels of subcooling. This would establish a "negative" radius of the cavity from Eq. (5-4a) on the assumption that the oxidation in the cavity would prevent further collapse to smaller radius. On subsequent reduction of pressure and increase in the degree of superheat, the radius, now taken as positive in Eq. (5-4a), should predict the superheat required for nucleation at the surface. The same formulation was explained[15] in terms of reentrant cavities (Fig. 5-1e), which eliminates the necessity of requiring oxidation to aid the process. Most test results fall below the predictions of this theory; however, Singer and Holtz[15] showed that this discrepancy was probably due to the presence of noncondensible gas. Their data showed that after extended periods of boiling to drive gas off the surface, the measured superheats agreed well with the theory. Surface finish had no effect on the results.

5.4 POSSIBLE MECHANISMS OF HEAT TRANSFER IN BOILING

As heat flux to a saturated liquid is increased beyond the incipient boiling point, the number of spots on the surface from which bubble columns rise increases. Photographs by Hsu and Graham suggest that between bubbles in pool boiling a thermal layer builds up and each bubble carries away the liquid in a region $2D_b$, thus pumping away the liquid superheat. In the region between the bubbles at the lower heat fluxes the heat transfer may be of the order of that associated with natural convection. Calculations based on this hypothesis were verified at low heat fluxes by Han and Griffith.[16]

As heat flux is increased, the nucleation site density increases as well as the columns of continuous vapor.[17] The zones between these vapor columns become "return passages," bringing liquid to the heating surface, where it is converted to vapor in the columns. As heat flux is increased further, this situation becomes unstable and the heating surface becomes starved for liquid, causing the transition to film boiling.

A typical history of the temperature at a cavity under a bubble is shown in Fig. 5-5a. At point A the bubble begins to grow on the surface. The major portion of the bubble grows in an essentially nonviscous way, except near the wall, where viscosity predominates. Measurements of Cooper and Lloyd[18] on toluene suggest that a thin liquid layer is left on the surface under the bubble (Fig. 5-5b). It is presumed that viscous force retards the motion of the liquid next to the wall as the bubble grows, thus leaving behind a liquid layer. At point B this layer has evaporated and the dry surface rises in temperature. At point C the bubble has departed and colder liquid covers the surface, reducing its temperature to point D. Then the surface continues to rise in temperature to point A, when a new bubble forms. Calculations of heat flux from the surface suggest that major contributions to the heat flux

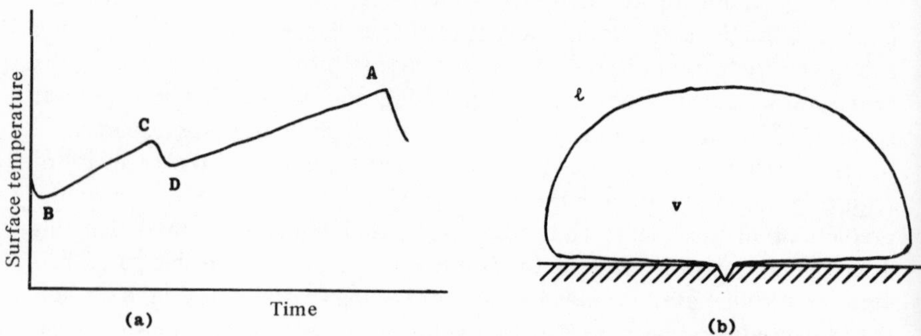

FIG. 5-5. Temperature history of a wall under a bubble.

occur while the thin liquid film evaporates under the bubble, *AB*, and when the colder liquid rushes into contact with the surface, *CDA*.

Cooper[19] showed that this microlayer is significant only at low pressure. At pressures above 1 atm for fluids such as water, hydrocarbons, and cryogens the bubble growth is thermally controlled (not dynamically controlled); then the microlayer does not greatly influence the growth and the heat transfer.

Boiling from a heated surface in highly subcooled liquids may be accompanied by a different dominant effect. Cumo[20] has observed photographically long "jets" of hot liquid streaming forth in front of bubbles forming at surfaces in subcooled liquid. Based on heat transfer measurements and photographs in highly subcooled liquids Brown[11] suggests that because of evaporation at the hot side and condensation at the cold side of the bubble, a small temperature difference exists around it. This results in a slightly lower surface tension at the outer edge of the bubble, producing significant flow of the interface which induces large flows of the surrounding liquid to form the observed jets. Here the bubble appears to induce large convection currents which are heated by the hot surface. This process has been called thermocapillarity.[11] It should have its greatest influence in boiling of subcooled liquids in forced convection, but is no doubt present in some degree at all times.

Thermocapillarity may tend to remove rapidly the thin layer of liquid under a bubble (Fig. 5-5b), reducing the time between points *A* and *B* (Fig. 5-5a). Also, the convection induced by thermocapillarity may require wall superheats higher than those predicted by Eq. (5-9), as observed in some test data.

Probably all of the preceding mechanisms occur simultaneously and participate to varying degrees in the heat transfer process under various conditions. The boundaries where different mechanisms predominate are not well established.

5.5 BUBBLE PARAMETERS

For liquids which wet the heating surface the size of bubbles at departure from the heating surface has been studied by a number of workers. Fritz[21] and Wark[22] equated boundary and surface tension to determine the following expression for departure diameter:

$$D_b = C_d\beta[2g_0\sigma/g(\rho_l - \rho_v]^{1/2} \tag{5-12}$$

where C_d was found experimentally to be 0.0148 for H_2O and H_2 bubbles.

Mikic *et al.*[23] and Lien and Griffith[24] studied bubble growth analytically and experimentally. Their results show clearly that all bubbles start their growth dynamically controlled and in later stages become thermally controlled, vaporization being governed by heat conduction. Mikic *et al.*[23] show that for a bubble growing at a wall at $T_w > T_{sat}$ into a liquid at $T_b < T_{sat}$

$$dR^+/dt^+ = \{t^+ + 1 - \theta[t^+/(t^+ + t_w^+)]^{1/2}\}^{1/2} - (t^+)^{1/2} \tag{5-13}$$

where

$$R^+ \equiv RA/B^2; \qquad t^+ = tA^2/B^2$$

$$A \equiv [b(T_w - T_{sat})h_{fg}\rho_v/T_{sat}\rho_l]^{1/2}; \qquad B \equiv [(12/\pi)(\mathrm{Ja}^2k_l/\rho_l c_l)]^{1/2}$$

$$\mathrm{Ja} \equiv (T_w - T_{sat})\rho_l c_l/h_{fg}\rho_v; \qquad \theta \equiv (T_w - T_b)/(T_w - T_{sat})$$

Here for the bubble growing at the wall $b = \pi/7$. Mikic and Rohsenow[25] show that the waiting time between bubbles can be approximated by

$$t_w = \frac{\rho_l c_l}{\pi k_l}\left\{\frac{(T_w - T_b)R_c}{T_w - T_{sat}\{1 + [2\sigma(fg)^{1/2}/R_c h_{fg}]\}}\right\}^2 \tag{5-14}$$

where R_c is the radius of the cavity. Mikic *et al.*[23] integrated Eq. (5-13) for various values of the parameters, neglecting R_c, which is the bubble radius at $t = 0$, compared with R throughout the growth; the results are shown graphically in Fig. 5-6.

For dynamically controlled growth $t^+ \ll 1$ and $t_w^+ \gg t^+$, Eq. (5-13) integrates to the following result:

$$R^+ = t^+ \tag{5-15}$$

For the thermally controlled region $t^+ \gg 1$, Eq. (5-13) and the curves of Fig. 5-6 are approximated by

$$R^+ = (t^+)^{1/2} - \theta[(t^+ + t_w^+)^{1/2} - (t_w^+)^{1/2}] \tag{5-16}$$

For a bubble growing in an initially uniformly superheated liquid $(t_w^+ \to \infty)$ from an initial radius greater than the critical radius given by Eq. (5-4a), Eq. (5-13) integrates to

$$R^+ = \tfrac{2}{3}[(t^+ + 1)^{3/2} - (t^+)^{3/2} - 1] \tag{5-17}$$

which is valid for both regimes of growth. Here $b = 2/3$ in the expression for A.

For the dynamically controlled region $t^+ \ll 1$, Eq. (5-17) becomes

$$R^+ = t^+ \tag{5-18}$$

which is the Rayleigh solution, and is identical to Eq. (5-15).

For the thermally controlled region $t^+ \gg 1$, Eq. (5-17) becomes

$$R^+ = (t^+)^{1/2} \tag{5-19}$$

which is the result obtained by Plesset and Zwick[26] and Scriven.[27]

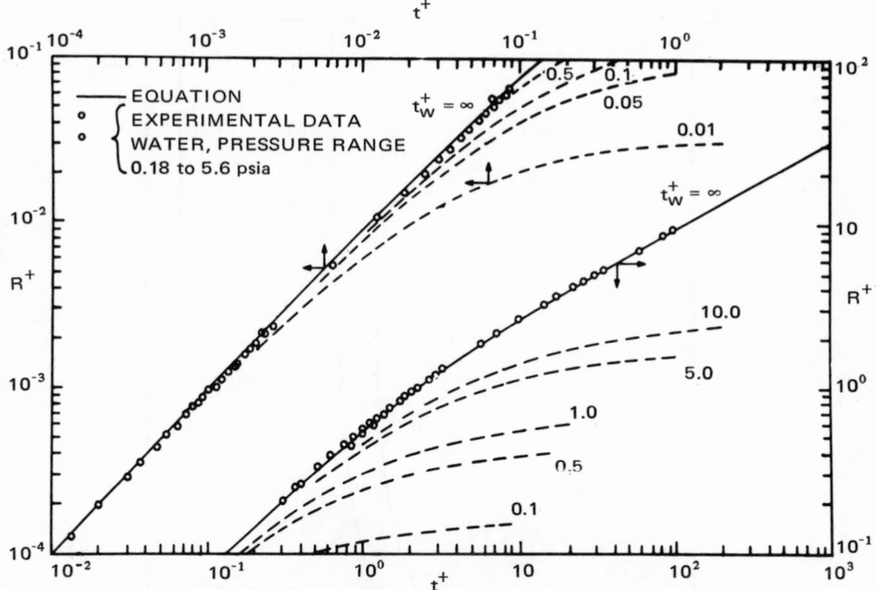

FIG. 5-6. Bubble growth curves; comparison with experimental data.[23]

Lien and Griffith[24] performed definitive experiments for bubble growth in superheated water over a pressure range of 0.18–5.6 psia, superheats in the range 15–28°F, and $58 < \mathrm{Ja} < 2690$. For pressures less than around 0.4 psia the duration of the bubble growth period remained in the dynamically controlled region, $R^+ = t^+$; for pressures above around 5 psia the dynamically controlled region existed for only a short time and practically all of the growth was governed by heat conduction. Between these pressures the early stage of growth was dynamically controlled and the later stage thermally controlled. The midrange of this region exists when $t^+ \approx 1$. The results agreed with the prediction of Eq. (5-17).

Experimental data for growth in nonuniform temperature fields for which waiting time t_w is recorded are limited but do agree[25] with the prediction of Eq. (5-15). Data for bubble growth on heated surfaces at reduced pressures do exist[28] but without observed values of t_w. These data fall below the prediction of Eq. (5-17), suggesting that Eq. (5-13) and Fig. 5-6 should apply and that the effect of waiting time on the data is significant.

The preceding solutions and results are for lower pressures where $\rho_v \ll \rho_l$. At higher pressures where bubble growth is thermally controlled Scriven showed the effect of going to higher pressures (significant values of ρ_v/ρ_l) as given by the curves of Fig. 5-7, where

$$R^+ = Ct^+ \tag{5-20}$$

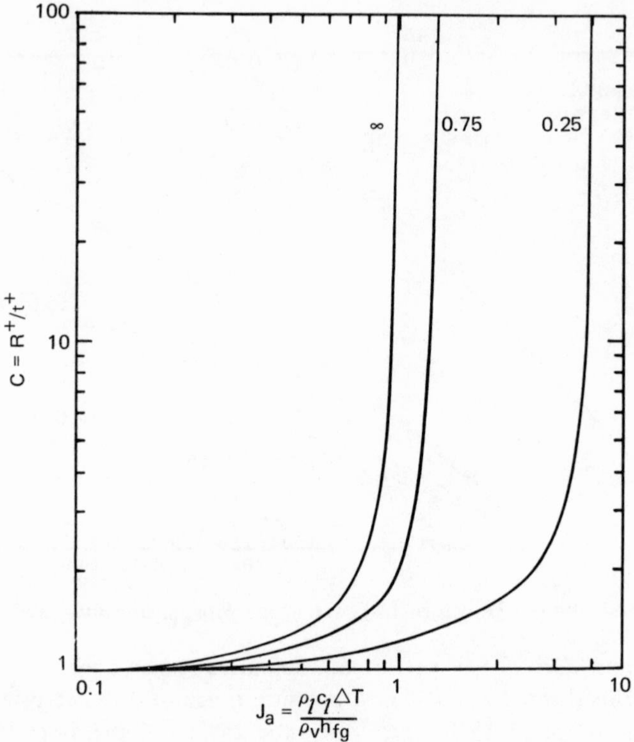

FIG. 5-7. Bubble growth at high pressures.[27] The labels on the curves give the values of ρ_v/ρ_l.

Bubble departure diameters as given by Eq. (5-12) do not agree well with data. Cole and Rohsenow[29] correlated the departure diameters D_b for various fluids as shown in Fig. 5-8. The straight line portion of the two curves where $p/p_{\text{crit}} < 0.2$ is given by the following equations

$$\text{water:} \qquad \text{Eo}^{1/2} = (1.5 \times 10^{-4})(\text{Ja}^*)^{5/4} \tag{5-21}$$

$$\text{other fluids:} \quad \text{Eo}^{1/2} = (4.65 \times 10^{-4})(\text{Ja}^*)^{5/4}$$

where

$$\text{Eo} \equiv g(\rho_l - \rho_v)D_b{}^2/g_0\sigma, \qquad \text{Ja}^* \equiv \rho_l c_l T_{\text{sat}}/\rho_v h_{fg}$$

With specific reference to cryogens, Eq. (5-21) shows good correlation with liquid nitrogen data.

The prediction of bubble frequency for growth at a heating surface is

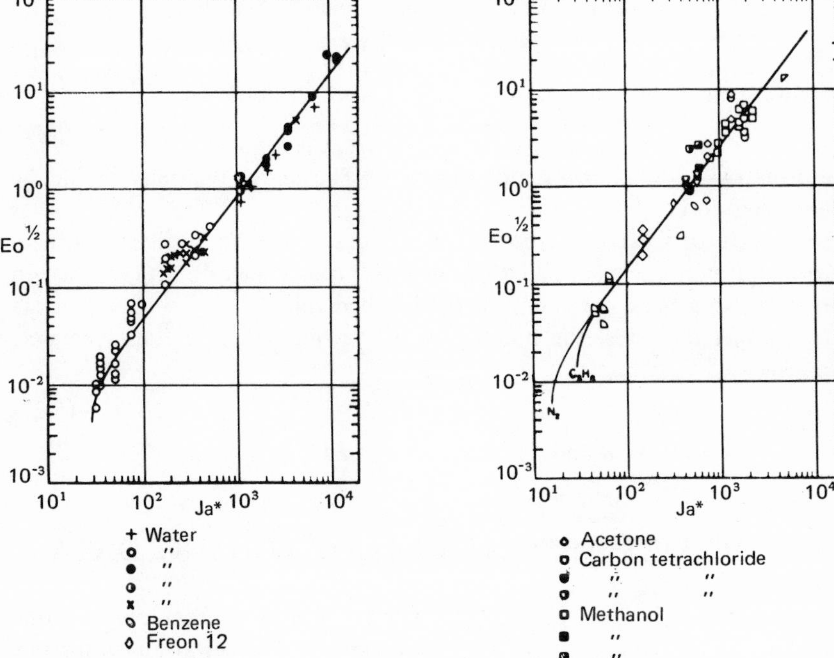

FIG. 5-8. Left: Departure diameter correlation for water.[29] Right: Departure diameter correlation for other liquids.[29]

not well established. Jakob and Linke suggested $fD_b = 920$ ft/hr for hydrogen and water bubbles. Later Zuber[30] and Peebles and Garber[31] obtained

$$fD_b = 1.18[t_c/(t_c + t_w)][\sigma g_c g(\rho_l - \rho_v)/\rho_l^2]^{1/4} \tag{5-22}$$

Cole[32] showed from data that $1.18[t_c/(t_c + t_w)]$ ranged between 0.15 and 1.4, raising serious doubt regarding the validity of the above equation.

Later Ivey[33] showed that the frequency–diameter relation depended on the regime of bubble growth:

dynamically controlled	$D_b f^2 = \text{const}$
thermally controlled	$D_b f^{1/2} = \text{const}$

In the intermediate region the exponent on f changes from 2 to $\frac{1}{2}$.

The analysis of Mikic *et al.*[32] for the thermally controlled regime leads to

$$D_b f^{1/2}/2\text{Ja}(\pi\alpha_t)^{1/2} = [t_c/(t_c + t_w)]^{1/2} + [t_w/(t_c + t_w)]^{1/2} - 1 \tag{5-23}$$

where t_c and t_w are bubble contact time and waiting time between bubbles, and $(t_c + t_w) = 1/f$. Over the range $0.2 < t_c/(t_c + t_w) < 0.8$ this becomes

$$D_b f^{1/2} = \tfrac{3}{4}(\pi\alpha_l)^{1/2}\text{Ja} \pm 10\% \tag{5-24}$$

In the dynamically controlled regime, Cole[34] suggested

$$D_b f^2 = \tfrac{4}{3}g(\rho_l - \rho_v)/C_{\text{drag}}\rho_l \tag{5-25}$$

which for steam at 1 atm, $C_{\text{drag}} \approx 1$, and $\rho_v \ll \rho_l$ reduces to

$$D_b f^2 = 1.32g \tag{5-26}$$

Information on f vs. D_b is not extensive. The above relations should be considered to be approximate.

Effects of forced convection on D_b and f are not well established. Koumoutsus *et al.*[35] showed D_b to decrease nearly linearly with increase in velocity. No general prediction method is available.

As q/A increases, the number of active nucleating cavities increases. Staniszewski[36] obtained

$$q \sim n^m \tag{5-27}$$

where $m = 1$ at low q and decreases to around $\frac{1}{2}$ at high q.

5.6 FACTORS AFFECTING POOL BOILING DATA

In this section we present the results of experimental observations showing the effect on pool boiling of changing the values of various properties and conditions. Except as noted, all data discussed are for liquids which wet the heating surface.

Pressure and temperature differences have a marked effect on all regimes

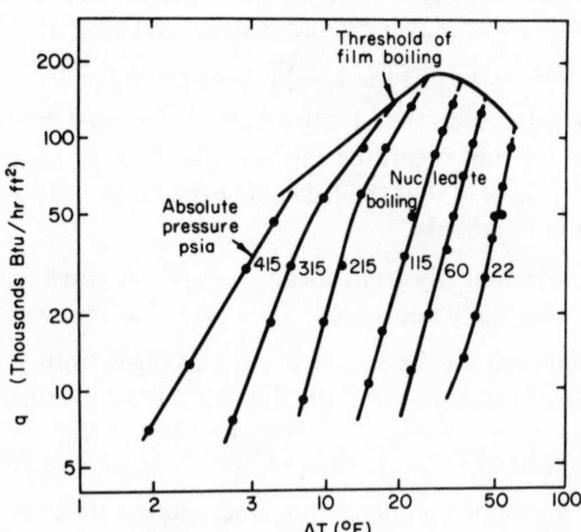

FIG. 5-9. Effect of pressure on pool boiling curve, pentane.[52]

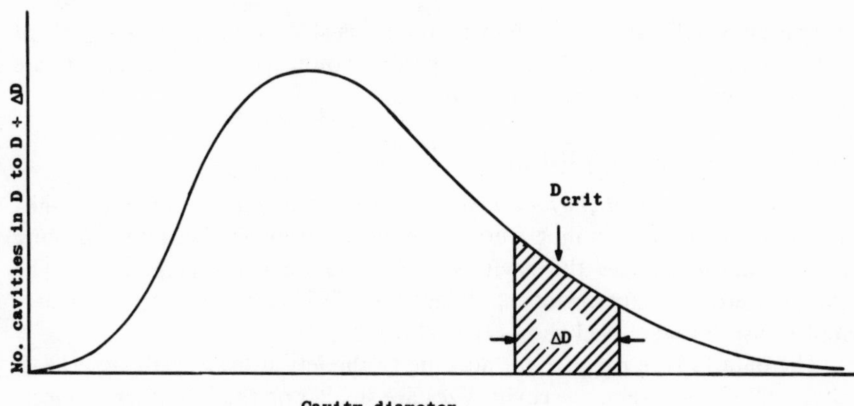

FIG. 5-10. Cavity size distribution.

of boiling. Figure 5-9 shows representative data for *n*-pentane. Similar data for neon are given in Ref. 37. Consider a cavity size distribution (number of cavities of a particular size range versus diameter of cavity) represented by the curve in Fig. 5-10. The first cavity to be activated has an approximate diameter as calculated for the point of tangency in Fig. 5-4. Then in accordance with Eq. (5-27), as q is increased, n must increase and more cavities must become activated as diameters spread on either side of D_{crit} (Fig. 5-10). As shown in Fig. 5-4, this requires greater wall superheat.

The slope of the q vs. ΔT curve expresses the change in ΔT necessary to increase the number of activated points sufficiently to accommodate the new heat flux. Although the slope is predominantly in the neighborhood of 3, observations are available with slopes as low as unity for contaminated surfaces and as high as approximately 25 for clean, polished surfaces. The actual slope depends upon the uniformity or distribution of the size and shape of the nucleating cavities (Fig. 5-10).

Griffith and Wallis[8] obtained data for a number of active nucleation sites vs. wall superheat and calculated the radius corresponding to the wall superheat from Eq. (5-5). For data on boiling water, methanol, and ethanol plotted on the same surface a single curve resulted for the number of active nucleation sites vs. this radius, suggesting the existence of a characteristic cavity size distribution (Fig. 5-10) for that surface.

5.6.1 Effect of Pressure

Pool boiling heat flux data plotted vs. wall superheat give curves as shown in Fig. 5-9. At higher pressures the data lie at lower superheat magnitudes. Of significance in explaining this shift to lower wall superheat is the fact that for all fluids, Eq. (5-4) predicts lower ΔT_{sat} required to activate given size cavities as pressure is increased.

5.6.2 Surface Condition

5.6.2.1 *Effect of Surface Finish:* As expected, surface finish can shift the position of the boiling curve markedly, probably because changing surface finish changes the cavity-size distribution curve (Fig. 5-10). The initial controlled experiments studying this effect were performed by Corty and Foust.[3] Figure 5-11 shows typical results.

The data for the rougher surfaces lie to the left at lower wall superheat, presumably because active cavity sizes are smaller on the smoother surfaces.

5.6.2.2 *Aging:* Experience indicates that aged surfaces have higher required ΔT for a given q. On metallic surfaces a scale or deposit may form from the boiling liquid or a film may form from oxidation or other chemical reaction. In either case the vapor-trapping cavity may shrink, necessitating higher superheat for activation.

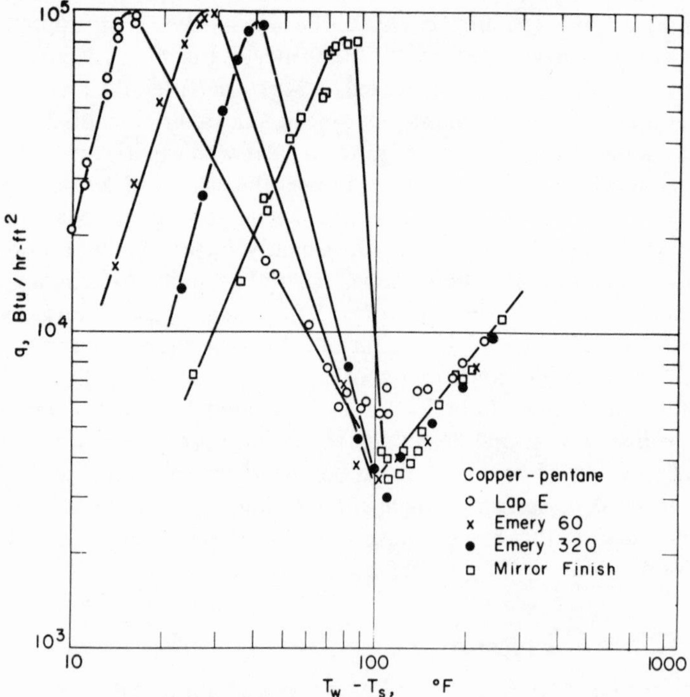

FIG. 5-11. Copper–pentane test results; effect of roughness.[68]

5.6.2.3 *Surface Coatings or Deposits:* In addition to the effect described above in connection with aging, there obviously will exist a temperature drop associated with conduction across the coating or deposit. Since coatings may be very thin and their properties not known, this additional temperature difference is usually included in ΔT when plotting boiling data. Including this constant, conduction resistance causes q vs. ΔT curves to be farther to the right and at lower slopes than those for clean surfaces. Curves for surfaces with various coatings are spread apart even farther than those of Fig. 5-11.

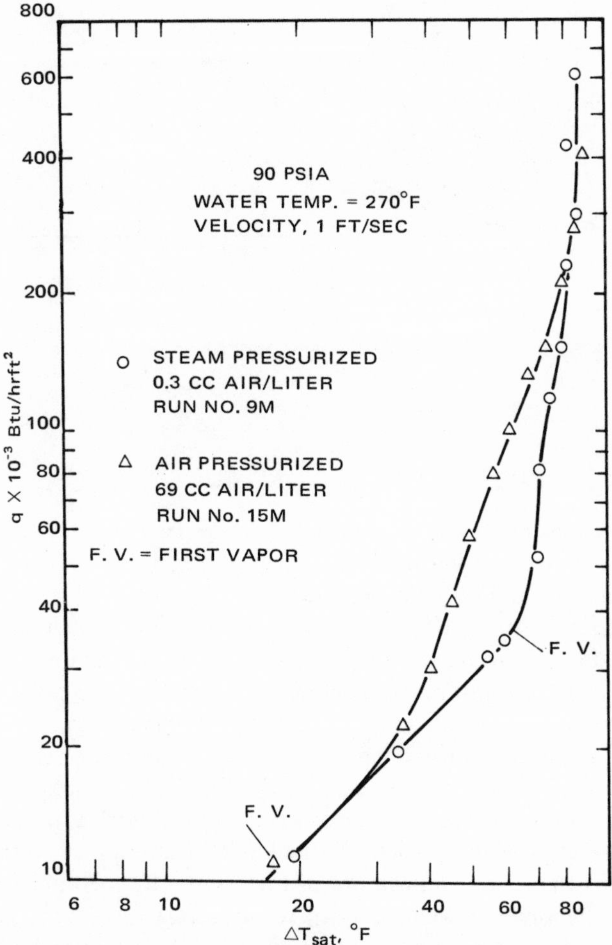

FIG. 5-12. Effect of dissolved gas on boiling curve.[70]

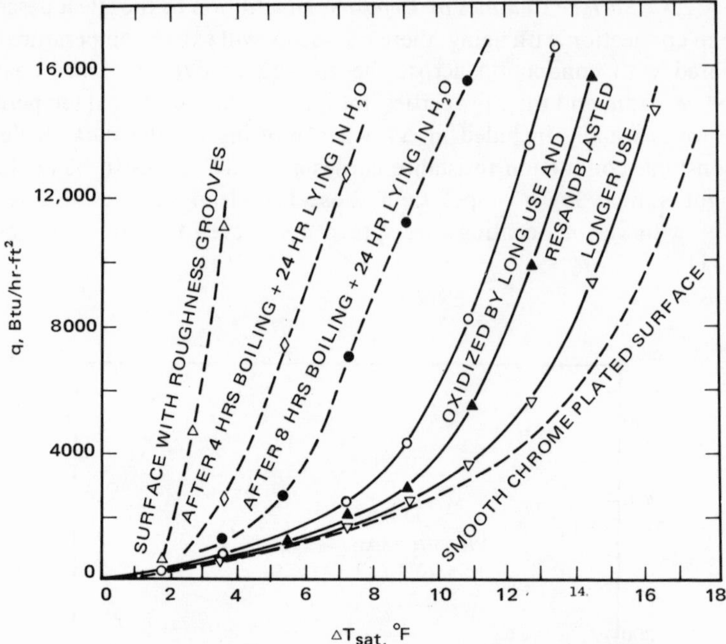

FIG. 5-13. Effect of surface sand-blasting on pool boiling.[71]

5.6.3 Effect of Gases

Noncondensible gases dissolved in the water tend to come out of solution at the hot surface. In general, they tend to move the q vs. ΔT curve to the left and to reduce the magnitude of q_{crit}.

Figure 5-12 shows the results when fresh liquid containing a gas is continually brought to the heating surface by forced convection. Gas bubbles may appear at temperatures well below the normal boiling point and additional convection, probably induced by thermocapillarity, may produce the higher heat fluxes as shown.

Figure 5-13 shows results for pool boiling. It is assumed that sand-blasting the surface causes air to be present in the nucleation cavities. As time goes on after heating is started, the boiling curve moves to the right to higher ΔT as the gas is pumped out of the cavities by the bubbles. Also an interesting phenomenon which appears to be associated with the presence of gas in the nucleating cavities is a kind of unstable boiling or bumping[7,38] which has been observed thus far only in boiling of liquid metals.

5.6.4 Hysteresis

The overshooting of the wall temperature on increase of q in natural convection and the significant drop in ΔT when boiling begins is known as the hysteresis effect. This overshooting is more pronounced with liquid metals, particularly alkali metals, which wet practically all solid surfaces extremely well. This tends to deactivate cavities as discussed in connection with Fig. 5-1c.

With liquid nonmetals this overshooting occurs less frequently since cavities such as those of Fig. 5-1d are usually present and remain active cavities even after a system is cooled. In nonmetals if some of the cavities become deactivated, the hysteresis overshoot can occur; however, on reducing the heat flux in the nucleate boiling region the data follow smoothly down the boiling curve to natural convection. Figure 5-14 demonstrates that this typical behavior is also observed with cryogenic fluids.

5.6.5 Size and Orientation of Heating Surface

There appears to be very little effect of size of wire or of orientation on the q vs. ΔT curve in the nucleate boiling region. A geometry effect, however, is pronounced in the nonboiling natural convection region, which is the

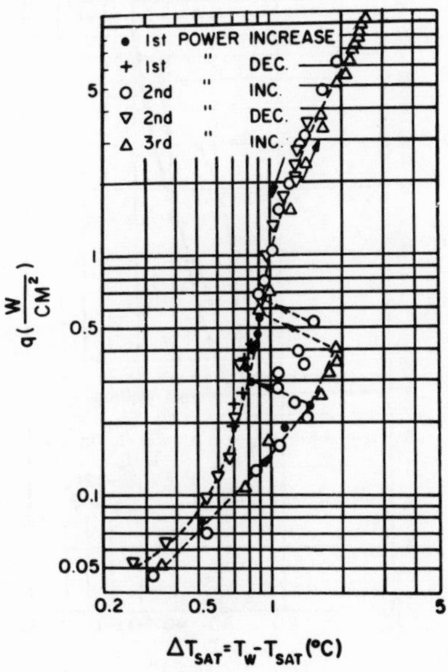

FIG. 5-14. Hysteresis at very low heat flux in nucleate pool boiling of liquid neon; absolute pressure 4 bars.[37]

left asymptote of the curve and is predicted by the normal single-phase natural convection data. In the film boiling regime geometry also influences the data, curves for smaller wire diameter lying to the left.

5.6.6 Agitation

The effect of agitation such as with a propeller is quite similar to the velocity effect shown in Fig. 4-7. Curves for greater rpm are higher up on the graph. This effect was studied by Austin[39] and more recently by Pramuk and Westwater.[40] Also, the critical heat flux increases with increasing rpm. No attempt to correlate this information is made, since the effect is greatly influenced by shape, size, and position of the agitation.

5.6.7 Subcooling

The effect of subcooling on the position of the boiling curve appears to depend on the geometry for convection. Figure 5-15 shows data for water[9] boiling on an 0.0645-in.-diameter stainless steel tube. Here the boiling curves lie farther to the right as subcooling is increased. The left asymptotes are for natural convection and should have higher q for greater subcooling at equal $(T_w - T_{sat})$. In contradiction with this are results[41] for a horizontal,

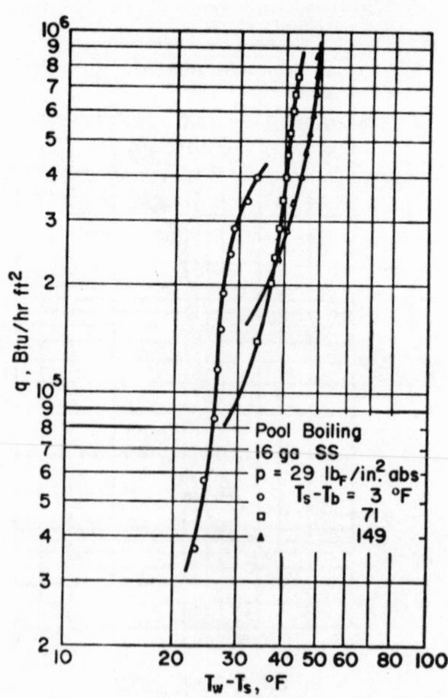

FIG. 5-15. Influence of subcooling on pool boiling.[9]

flat, nichrome heater, 4.25 in. square and 0.0165 in. thick. These show the boiling curves lie farther to the left as subcooling increases. The differences between the results of Refs. 9 and 41 may be due to different convection geometries or they may be due to different nucleating conditions of the surface.

5.6.8 Non-Wet Surfaces

When the liquid does not wet the heating surface, very large vapor bubbles form and cover much larger portions of the heating surface.[42,43] The net result is that very greatly increased magnitudes of $(T_w - T_{sat})$ are required to transfer a given q. The critical heat flux is also reduced by factors of 10–20.

5.6.9 Gravitational Field

At the NASA laboratories, a ribbon heater immersed in a beaker of water was photographed during free-fall conditions, and the results were reported by Siegel and Usiskin.[44] In each case the heat fluxes were in the nucleate boiling range under a normal gravitational field. In the free-fall condition with the lower heat fluxes, bubbles grew while remaining attached to the ribbon. At the higher heat fluxes a very large vapor volume formed around the ribbon. This suggests that nucleate boiling is essentially non-existent under zero-g conditions.

Subsequently, Siegel and Usiskin added a small amount of friction to the free-fall system, raising g to approximately 0.09. They reported verbally that under these conditions nucleate boiling appeared to continue throughout the fall. This indicates that only a small g field is needed to maintain nucleate boiling.

Merte and Clark[45] reported the results of tests on a heated surface at the bottom of a pool. The system was placed in a centrifuge and rotated so that the resultant acceleration field was normal to the surface. Their boiling tests covered a range of 1–21 g, with the results replotted as shown in Fig. 5-16. There seems to be very little effect on the position of the q vs. $T_w - T_{sat}$ curves at the higher heat fluxes. The displacement of the curves at the lower heat fluxes is probably due to the effect of superimposed natural convection effects.

5.7 COMPOSITE OF BOILING DATA

Temperature differences $T_w - T_{sat}$ associated with nucleate boiling are small 1–100°F (0.5–55°C). In many applications this resistance is not of primary importance; hence an order of magnitude for this temperature

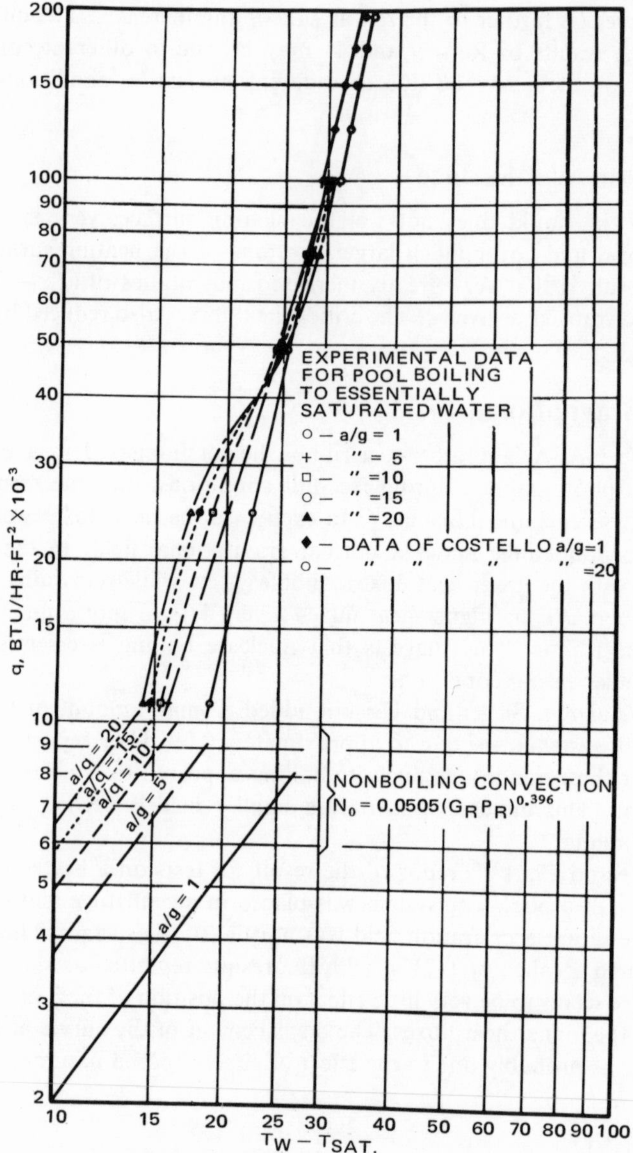

FIG. 5-16. Influence of system acceleration normal to heating surface on convection and pool boiling.[45]

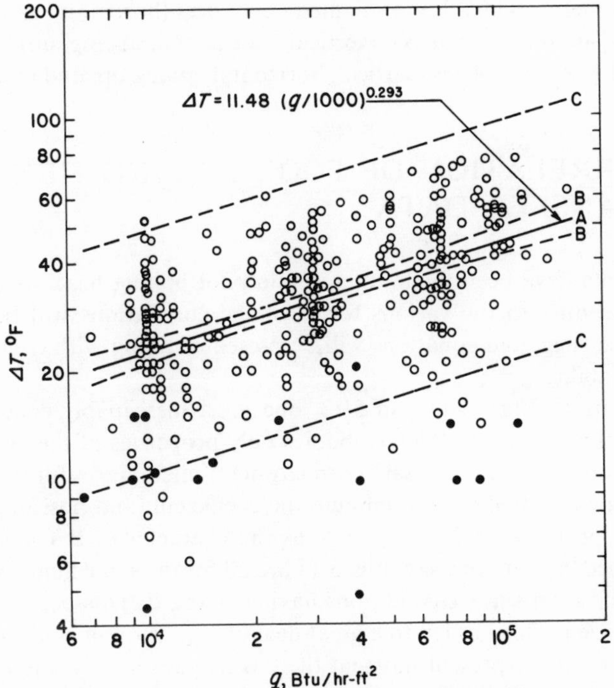

FIG. 5-17. Nucleate pool boiling of organic liquids (circles) and cryogens (solids) at 1 atm: *A*, regression line; *B*, 95% confidence limits on regression line; *C*, 95% probability limits on ΔT predicted by regression.[46]

difference is of interest. Figure 5-17 shows a composite of nucleate boiling data for various organic fluids and surfaces.[46] The conditions for the 276 data points of Fig. 5-17 are combinations of the following:

1. Liquid: Acetone, benzene, *i*-butanol, *n*-butanol, carbon disulfide, carbon tetrachloride, diethyl ether, diphenyl, ethanol, ethyl acetate, Freon-12_R, Freon-113_R, heptane, *n*-hexane, cryogens, methanol, methyl choloroform, *n*-pentane, *i*-propanol, stryrene, *m*-terphenyl, *o*-terphenyl, toluene.
2. Surface: Brass, chromium plate, copper, gold, Inconel, nickel plate, platinum, stainless steel, vitreous enamel, zinc (crystal).
3. Surface treatment: None; No. 36, 60, 120 (lapped), 150, 200, and 320 grit polishes; mirror finish; acid- and steel-wool-cleaned; annealed and unannealed; fresh and aged.

4. Geometry: 0.005–1.5-in.-diameter cylinders (horizontal and vertical); $\frac{3}{4}$–$3\frac{3}{4}$-in.-diameter disks (vertical, horizontal, facing up); $\frac{3}{4} \times \frac{3}{4}$-in. and $\frac{3}{4} \times 4$-in. plates (vertical, horizontal, facing up, and facing down)

5.8 CORRELATION OF POOL BOILING HEAT TRANSFER DATA

Attempts have been made to correlate pool boiling heat transfer data. The logic leading to the various forms of the correlations will be omitted here and the suggested equations will be presented along with comments on their applicability.

Referring to Figs. 5-11 and 5-13, one must inescapably conclude that any correlation equation which embodies only properties of the fluid (liquid or vapor) cannot be a "universal" correlation for all fluids or, for that matter, for any particular fluid. As a minimum the coefficient (and possibly even the exponents) must change in magnitude as the character of the solid surface changes, since the data for, say, Fig. 5-11 are all for the same fluid conditions, with only the solid surface conditions having changed. This point cannot be overemphasized. This has led to a great deal of confusion in work on boiling heat transfer. At the present moment there is no satisfactory way to include quantitatively the effect of the solid surface in any correlation equation; in spite of this, some researchers present correlation equations with a fixed magnitude for the coefficient. The best that can be accomplished is to correlate the effect of pressure for a given fluid and solid surface.

Many of the proposed correlations were developed by analyzing a simplified model of boiling leading to some dimensionless groups of quantities. Various forms of bubble Nusselt numbers, bubble Reynolds numbers, and Prandtl number appear in these equations. One of the early correlations[47] employed such groups with the characteristic dimension D given by Eq. (5-12). The equation proposed is

$$\frac{c_l(T_w - T_{\mathrm{sat}})}{h_{fg}} = C_{\mathrm{sf}}\left\{\frac{q}{\mu_l h_{fg}}\left[\frac{g_0\sigma}{g(\rho_l - \rho_v)}\right]^{1/2}\right\}^r\left(\frac{c_l\mu_l}{k_l}\right)^s \tag{5-28}$$

where C_{sf} should be a function of the particular fluid–heating surface combination. From Fig. 5-18a, 5-18b, or 5-19 the exponent $r = 0.33$. A cross plot of $c_l\,\Delta T/h_{fg}$ vs. Pr for constant values of the ordinate shows $s = 1.0$ for water, but $s = 1.7$ for all other fluids. The final correlation is shown in Fig. 5-18c, which results in $C_{\mathrm{sf}} = 0.013$ with a spread of approximately $\pm 20\%$. This process was repeated for other data with the results as shown in Table 5-1.

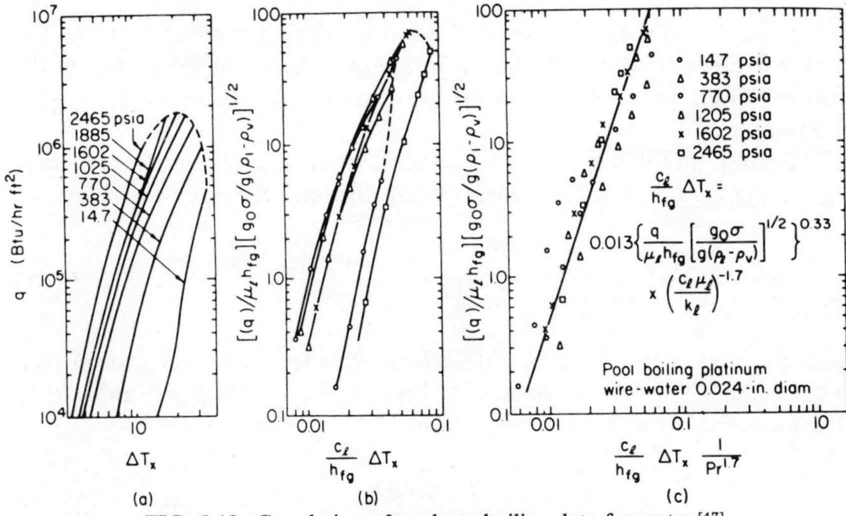

FIG. 5-18. Correlation of nucleate-boiling data for water.[47]

It should be emphasized that accurate values of fluid properties are essential in obtaining a correlation or in using Eq. (5-28). Also, the heating surface should be clean. The presence of a contamination or deposit on the heating surface can shift the relative position and slope of the curve, thus changing r and s in addition to C_{sf}. It should also be noted that Eq. (5-28) includes a $g^{1/6}$ term, which came from the expression for D in Eq. (5-12). As noted in Fig. 5-15, there appears to be no effect of g on the position of the boiling curve; therefore Eq. (5-28) should be used with $|g| = |g_0|$.

TABLE 5-1
Correlation Equation (5-28) with $r = 0.33$

Surface–fluid combination	C_{sf}	s
Water–nickel[48]	0.006	1.0
Water–platinum[49]	0.013	1.0
Water–copper[49]	0.013	1.0
Water–brass[51]	0.006	1.0
CCl_4–copper[51]	0.013	1.7
Benzene–chromium[52]	0.101	1.7
n-Pentane–chromium[52]	0.015	1.7
Ethyl alcohol–chromium[52]	0.0027	1.7
Isopropyl alcohol–copper[51]	0.0025	1.7
35% K_2CO_3–copper[51]	0.0054	1.7
50% K_2CO_3–copper[51]	0.0027	1.7
n-Butyl alcohol–copper[51]	0.0030	1.7

Later Forster and Zuber[53] used similar dimensionless groups, but with $\dot{R}R$ of Eq. (5-21) as the velocity times characteristic dimension in the Reynolds number and $2\sigma/\Delta p$ as the characteristic dimension in the Nusselt number.

Forster and Greif[54] modified the Forster–Zuber equation by not linearizing the Δp vs. ΔT relationship and suggested the following equation:

$$\frac{q}{h_{fg}\rho_v}\left(\frac{2\sigma}{\alpha\Delta p}\right)^{1/2}\left(\frac{\rho_l}{\Delta p}\right)^{1/4} = C_2\left\{\frac{\rho_l}{\mu}\left[\frac{c\rho_l(\pi\alpha)^{1/2}T_{\text{sat}}}{(h_{fg}\rho_v)^2}\,\Delta p\right]^2\right\}^{5/8}\text{Pr}^{1/3} \tag{5-29}$$

where C_2 was suggested to be 0.0012 from data for water at 1 and 50 atm, *n*-butyl alcohol at 3.4 atm, analine at 2.4 atm, and mercury at 1 and 3 atm. The following form of Eq. (5-29) approximates the data:

$$q = K_{\text{sf}}\left(\frac{\alpha c\rho_l T_{\text{sat}}}{h_{fg}\rho_v\sigma^{1/2}}\right)\left[\frac{cT_{\text{sat}}(\alpha)^{1/2}}{(h_{fg}L\rho_v)^2}\right]^{1/4}\left(\frac{\rho_l}{\mu}\right)^{5/8}\text{Pr}^{1/3}(\Delta p)^2 \tag{5-30}$$

For the limited amount of data K_{sf} was suggested to be 43×10^{-6}. Recently this equation was compared with other data. For the ethanol–chromium data of Cichelli and Bonilla[52] it was found that K_{sf} should be 82×10^{-6} and for the water–platinum data of Addoms[49] $K_{\text{sf}} = 142 \times 10^{-6}$. Obviously the coefficient does indeed change in magnitude for various surface–fluid combinations. Equation (5-30) needs more extensive comparison with data.

Equation (5-30) is identical to the earlier Forster–Zuber equation, except that Δp and ΔT are related by Eq. (5-6). Also, Eq. (5-30) yields a varying exponent of q/A vs. ΔT, increasing with ΔT in the range from 2 to 4. In many cases the curve cuts across natural boiling data at a lower slope.

Other correlations have been suggested by Gilmour,[55] McNeilly,[56] and Levy.[57] The Gilmour correlation contains a size effect not verified by experiment. The Levy procedure is dimensional and employs an empirical curve around which data scatter by a factor of five or more, which is about the same variation observed in C_{sf} of Table 5-1. (In the Levy paper as originally published the ordinate of Fig. 4 should read multiplied by 10^{-5} instead of 10^{-6}.)

Various Russian workers have suggested correlation equations in the following form[58]:

$$\text{Nu}_* = A\text{Pr}^{n_1}\text{Pe}_*^{n_2}K_p^{n_3}K_t^{n_4}\text{Ar}_*^{n_5} \tag{5-31}$$

with the following sets of parameters:

Authors	A	n_1	n_2	n_3	n_4	n_5
M. A. Kichigan and N. Y. Tobilevich	1.04×10^{-4}	0	0.7	0.7	0	0.125
S. S. Kutateladze	7.0×10^{-4}	−0.35	0.7	0.7	0	0
V. M. Borishanskiy and F. P. Minchenko	8.7×10^{-4}	0	0.7	0.7	0	0
G. N. Kruzhilin and Ye. K. Averin	0.082	−0.5	0.7	0	0.377	0
D. A. Labuntsov	0.125	−0.32	0.65	0	0.35	0

In Eq. (5-31)

$$\mathrm{Nu}_* \equiv \frac{h}{k}\left[\frac{g_0\sigma}{g(\rho_l - \rho_v)}\right]^{1/2}, \qquad \mathrm{Pe}_* \equiv \frac{q}{\alpha\rho_v h_{fg}}$$

$$K_p \equiv \frac{p}{[g\sigma(\rho_l - \rho_v)/g_0]^{1/2}}, \qquad K_t \equiv \frac{(\rho_v h_{fg})^2}{Jc_l T_{\mathrm{sat}}\rho_l[g\sigma(\rho_l - \rho_v)/g_0]^{1/2}} \tag{5-32}$$

$$\mathrm{Ar}_* \equiv \frac{g}{v^2}\left(\frac{g_0\sigma}{g(\rho_l - \rho_v)}\right)^{3/2}\left(1 - \frac{\rho_v}{\rho_l}\right)$$

It should be emphasized again that the coefficient A in Eq. (5-31) is not a constant, but varies with the surface fluid combination.

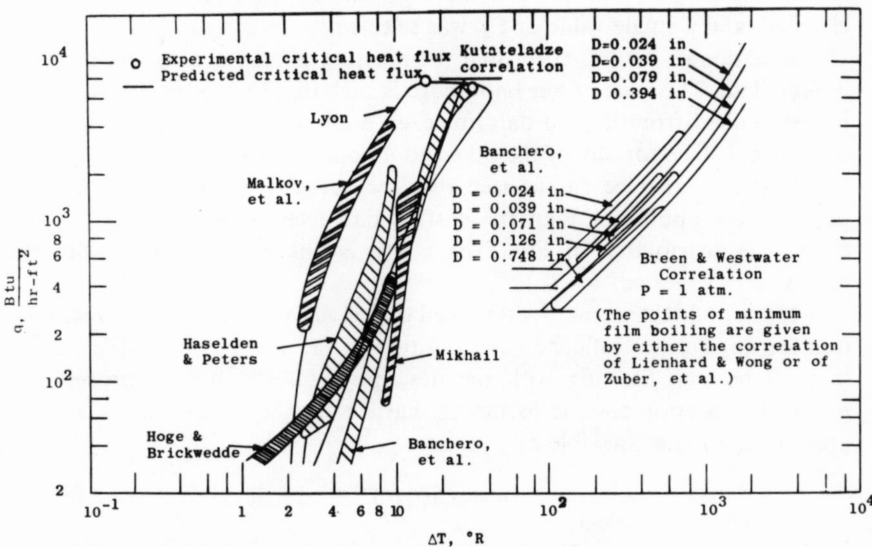

FIG. 5-19. Experimental nucleate and film pool boiling of oxygen at 1 atm compared with the predictive correlations of Kutateladze, and Breen and Westwater.[63]

Cryogenic fluids—liquid oxygen, nitrogen, hydrogen, helium, etc.—in pool boiling behave like noncryogenic fluids. The position of the q vs. $(T_w - T_{sat})$ curve depends strongly on the nature of the solid surface. A survey of such data is presented by Zuber and Fried,[59] Richards *et al.*,[60] Clark,[61] Seader *et al.*,[62] and Brentari and Smith.[63] Figure 5-19 shows a composite plot of a wide variety of pool boiling and also film boiling data for oxygen. Brentari and Smith[63] attempted to correlate all of these data with a modified Kutateladze equation

$$q = 4.87 \times 10^{-11}\left[\frac{C_{pl}}{h_{fg}\rho_v}\right]^{1/5}\left[\frac{k_l\rho_l^{1.282}p^{1.75}}{\sigma^{0.906}\mu_l^{0.626}}\right](T_w - T_{sat})^{2.5} \tag{5-33}$$

which represents the solid line through the central portion of the data. Quite obviously the coefficient of the equation must vary by a factor of 10 or so to accommodate all of the data. Similar composite data for nitrogen, hydrogen, and helium[63] exist.

Clark[61] shows that a modified form of Eq. (5-28) also correlates the boiling data of cryogenic fluids:

$$\frac{q}{h_{fg}\mu_l}\left[\frac{g_0\sigma}{g(\rho_l - \rho_v)}\right]^{1/2} = 3.25 \times 10^5\left[\frac{C_{pl}\Delta T}{h_{fg}}\left(\frac{T/T_c}{\mathrm{Pr}_l}\right)^{1/8}\right]^{2.89} \tag{5-34}$$

Here Eq. (5-28) was modified by including an additional pressure effect empirically in the form of T/T_c, the exponent on the Prandtl number was changed, and a single value of C_{sf} was selected.

Equation (5-28) has correlated the pressure effect for a variety of pool boiling data. The state of our knowledge is such that the coefficient C_{sf} must be determined from limited data for each fluid–surface combination. This, of course, is true for any of the other pool boiling correlations.

It should also be emphasized that the actual metal of the surface is perhaps less important than the surface character as represented by the cavity-size distribution of Fig. 5-10, which is unknown for practically all of the surfaces tested.

Mikic and Rohsenow[64] attempted to show how the cavity-size distribution for any surface influences the position of the q/A vs. ΔT boiling curve in pool boiling. Starting with the description of the boiling process as outlined in Section 5.4 and assuming that the number of cavities of radius greater than r is expressible by

$$n = C_1(r_s/r)^m \tag{5-35}$$

where r_s is the radius of the largest cavity present and C_1, r_s, and m are determined from cavity-size distribution measurements, Brown and Bergles[65] obtained the following expression:

$$q = (A_{nc}/A)q_{nc} + q_b \tag{5-36}$$

where A_{nc}/A is the fraction of the area where bubbles are not being formed, q_{nc} is the natural convection heat transfer at this area, and q_b is given by

$$\frac{q_b}{\mu_l h_{fg}}\left[\frac{g_0\sigma}{g(\rho_l - \rho_v)}\right]^{1/2} = B(\varphi\,\Delta T)^{m+1} \tag{5-37}$$

where

$$B \equiv \left(\frac{r_s \mathrm{J}}{2}\right)^m \frac{2\sqrt{\pi}\, g_0^{11/8}}{g^{9/8}}\, C_2^{5/3} C_3^{3/2} C_1 \tag{5-38}$$

$$\phi^{m+1} \equiv \frac{k_l^{1/2}\rho_l^{17/8} c_l^{19/8} h_{fg}^{(m-23/8)} \rho_v^{(m-15/8)}}{\mu_l[(\rho_l - \rho_v)]^{9/8} \sigma^{(m-11/8)} T_s^{(m-15/8)}} \tag{5-39}$$

Here $C_2 = 0.00015$ for water and 0.000465 for other fluids [Eq. (5-21)] and $C_3 = 0.6$ [Eq. (5-22)]; C_1, r_s, and m are to be determined from the cavity size distribution, Eq. (5-35). Note that B is solely a function of cavity size distribution and φ is a function of fluid properties, except for the exponent m. Further if the cavity size distribution has a slope of m, then the q vs. ΔT curve should have a slope of $m + 1$.

For most fluids $q_{nc} \ll q_b$ in Eq. (5-36) and may be neglected.

The cavity size distribution for most surfaces for which data are available has not been measured. In these cases the boiling data may be used to determine m and B. Mikic and Rohsenow[64] show that Eq. (5-38) used in this way does correlate existing boiling data.

5.9 NOMENCLATURE

A	= area of heating surface
A_b	= surface area of a bubble
C_{sf}	= coefficient of Eq. (5-28), which depends on the nature of the heating-surface–fluid combination
c_l	= specific heat of saturated liquid, Btu/lb$_m$-°F
D	= tube diameter
D_b	= diameter of the bubble as it leaves the heating surface, ft
D_e	= equivalent diameter = 4 (flow area)/perimeter
f	= frequency of bubble formation, hr^{-1}
G	= mass velocity in a channel or tube, lb$_m$/hr-ft^2
G_b	= mass velocity of bubbles at their departure from the heating surface, lb$_m$/hr-ft^2

g = acceleration of gravity
g_0 = conversion factor, 4.17×10^8 lb_m-ft/hr^2-lb_f
H = enthalpy, Btu/lb_m
h_{fg} = latent heat of evaporation, Btu/lb_m
h = $(q/A)/\Delta T$, film coefficient of heat transfer, Btu/hr-ft^2-°F
J = 778 ft-lb/Btu
Ja = Jakob number, $\rho_l c_l(T_w - T_{sat})/\rho_v h_{fg}$
k_l = thermal conductivity of saturated liquid, Btu/hr-ft-°F
L = length
M = molecular weight
Nu = Nusselt number = hd/k
Nu_b = bubble Nusselt number
n = number of points of origin of bubble columns per unit area of heating surface
Pr_l = Prandtl number $C_l\mu_l/k_l$
P_c = critical pressure
p_l = pressure on liquid side of interface of radius r
p_v = pressure on vapor side of interface of radius r
q = heat transfer rate per unit heating surface area, Btu/hr-ft^2
q_b = heat transfer rate to bubble per unit heating surface while bubble remains attached to the surface, Btu/hr-ft^2
R = gas constant, also bubble radius
Re_b = bubble Reynolds number
r = radial distance
S = entropy, Btu/lb_m°F
T_b = temperature of subcooled liquid, °F
T_c = critical thermodynamic temperature
T_l = temperature of liquid, °F
T_{sat} = saturation temperature, °F
T_w = temperature of heating surface, °F
t = time
v = volume, ft^3/lb_m
W,w = flow rate, lb_m/hr
z = depth

Greek Letters

α = thermal diffusivity, $k/\rho c_p$
β = bubble contact angle
ΔT_{sat} = $T - T_{sat}$
$\Delta T_{subcool}$ = $T_{sat} - T_l$

μ_1 = viscosity of saturated liquid, $lb_m/ft\text{-}hr$
ρ_l = density of saturated liquid, lb_m/ft^3
ρ_v = density of saturated vapor, lb_m/ft^3
σ, σ_{lv} = surface tension of liquid–vapor interface, lb_f/ft
σ_{sl} = surface tension of solid–liquid interface, lb_f/ft
σ_{vs} = surface tension of vapor–solid interface, lb_f/ft
v = kinematic viscosity, μ/ρ

5.9 REFERENCES

1. S. W. Gouse, "An Index to Two-Phase Gas-Liquid Flow Literature," MIT Report No. 9, MIT Press, Cambridge, Mass. (1966).
2. J. W. Westwater and P. H. Strenge, *Chem. Eng. Progr. Symp. Series* **29**, 95 (1959).
3. C. Corty and A. S. Foust, *Chem. Eng. Progr. Symp. Series* **17**, 51 (1955).
4. S. G. Bankoff, in *Proc. of Heat Transfer and Fluid Mechanics Institute*, Stanford Univ. Press (1956).
5. W. M. Rohsenow, *Ind. Eng. Chem.* **58**, 1 (1966).
6. P. Griffith, in *Symp. on Boiling Heat Transfer*, *Proc. Inst. Mech. Engrs.* **180** Part 1 and 3C (1965–1966).
7. I. Shai, MIT Heat Transfer Lab., Rept. 76303-45 (January 1967).
8. P. Griffith and J. D. Wallis, *Chem. Eng. Progr. Symp. Series* **30**, 49 (1960).
9. A. E. Bergles and W. M. Rohsenow, *Trans. ASME, J. Heat Transfer* **86**, 365 (1964).
10. Y. Y. Hsu and R. W. Graham, NASA Tech. Note, TNP-594 (May 1961).
11. W. Brown, Sc.D. Thesis, MIT Heat Transfer Lab., Mech. Eng. Dept. (January 1967).
12. W. Frost and G. S. Dzakowic, ASME paper 67-HT-61 (1967).
13. D. B. R. Kenning and M. G. Cooper, in *Boiling Symp., Inst. Mech. Engs. (London)*, **180C** (1965).
14. C. W. Deane IV, and W. M. Rohsenow, MIT Heat Transfer Lab., Rept. No. DSR 76303-65 (October 1969).
15. R. M. Singer and R. E. Holtz, paper at Intern. 4th Heat Trans. Conf., Versailles, France (September 1970).
16. C. Y. Han and P. Griffith, MIT Heat Transfer Lab., Rept. No. 19 (1962).
17. R. Moissis and P. J. Berenson, *Trans. ASME, J Heat Transfer* **85C**, 221 (1963).
18. M. G. Cooper and J. P. Lloyd, *3rd Intern. Heat Trans. Conf.* (1966), p. 193; also *Intern. J. Heat Mass Transfer* **12**, 895 (1969).
19. M. G. Cooper, *Intern. J. Heat Mass Transfer* **12**, 915 (1969).
20. M. Cumo, C.N.E.N., C.S.N., Rome, Italy, personal communication.
21. W. Fritz, *Phys. Z.* **36**, 379 (1935).
22. J. W. Wark, *J. Phys. Chem.* **37**, 623 (1933).
23. B. B. Mikic, W. M. Rohsenow, and P. G. Griffith, *Intern. J. Heat Mass Transfer*, **13**, 657 (1970).
24. Y. Lien and P. Griffith, *PhD. Thesis, ME Dept., MIT*, Jan. 1969.
25. B. B. Mikic and W. M. Rohsenow, in *Progress in Heat and Mass Transfer*, Vol. II, Pergamon Press, London (1969), p. 283.
26. M. S. Plesset and J. A. Zwick, *J. Appl. Phys.* **25**, 493 (1954).

27. L. E. Scriven, *Chem. Eng. Sci.* **10**, p. 1–13 (1959).
28. R. Cole and H. L. Shulman, *Intern. J. Heat Mass Transfer* **9** (12), 1377 (1966).
29. R. Cole and W. M. Rohsenow, *Chem. Eng. Progr. Symp. Series* **65**, 92, 211 (1969).
30. N. Zuber, USAEC Rept. AECU-4439 (1959).
31. F. N. Peebles and H. J. Garber, *Chem. Eng. Progr.* **49**(2), 88 (1953).
32. R. Cole, *AIChE J.*, **10**(7), 779 (1967).
33. H. J. Ivey, *Intern. J. Heat Mass Transfer* **10**(8), 1023 (1967).
34. R. Cole, *AIChE J.*, **6**(4), 533 (1960).
35. N. Koumoutsus, R. Moissis, and A. Spyridonos, *Trans. ASME, J. Heat Transfer* **90C**(2), 223 (1968).
36. B. E. Staniszewski, Technical Rept. No. 16, DSR 7673, Office of Naval Research Contract Nonr-1841 (39), MIT Heat Transfer Lab. (August 1959).
37. J. M. Astruc, P. Perroud, A. Lacaze, and L. Weil, in *Advances in Cryogenic Engineering*, Vol. 12, Plenum Press, New York (1967), p. 387.
38. P. J. Marto and W. M. Rohsenow, *Trans. ASME, J. Heat Transfer* **88C**(2), 183 (1966).
39. L. Austin, *Mitt. Forsch.* **7**, 75 (1903).
40. F. S. Pramuk and J. W. Westwater, *Chem. Eng. Progr. Symp. Ser.* **52**(18), 79 (1956).
41. E. E. Duke and V. E. Shrock, *Fluid Mech. Heat Trans. Inst.*, **1961**, (June) p. 130.
42. M. Jakob and W. Fritz, *Forsch. Gebiete Ing.* **2**, 434 (1931).
43. E. K. Averin, AERE Lib/Trans. 562, from *Izv. Akad. Nauk SSSR (OTN)* **1954**(3), 116.
44. R. Siegel and C. Usiskin, *Trans. ASME, J. Heat Transfer* **81**, 3 (1959).
45. H. J. Merte Jr. and J. A. Clark, Univ. Michigan Rept. No. 2646-21-T, Tech. Rept. No. 3 (November 1959).
46. R. J. Armstrong, *Intern. J. Heat Mass Trans.* **9**, 1148 (1966).
47. W. M. Rohsenow, *Trans. ASME* 969 (1952).
48. W. M. Rohsenow and J. A. Clark, Heat Transfer Fluid Mech. Inst., Stanford, California (1951).
49. J. M. Addoms, D.Sc. Thesis, Chem. Eng. Dept., MIT (June 1948).
50. D. S. Cryder and A. C. Finalborgo, *Trans. AIChE* **33**, 346 (1937).
51. E. L. Piret and H. S. Isbin, AIChE Heat Transfer Symp., St. Louis (December 1953).
52. M. T. Cichelli and C. F. Bonilla, *Trans. AICHE* **41**, 755 (1945).
53. K. Forster and N. Zuber, *AIChE J.* Vol. 1, 531, 1955.
54. K. Forster and R. Greif, Progr. Rept. No. 7, Dept. of Eng., UCLA, Los Angeles, California (1958).
55. C. H. Gilmour, *Chem. Eng. Progr.* **54**(10), 77 (1958).
56. M. J. McNeilly, *J. Imperial College Chem. Eng. Soc.* **7**, 18 (1953).
57. S. Levy, *J. Heat Transfer* **81**, 37 (1959).
58. S. S. Kutateladze, *Fundamentals of Heat Transfer*, Academic Press, New York (1963).
59. N. Zuber and E. Fried, Am. Rocket Soc. Propellants, Combusion and Liquid Rocket Conf. (April 1961).
60. R. J. Richards, W. G. Steward, and R. B. Jacobs, NBS Tech. Note 122, Boulder, Colorado (October 1961).
61. J. A. Clark, in *Cryogenic Technology* (R. W. Vance, ed.), John Wiley, New York (1963), Chapt. 5.; also *Advances in Heat Transfer*, Vol. 5 (Hartnett and Irving eds.), Academic Press, New York (1968), p. 325.
62. J. D. Seader, W. S. Miller, and L. A. Kalvinskas, NASA CR-243 (1965).
63. E. G. Brentari and R. V. Smith, in *Advances in Cryogenic Engineering*, Vol. 10, Plenum Press, New York (1965), p. 325.

64. B. B. Mikic and W. M. Rohsenow, *Trans. ASME, J. Heat Transfer* **91** (2), 245 (1969).
65. W. T. Brown Jr. and A. E. Bergles, Ph.D. Dissertation, MIT (January 1967).
66. J. C. Chen, *Trans. ASME, J. Heat Transfer* **90C**, 303 (1968).
67. H. J. Ivey and D. J. Morris, UK Rept. AEEW-R-137, Winfrith (1962).
68. P. Berenson, D. Sc. Thesis, Mech. Eng. Dept. MIT, (February 1960); also Tech Rept. No. 17, MIT Heat Transfer Lab. (March 1960).
69. P. J. Marto and W. M. Rohsenow, *Trans. ASME* **88C**(2), 196 (1966).
70. W. H. McAdams, W. E. Kennel, C. S. Minden, C. Rudolf, and J. E. Dow, *Ind. Eng. Chem.* **41**, 1945 (1959).
71. M. Jakob, *Mech. Eng.* **58**, 643 (1936).

CRITICAL HEAT FLUX 6

W. FROST and R. VON RETH

6.1 INTRODUCTION

The critical heat flux is the heat flux at which a transition from nucleate to film boiling occurs. It is of interest to designers because a marked reduction in heat transfer results from the transition, which may be rather sudden and which is accompanied by a substantial rise in surface temperature. Knowledge of the critical heat flux dictates whether a system is designed with a nucleate boiling heat transfer coefficient (Chapter 5), representing a low thermal resistance, or with a film boiling heat transfer coefficient (Chapter 7), representing a high thermal resistance. Although not frequently encountered with cryogenic fluids nor with more volatile common refrigerants, the temperature rise associated with the transition to film boiling can, with fluids such as water, be sufficiently great to melt the heating surface and destroy the system.

The critical heat flux is encountered under a number of flow conditions. These may be classified as follows:

1. Natural convection boiling critical heat flux: (a) Unconfined as in a large pool. (b) Confined as in a natural circulating thermosiphon.
2. Forced convection boiling critical heat flux: (a) Internal flows. (i) axial flow in round ducts; (ii) axial flow in ducts of other geometric cross sections; (iii) swirl flow. (b) External flow. (c) Rod bundle flows.

In any of these flow regimes, the liquid may be either saturated or subcooled. The axial forced flow systems are often classified also as having uniform or nonuniform heating sections.

W. FROST University of Tennessee Space Institute, Tullahoma, Tennessee.
R. VON RETH Messerschmitt–Bolkow–Blohm.

Topics 2a(iii) and 2c will not be dealt with to any extent in this chapter, nor will nonuniform heating sections.

These notes emphasize presently existing correlating techniques which can be applied in design. No derivations of theories nor arguments for or against proposed hypotheses of the burnout mechanism are given. Experimental findings reported in the literature are presented if they shed light on correct application of the proposed correlations.

Although the authors have attempted to assemble as much of the reported experimental evidence for cryogenic and low-temperature liquids as possible, the available data are scarce. Hence, it is frequently necessary to present correlations or arguments verified only for water or for organic liquids. Fortunately, where experimental data for cryogens are compared with correlations developed for other fluids, the agreement is relatively good. This lends a certain degree of confidence to the cryogenic engineer when correlations or extrapolations of data for other than cryogenic fluids are applied to design analysis.

6.2 NATURAL CONVECTION BOILING, CRITICAL HEAT FLUX

6.2.1 Unconfined Natural Convection

6.2.1.1 *Saturated:* The transition from nucleate to film boiling in a large pool of liquid occurs when the vapor generation from the heated surface becomes so profuse that the mechanism of vapor removal fails and the heated surface becomes separated from the liquid by a continuous film of vapor. Disagreement exists on the exact physical phenomenon by which this occurs. Early studies[1–7] hypothesized the problem as being purely hydrodynamic in nature and practically all of the currently used critical heat flux correlations are developed on this assumption.

Seader *et al.*[8] have found that nearly all the theories proposed can be rearranged in terms of a superficial vapor velocity,

$$\varphi_1 = q_{\max}/\rho_v h_{fg} \tag{6-1}$$

where φ_1 has the form given in Table 6-1 and the accompanying Figs. 6-1 and 6-2 taken from this reference. Note that φ_1 has the dimensions of length per unit time and consistent units are required in evaluating it. Physical models employed in deriving the listed correlations differ slightly; however, all generally envision the number of nucleation sites becoming very numerous as the critical heat flux is approached and the frequency of bubble formation

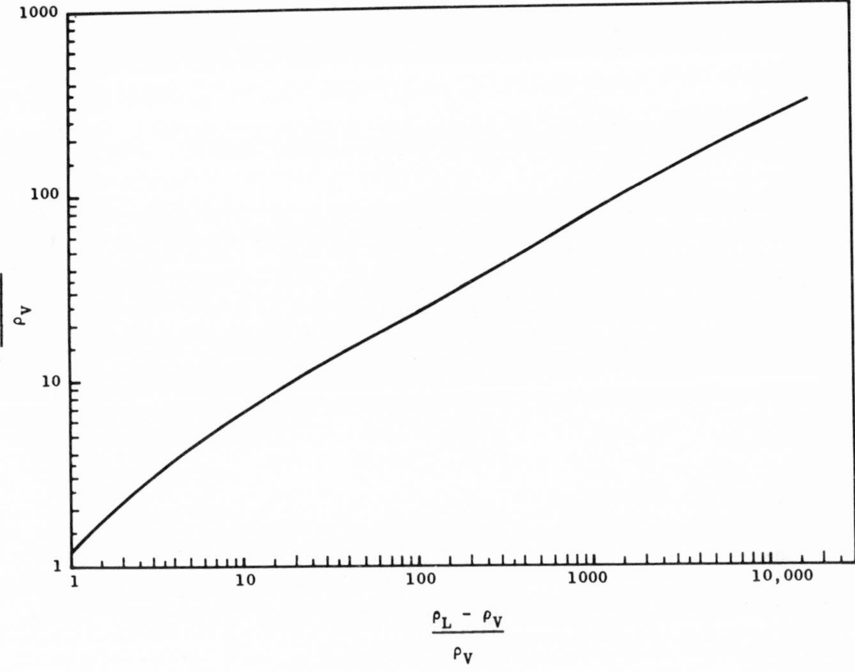

FIG. 6-1. Correlation parameter of Addoms and Noyes.

sufficiently great that departing bubbles coalesce with their predecessor forming columns of vapor closely spaced on the heated surface. The correlation of Rohsenow and Griffith[1] [Eq. (1) in Table 6-1] is a semianalytical development which assumes that coalescence of these vapor columns results in burnout. Kutateladze[2] based his prediction techniques on dimensional analysis supported with liquid–vapor instability arguments, and Zuber[3] theoretically arrived at an almost identical correlation by analyzing the stability between a two-dimensional pattern of vapor jets flowing upward and liquid jets flowing downward. The development of other correlations uses similar approaches and may be found in the appropriate references.

Figures 6-3–6-5 from Ref. 8 compare all the correlations with a wide range of peak heat flux data for hydrogen, nitrogen, and oxygen, respectively. The hydrogen data of Fig. 6-3 show considerable scatter. Equations (2), (4)–(7), (9), (10b), and (11) in Table 6-1 lie within the scatter at low pressures, but all the equations over predict the peak heat flux at high pressures. The nitrogen data also show scatter on the order of $\pm 30\%$ at intermediate pressures and $\pm 75\%$ at atmospheric pressure. Equations (2), (4)–(6), (9),

TABLE 6-1

Summary of Pool Nucleate Boiling Maximum Heat Flux Theories[a]

Ref.	ϕ_1	Equation No.
Addoms[4]	$[g(k_l/\rho_l C_l)]^{1/3}\{\varphi[(\rho_l - \rho_v)/\rho_v]\}$ (see Fig. 6-1)	(1)
Rohsenow and Griffith[1]	$143[(\rho_l - \rho_v)/\rho_v]^{0.6}$, ft/hr	(2)
Griffith[9]	$\{[g(\rho_l - \rho_v)/\mu_l](k_l/\rho_l C_l)^2\}^{1/3}[\varphi(P/P_c)]$ (see Fig. 6-2)	(3)
Zuber and Tribus[10]	$K[\sigma g g_c(\rho_l - \rho_v)/\rho_v^2]^{1/4}[\rho_l/(\rho_l + \rho_v)]^{1/2}$, $0.12 \leqslant K \leqslant 0.16$	(4)
Kutateladze[2]	$K[\sigma g g_c(\rho_l - \rho_v)/\rho_v^2]^{1/4}$, $0.095 \leqslant K \leqslant 0.20$	(5)
Borishanskii[2]	$[\sigma g g_c(\rho_l - \rho_v)/\rho_v^2]^{1/4}$ $\{0.13 + 4\{\mu_l^2[g(\rho_l - \rho_v)]/\rho_l(\sigma g_c)^{3/2}\}^{1/2}\}^{0.4}$	(6)
Noyes[6]	$0.144(g g_c \sigma/\rho_l)^{1/4}[(\rho_l - \rho_v)/\rho_v]^{1/2}(C_l\mu_l/k_l)^{-0.245}$	(7)
Noyes[6] (alternate correlation)	$[g(k_l/\rho_l C_l)]^{1/3}(C_l\mu_l/k_l a^b)^{1/2}[\varphi(\rho_l - \rho_v)/\rho_v]$ (see Fig. 6-1)	(8)
Chang and Snyder[5]	$0.145[\sigma g g_c(\rho_l - \rho_v)/\rho_v^2]^{1/4}[(\rho_l + \rho_v)/\rho_l]^{1/2}$	(9)
Chang[11]	$K[\sigma g g_c(\rho_l - \rho_v)/\rho_v^2]^{1/4}$, $K = 0.098$ vertical	(10a)
	$K = 0.13$ horizontal	(10b)
Moissis and Berenson[7]	$0.18\ [\sigma g g_c(\rho_l - \rho_v)]^{1/4}[(\rho_l + \rho_v)/\rho_l\rho_v]^{1/2}$ $\times\ [1 + 2(\rho_v/\rho_l)^{1/2} + (\rho_v/\rho_l)]^{-1}$	(11)

[a] $q_{max}/h_{fg}\rho_v = \phi_1$ (consistent units except as noted).

TABLE 6-2

Reference Values for Saturated Pool Boiling Burnout (*Evaluated at $P_1 = 0.05$ and $a/g = 1$*)

	Range of $q_{max,ref}$, W/cm²	
Ammonia	117.7	153.9
Argon	24.9	32.5
Carbon dioxide	48.3	63.1
Carbon tetrachloride	25.7	33.6
Ethanol	71.3	93.0
Freon 12	26.6	34.8
Helium	0.35	0.45
Parahydrogen	7.03	9.21
Kerosene (JP-4)	31.0	40.6
Neon	10.5	13.8
Nitrogen	18.6	24.3
Oxygen	39.7	38.9
n-Pentane	27.2	35.6
Propane	21.4	27.9
Water	268.2	350.7
Methane	27.8	36.2
Ethane	31.7	41.5
Ethylene	30.1	39.3

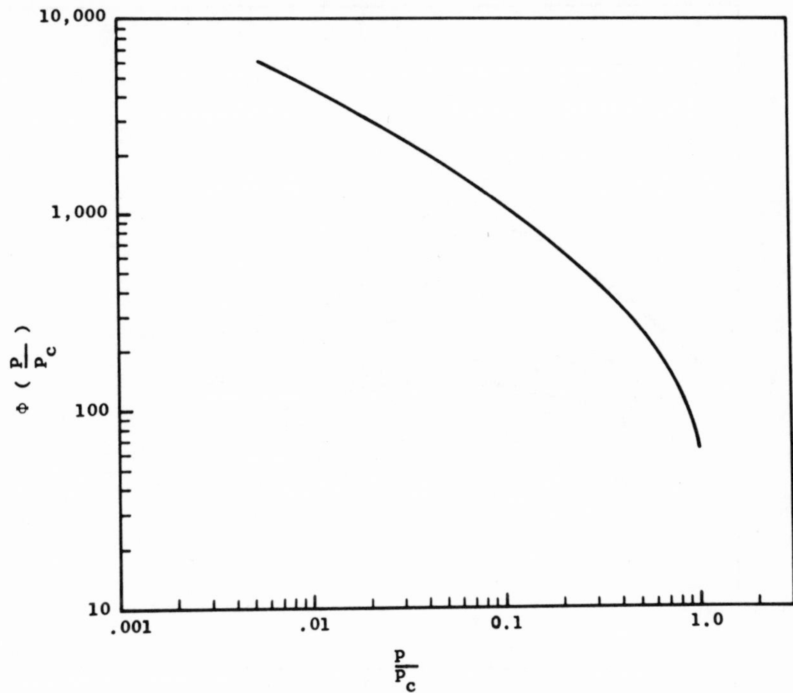

FIG. 6-2. Correlation parameter of Griffith.

and (10b) in Table 6-1 lie within the scatter at all pressures for which data are presented. The oxygen data are not sufficient to show scatter and the sparse data show reasonably good agreement with Eq. (5) in Table 6-1.

Figure 6-6 compares the Kutateladze correlation [Eq. (5), Table 6-1] with experimental data on oxygen, nitrogen, hydrogen, and helium.[9] The dashed portion of the curve represents conditions above a reduced pressure of 0.6, for which the correlation begins to depart from the data. Brentari *et al.*[12] and Lyon *et al.*[13] indicate that the available maximum heat flux prediction techniques for cryogens are marginally successful at reduced pressures above 0.6 and become virtually useless above a reduced pressure of 0.8. Much of the discrepancy in the burnout data which has resulted in a multitude of critical heat flux correlation has now been traced to heater geometry and heater surface treatment. These effects have not as yet been incorporated into a satisfactory general prediction technique and consequently no one particular equation in Table 6-1 developed on the basis of a primarily hydrodynamic instability analysis can be singled out as better than

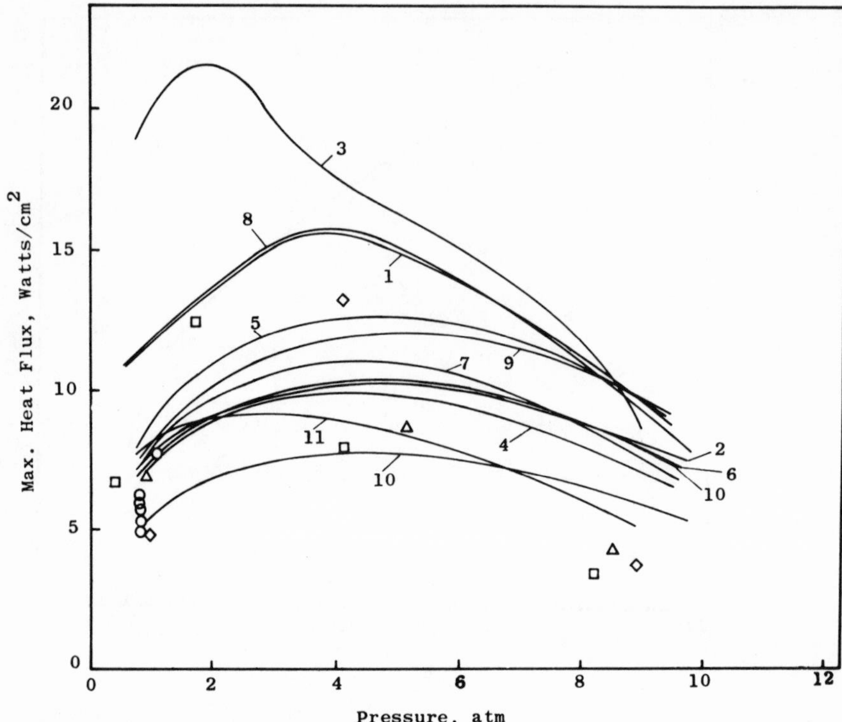

FIG. 6-3. Comparison of theories for pool nucleate boiling maximum heat flux for hydrogen. (1) Addoms; (2) Rohsenow and Griffith; (3) Griffith; (4) Zuber and Tribus; (5) Kutateladze; (6) Borishanskii; (7) Noyes; (8) Noyes (alternate); (9) Chang and Snyder; (10) Chang (horizontal); (10) Chang (vertical); (11) Moissis and Berenson.

the others. Many of them, however, do predict the experimentally determined dependence on pressure P through the physical fluid properties, and on the local acceleration of gravity g.

Chichelli and Bonilla[14] and Kutateladze and Borishanskii[15] initially investigated the critical heat flux for water and for organic liquids as a function of reduced pressure. A maximum value of the critical heat flux occurs at reduced pressures on the order of 0.25–0.33.

Frost and Dzakowic[16] found that the Kutateladze correlation along with a number of the others listed in Table 6-1 can be represented with a single curve for a large number of fluids by plotting $q_{max}/q_{max,ref}$ versus reduced pressure P_r. The reference value was arbitrarily chosen at $P_r = 0.05$ and is tabulated for a number of fluids in Table 6-2. Figure 6-7 compares Eq. (5), Table 6-1, plotted in this manner with data for ten different fluids. The scatter is appreciable.

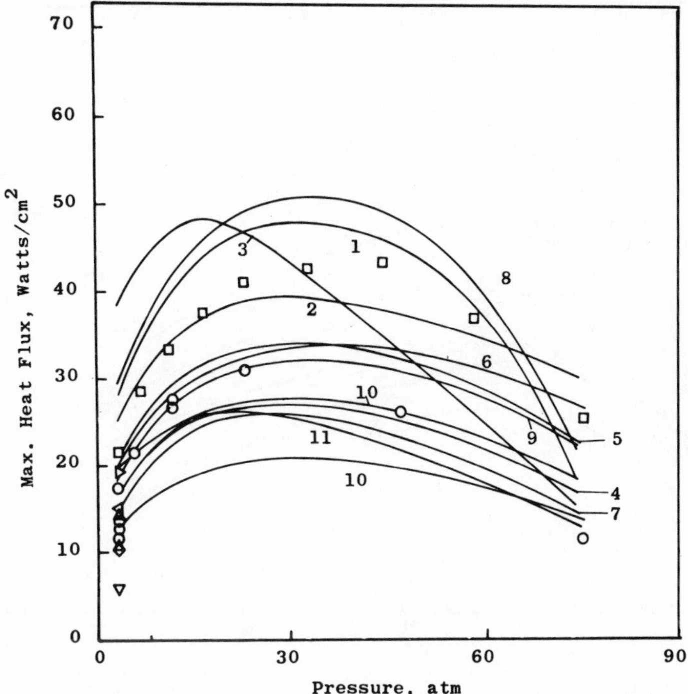

FIG. 6-4. Comparison of theories for pool nucleate boiling maximum heat flux for nitrogen. (1) Addoms; (2) Rohsenow and Griffith; (3) Griffith; (4) Zuber and Tribus; (5) Kutateladze; (6) Borishanskii; (7) Noyes; (8) Noyes (alternate); (9) Chang and Snyder; (10) Chang (horizontal); (10) Chang (vertical); (11) Moissis and Berenson.

Sciance *et al.*,[17] working with methane, set out to obtain accurate saturated pool boiling data over a range from atmospheric pressure to the critical pressure. They concluded that Noyes' equation [Eq. (7), Table 6-1] correlated their data so well that it was not appropriate to change the constants (Fig. 6-8). Figure 6-9 compares data of Fig. 6-7 with Noyes' equation [Eq. (7), Table 6-1] plotted similarly but with the Prandtl number ratio included in the expression on the ordinate.

Borishanskii[18] and Cobb and Park[19] also plotted the critical heat flux on reduced coordinates. They used actual measured heat fluxes for reference values, as contrasted to Figs. 6-7 and 6-9 where reference values were calculated from equations (Table 6-2). Better agreement with the data is achieved using measured reference values as illustrated by Fig. 6-10 from Cobb and Park.[19] This is because the critical heat flux is a function of system geometry and heat transfer surface chemistry as well as reduced

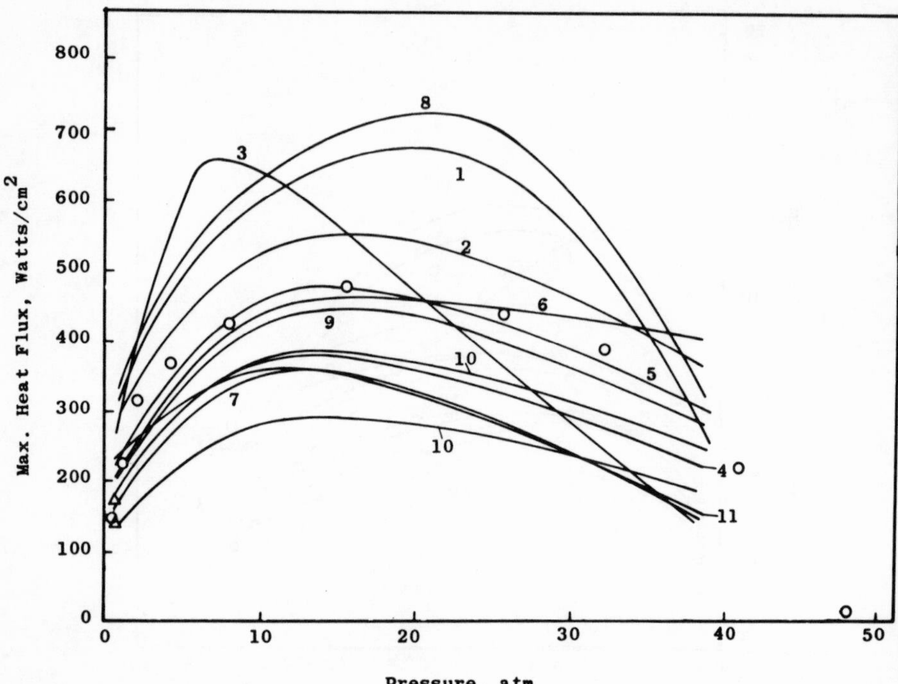

FIG. 6-5. Comparison of theories for pool nucleate boiling maximum heat flux for oxygen. (1) Addoms; (2) Rohsenow and Griffith; (3) Griffith; (4) Zuber and Tribus; (5) Kutateladze; (6) Borishanskii; (7) Noyes; (8) Noyes (alternate); (9) Chang and Snyder; (10) Chang (horizontal); (10) Chang (vertical); (11) Moissis and Berenson.

pressure (as will be discussed in the following section). Both these effects are factored out when the ratio is taken.

Cobb and Park[19] give the least square fit of their data as

$$q_{\max}/q_{\max,\mathrm{ref}} = 1.70 - 3.90T_r - 0.048T_r^2 + 2.41T_r^3 + 7.58T_r^4 + 5.20T_r^5 + 12.88T_r^6 \tag{6-2}$$

which has an average deviation of 12.6%. The only difficulty with using this correlation is that an experimental value of the peak heat flux is required before any design calculation can be made.

The variation of critical heat flux with local acceleration is currently of considerable interest in view of space travel. Usiskin and Siegel[20] and Merte and Clark[21] have experimentally investigated reduced- and zero-gravity effects on boiling of water and of liquid nitrogen. The data substantiate the $\frac{1}{4}$-power dependence on g expressed by the majority of the

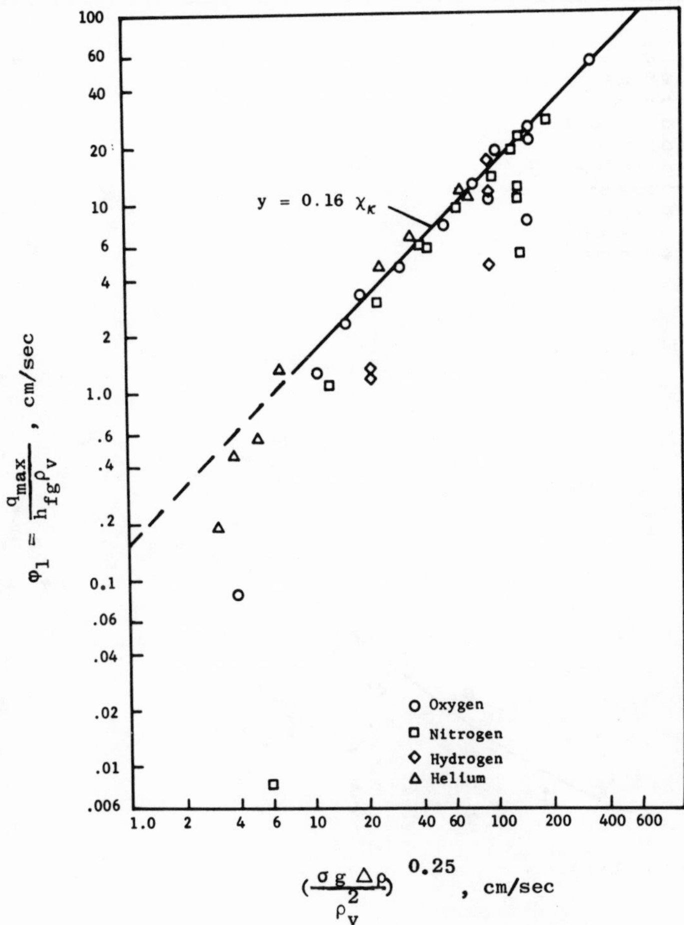

FIG. 6-6. Comparison of the maximum nucleate heat transfer fluxes with the Kutateladze maximum correlation.

burnout correlations down to values of a/g on the order of 0.1. At a/g less than 0.1 the $\frac{1}{4}$-power dependence seems no longer valid. Figure 6-11 and Fig. 6-12 show data for oxygen, nitrogen, hydrogen, and water from Refs. 13 and 20–22 as plotted by Clark.[23] The data indicate that the critical heat flux does not reach a value of zero at zero g, but has a finite value which is still a substantial fraction of the critical heat flux experienced at a gravity of 1 g. A recent study by Kirchenko and Dolgoi[24] modeled weightlessness using a flat layer of liquid in an inclined thin container. They achieved a

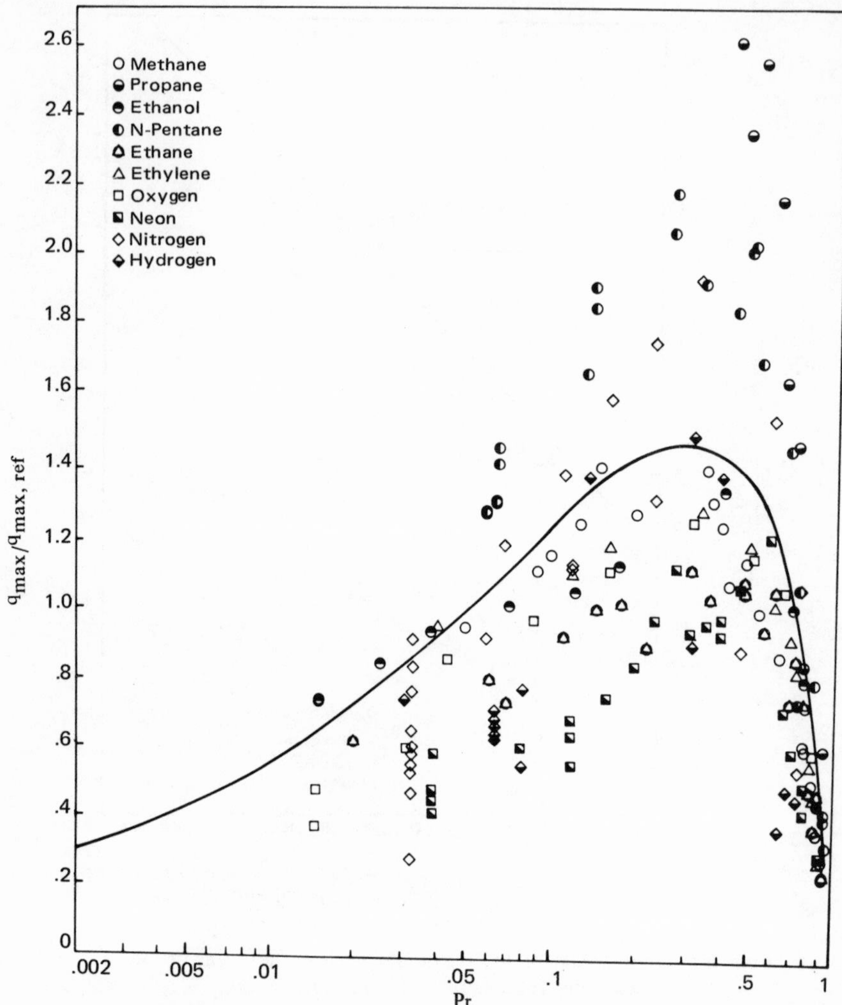

FIG. 6-7. Comparison of Kutateladze maximum heat flux correlation in reduced coordinates with data.

0.26 ± 0.1 power dependence for water, ethyl ether, and ethyl alcohol up to an a/g of 0.2. At zero g they found the peak flux to be 0.38 times that at 1 g.

Peak heat flux measurements for water with increased local acceleration have been made by Costello and Adams,[25] Ivey,[26] and Beasant and Jones.[27] Costello and Adams[25] reported the critical heat flux is proportional to

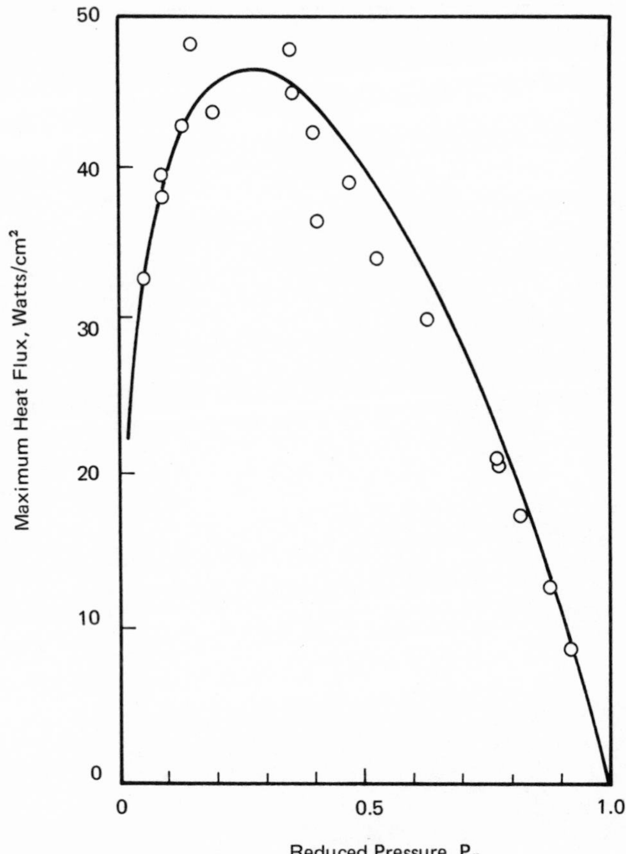

FIG. 6-8. Comparison of Noyes maximum heat flux correlation in reduced coordinates with methane data.[17]

$(a/g)^{0.15}$ for $1 < a/g < 10$ and to $(a/g)^{1/4}$ for $10 < a/g < 100$ and suggest that the value of the exponent is sensitive to surface characteristics. Ivey[26] found a value of 0.273 for the exponent in the range $1 < a/g < 160$, and Beasant and Jones[27] found the exponent on a/g to be 0.27, 0.25, 0.22, and 0.14 corresponding to pressures of 1.0, 3.4, 10.2, and 20.4 atm, respectively. In general, however, variation of the heat flux with $(a/g)^{1/4}$ should give reasonable predictions for engineering purposes in the range of $a/g > 0.1$.

Costello and Adams[28] also investigated the orientation and geometry of the heater with respect to the acceleration vector. They concluded that little effect on the peak heat flux occurs due to orientation of the acceleration vector except where the vector is directed wholly or partially toward a flat

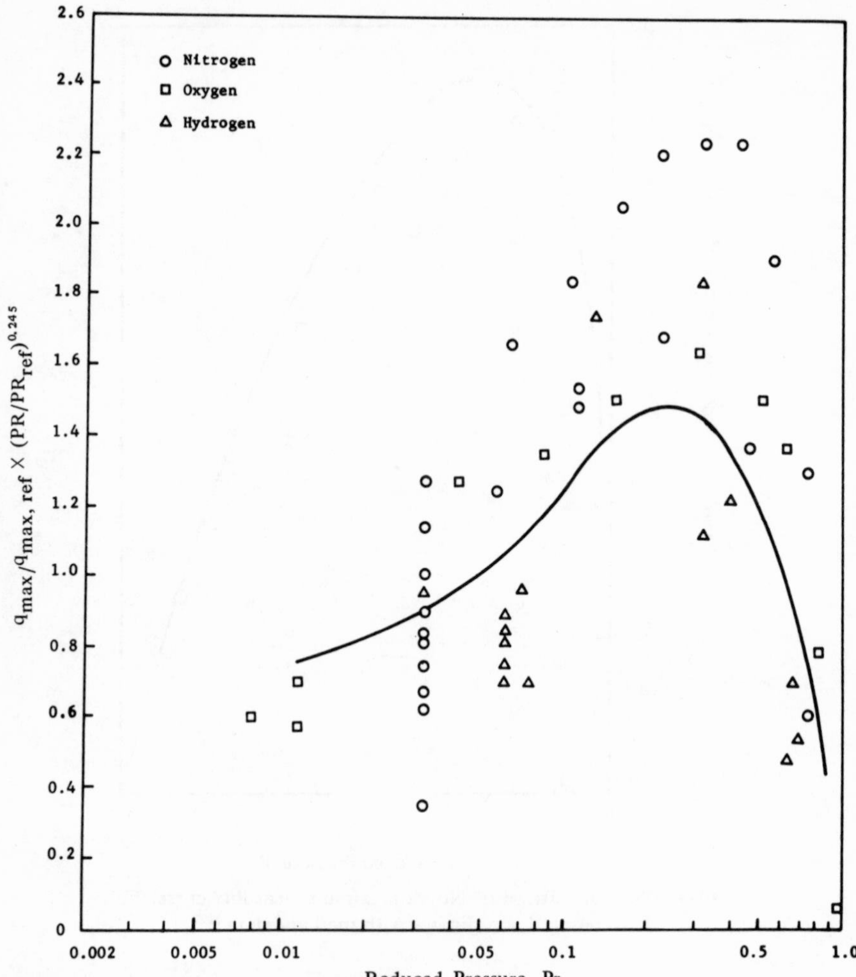

FIG. 6-9. Comparison of Noyes maximum heat flux correlation in reduced coordinates with nitrogen, oxygen and hydrogen data.

or cylindrical surface. For semicylinders, if the vector is directed normally toward the axis, the peak heat flux increases with a/g to powers as great as 0.38. For flat plates, the bubbles are held against the heater and the peak heat flux actually declines with a/g (Fig. 6-13). A directional influence of acceleration on peak heat flux is also illustrated in Fig. 6-12. Negative accelerations of $-0.3\,g$ reduce the critical heat flux in oxygen by a factor of approximately 0.05 from that attained at normal gravity.

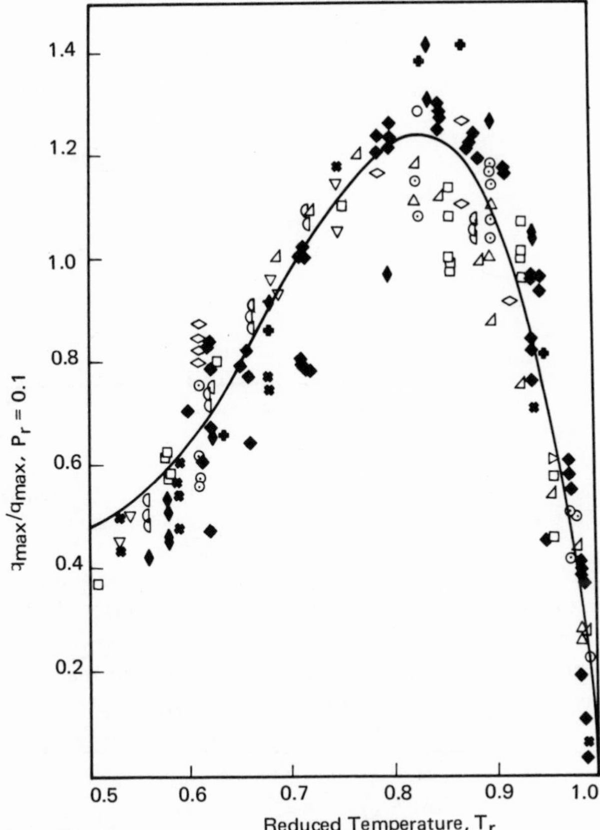

FIG. 6-10. Empirical correlation by Cobb and Park.[19]

Considering again the correlations given in Table 6-1, it is seen that a number of the equations give a range of values within which the empirical constant K is expected to lie. This uncertainty interval for K is not necessarily due to experimental inaccuracies. Zuber's[3] theoretical analysis reveals a spectrum of wavelengths any of which could result in the liquid–vapor interface in the two-phase boundary layer becoming unstable, causing burnout to occur. Accordingly, the critical heat flux may span an uncertainty band of approximately $\pm 14\%$.

Gambill[29] further investigated this uncertainty band experimentally. He concluded that an intrinsic randomness on the order of $\pm 9.3\%$ exists in the critical heat flux determined under saturated pool boiling conditions with minimum surface variability. He emphasized the minimum surface

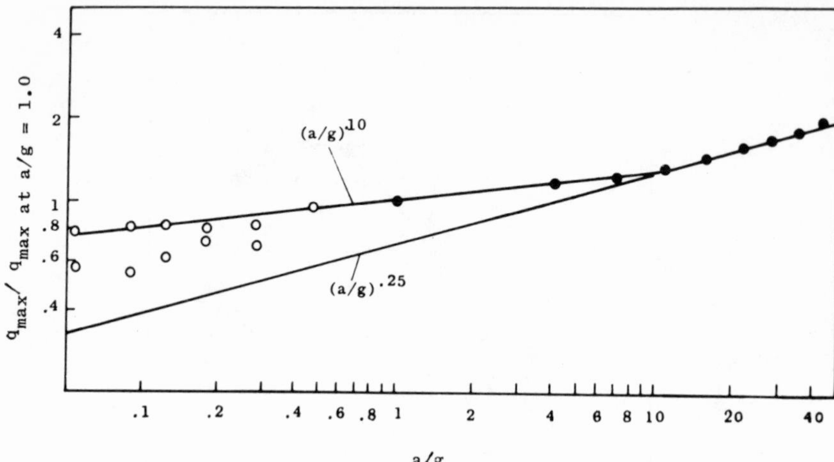

FIG. 6-11. Influence of acceleration on maximum heat flux. The open circles give the data of Usiskin and Siegel, the solid circles the data of Costello and Adams.

variability since he additionally observed that surface conditions constitute an important influence on the peak heat flux which purely hydrodynamic theories of burnout do not take into account.

Other experimental evidence of the variation in the magnitude of the peak heat flux due to surface conditions or deposits is available. Costello

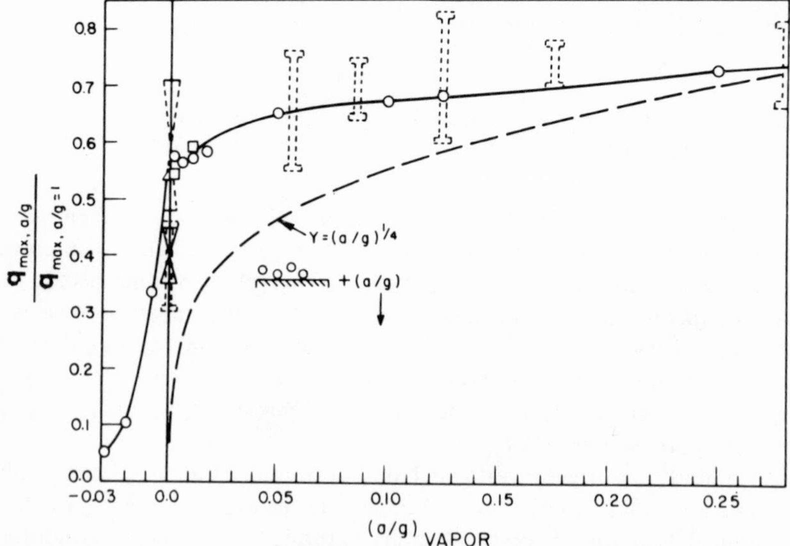

FIG. 6-12. Normalized maximum heat flux at low zero and negative accelerations.[23]

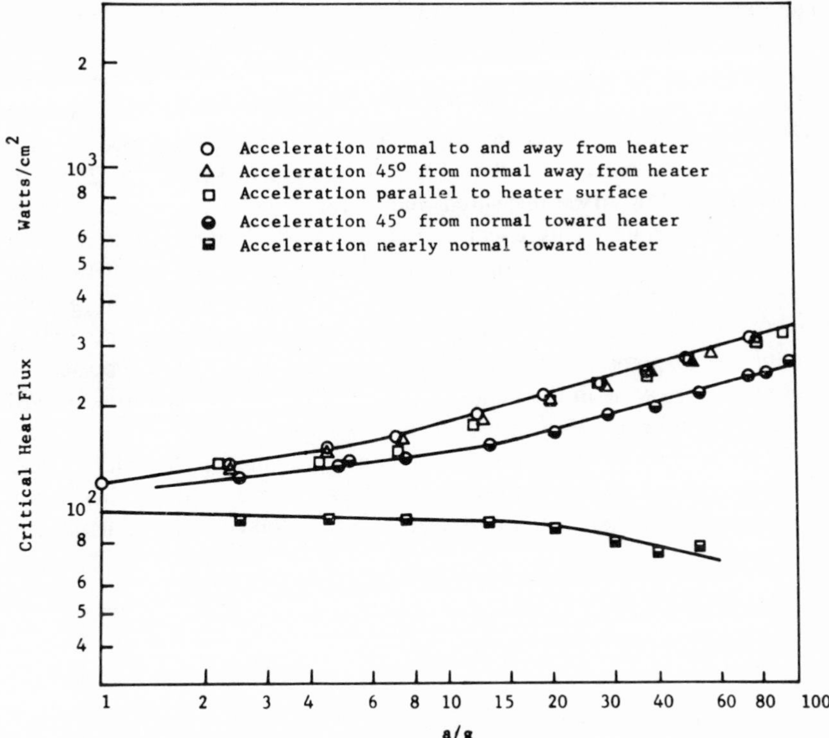

FIG. 6-13. Influence of surface orientation with varying acceleration on maximum heat flux.[25]

and Frea[30] found that coatings, probably of calcium composition, formed by boiling heaters in tap water for extended periods improved surface wettability and allowed the attainment of extremely high heat fluxes. Farber and Scorah[31] oxidized wire surfaces by first heating them red hot in film boiling. The process was called "Farberizing" and peak heat fluxes 2–3 times normal values were obtained by this surface treatment. Wires 0.040-in. in diameter made of nickel, tungsten, Chromel A, and Chromel C were used. Ivey and Morris[32] observed higher critical heat fluxes on oxidized aluminum surfaces than on clean aluminum surfaces. However, for wires of metals that are dissimilar but not easily oxidized such as nickel, platinum, silver, stainless steel, and Chromel, the burnout data were all within a 20% error band. Bernath[33] also reports higher critical heat fluxes on oxidized aluminum surfaces. He explained this effect as resulting from the several-fold more effective absorption of water molecules by hydratic aluminum oxide than by other metal oxides. Seader *et al.*[8] report that Class *et al.*[34] obtained a 50–100% increase in the peak heat flux for liquid nitrogen by coating the heater surface with silicone grease. This observation contradicts the experience of those authors who find the peak heat flux to be drastically reduced on oil- or grease-covered heaters in boiling water. Also, Gaertner[35] reported that the peak heat flux in water on a platinum wire was reduced from 155.2 W/cm^2 to a value too low to be measured due to the silicone grease accidentally covering the surface. This effect is related to the liquid–surface contact angle. For wetting surfaces (contact angle less than 90°) the liquid tends to adhere to the surface, whereas for nonwetting surfaces (contact angle greater than 90°) the vapor tends to spread out over the surface, causing premature burnout. An explanation of the 50–100% higher heat fluxes reported by Class *et al.*[34] may lie in the adhesive energy of liquid nitrogen to a greased surface. This information is, however, not available.

Other studies of the influence of contact angle on the critical heat flux have been made. Walker[36] investigated the critical heat flux for water on Nichrome, stainless steel, and Teflon-coated stainless steel

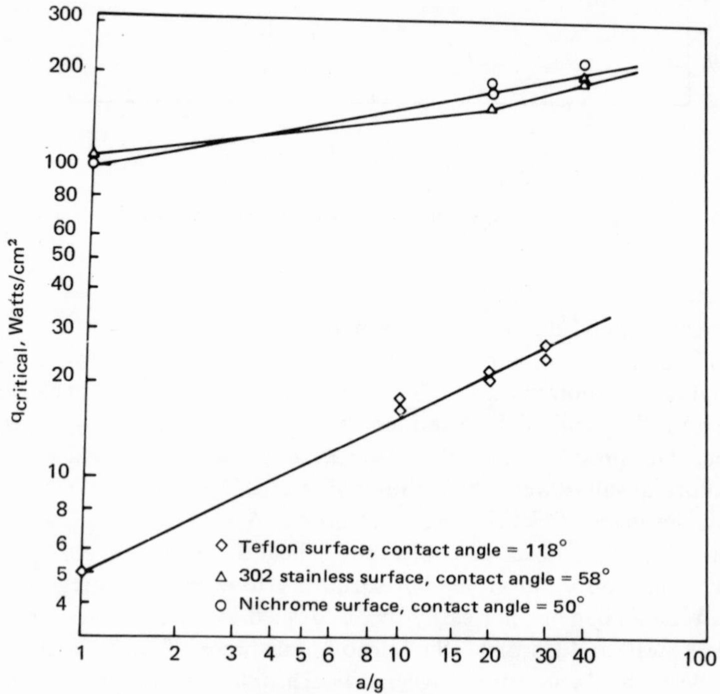

FIG. 6-14. Influence of contact angle with varying acceleration on maximum heat flux.[36]

heaters at various values of a/g (Fig. 6-14). He concluded that the critical heat flux is dependent upon the heater material employed, and increases as the surface wettability increases. Gaertner[35] also studied the influence of surface chemistry on burnout. For distilled water boiling on a nonwetting fluorocarbon-coated steel surface, the critical heat flux occurred at 1% of the normal value of the burnout heat flux for metallic surfaces (essentially at the incipient boiling heat flux).

Owens[37] observed burnout values of 111 W/cm^2 for pure water and 164 W/cm^2 for tap water on a carbon tetrachloride-cleaned surface and 142 W/cm^2 for pure water on an acetone-cleaned surface. In a subsequent test, pure water on the same surface cleaned with carbon tetrachloride and diethyl ether (similar solvents) gave burnout values in close agreement, but for tap water gave a burnout value which was 29% higher. Owens reported that water was found to spread on acetone, but not on 17 other organic liquids, indicating further a contact angle effect. Also, the tap water formed a deposit on the heated surface, which is probably similar to that reported by Costello and Frea,[30] which, they showed, increased surface wettability. Berenson[38] reported a burnout heat flux of 28 W/cm^2 for pentane on a clean copper surface, where the contact angle is on the order of 10°, and 31 W/cm^2 on an oxidized copper surface, where the contact angle is approximately zero.

Lyon,[39] in boiling tests with unpurified liquid oxygen and oxygen-rich mixtures, observed a continuous deposition of minute amounts of solid (suspected to be carbon aldehydes) on the heat transfer surface and found

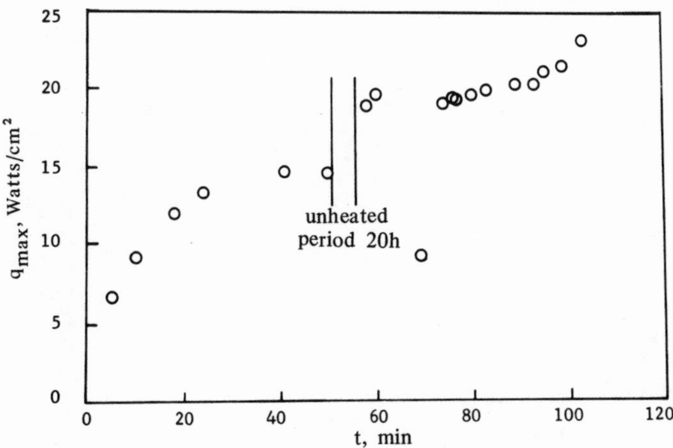

FIG. 6-15. Increase in maximum heat flux in oxygen with time due to deposit formation.

TABLE 6-3a

Peak Nucleate Boiling Fluxes for Liquid Nitrogen on Vertical Cylindrical Surfaces at 1 atm[39]

Surface	Peak nucleate boiling flux, W/cm^2			
	Test block 1	Test block 3	Test block 4	Test block 5
Gold	1.28	1.28	1.28	1.29
Copper	14.2[a]	14.4	14.1	13.4
Cu_2S	—	15.5[b]	12.6[d]	14.4[e]
	—	14.3[c]	—	—
Phenylacetonitrile	—	16.2	—	—
Ethyl alcohol (95%)[g]	—	16.4	—	15.4
Trioctylphosphate	—	—	15.1	15.0
KEL-F Oil No. 3	—	—	—	15.6
Fluorochemical N-43[f]	—	—	—	14.8
White mineral oil (tech. grade)	—	—	—	15.7
H_3PO_4 (85%)[g]	—	—	—	12.7
H_2SO_4 (96%)[g]	—	—	—	12.6

[a] Obtained from quenching curves.
[b] Initial value; surface not washed with ethyl alcohol.
[c] After surface exposed to LN_2 for about 8 hr.
[d] Standard surface preparation.
[e] Surface not washed with ethyl alcohol.
[f] Minnesota Mining and Mfg. Co.
[g] Concentration at time of application.

TABLE 6-3b

Peak Nucleate Boiling Fluxes for Liquid Oxygen on Various Vertical Cylindrical Surfaces at 1 atm[39]

Surface	Peak nucleate boiling flux, W/cm^2				
	Test block 1	Test block 3	Test block 4	Test block 5	Test block 6
Gold	18.5	17.9	16.9	17.5	—
Copper	17.9[a]	—	18.8	18.3	17.3
Cu_2S	—	—	—	18.5	—
Ethyl alcohol (95%)[c]	—	13.8[b]	—	19.6[b]	—
	—	15.5	—	21.9	—
Trioctylphosphate	—	—	20.2	19.8	21.5
KEL-F oil No. 3	—	—	—	20.3	—
White mineral oil	—	—	—	19.2	—
H_3PO_4 (85%)[c]	—	—	—	16.2	—
H_2SO_4 (96%)[c]	—	—	—	15.8	—

[a] Obtained from quenching curves.
[b] Value increased with each determination.
[c] Concentration at time of application.

the peak heat flux to increase steadily as the experiment continued. Figure 6-15 shows some of the reported results. To substantiate further the influence of surface texture, a number of specially treated surfaces were prepared. Peak heat flux tests were made on these surfaces and the results taken from this reference are tabulated in Table 6-3a and 6-3b. A definite difference in peak heat flux is observed with liquid nitrogen on copper and on electroplated gold. With liquid oxygen the difference, if any, is obscured by the scatter. The remaining data also show approximately 25% scatter in peak heat flux with variation in surface chemistry.

Although the majority of the above surface preparation and contamination effects were observed for fluids other than cryogens and cannot be extrapolated due to the unknown fluid–surface interaction, it is apparent that many of the observed discrepancies in experimental peak heat fluxes may be attributed to these effects and the designer must remain aware of them. Some researchers[29,30–37] conclude that variation in the critical heat flux due to different surface treatment cannot be reconciled with purely hydrodynamic theory. On the other hand, some[35] suggest that present hydrodynamic instability analysis must be refined to take such effects into account.

Additional experimentally observed effects not presently accounted for in the hydrodynamic instability analysis are heater geometry, size, and orientation. Bernath[33] compared burnout data for horizontal and vertical wires reported in the literature. A ratio of vertical-to-horizontal burnout flux of 0.76 was observed for approximately equivalent experimental conditions. The shortness of the vertical heater was significant. For vertical heaters boiling in water at a pressure of 93 psia and a subcooling of 97°C, heaters shorter than 8 in. attained higher heat fluxes. Successively shorter heaters burned out at heat fluxes asymptotically approaching the value for

TABLE 6-4

Peak Nucleate Boiling Fluxes for N_2 and O_2 on Test Block 5 (Gold Surfaces) for Vertical and Horizontal Orientation[39]

Fluid	Gold surface	Peak nucleate boiling flux, Btu/hr-ft^2	
		Vertical	Horizontal
Nitrogen	Polished	13.6	11.5
Nitrogen	Polished	12.8	10.7
Oxygen	Polished	17.8	13.3
Oxygen	Rough	16.9	15.4
Nitrogen	Rough	12.8	11.8

horizontal surfaces. Increasing the length of the vertical heaters over 8 in. showed no further variation in peak heat flux. Chang's[11] theoretical analysis supports this observation; however, it does not agree with the results of Lyon,[39] which show a higher maximum heat flux in nitrogen and oxygen with vertical cylinders than horizontal cylinders (Table 6-4). Kutateladze[2] also reported that the value of the peak heat flux is lower for a horizontal than a vertical plate, but gave no reference to experimental evidence. At present the authors are unaware of any satisfactory explanation for this disagreement.

Ishigai *et al.*[40] investigated boiling heat transfer to water from a flat surface facing downward. The heater surface was the circular, flat end of a copper cylinder of diameter d which was encased in an annular insulating material of outside diameter D. The peak heat flux increased with increasing values of d/D, and extrapolated to a peak heat flux value of 197 W/cm² at $d/D = 1.0$. This is 40% higher than the average of 12 different reported values for horizontal wires (140 W/cm²). The highest actual measured burnout flux was, however, 29% lower than the aforementioned average value for horizontal wires. This is in contrast to the results of Refs. 13 and 28, which simulated downward-facing plates by reversing the acceleration vector and obtained appreciable reduction in peak heat flux.

The size of the heating surface also influences the critical heat flux. Ivey and Morris[41] plotted the variation of critical heat flux with diameter reported by five different investigators (Fig. 6-16). The critical heat flux decreases with decreasing diameter to approximately 2×10^{-2} in., where the authors[41] detected a discontinuity in the curve attributable to a change in the mechanism of the critical heat flux. Further decrease in diameter

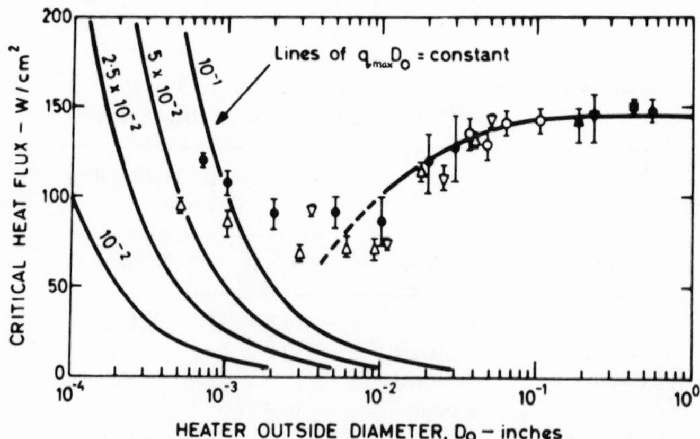

FIG. 6-16. Variation of maximum heat flux with wire diameter.[41]

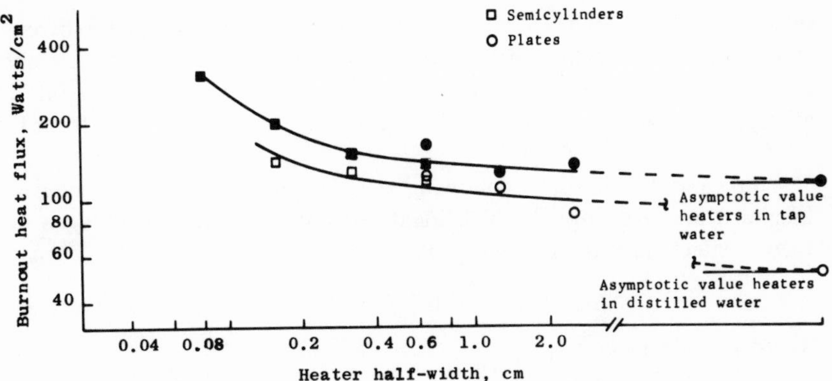

FIG. 6-17. Maximum heat flux variation with heater width.[44]

below a value of 10^{-2} in. results in an increase in the critical heat flux. They concluded that burnout occurs by a multiple-bubble blanketing process for wire diameters $D_0 > 2 \times 10^{-2}$ in. and by a single-bubble blanketing process for wire diameters $D_0 < 10^{-2}$ in. For the smaller region they propose

$$q_{max} D_0 = \text{const} \tag{6-3}$$

Some of these curves are shown in Fig. 6-16.

Ivey and Morris[41] reported that Pitts[42] has developed a model for predicting the variation of the critical heat flux with reduction in wire diameter for the range 10^{-1} to 2×10^{-2} in.

Other investigators reported a somewhat different variation in burnout with heater size. Frea[43] showed an increase in the critical heat flux with decreasing heater diameter. Costello *et al.*,[44] working with flat plates $\frac{1}{2}$–2 in. wide, obtained a substantial increase in peak heat flux with decreasing width. Their results along with those of Frea[43] are plotted in Fig. 6-17. Also illustrated is the influence of a surface deposit formed from first boiling the heater in tap water. These authors[44] postulated that burnout occurs when a critical amount of energy limited by wettability must be removed from the surface by latent transport. Superimposed on the latent heat transport is a component of energy removed by an induced convective effect. The convective component varies with heater diameter as shown in Fig. 6-17, where the asymptotic value of the critical heat flux as the diameter approaches infinity is taken as the critical latent heat transport. The asymptote was obtained by simulating an infinite flat plate heater by enclosing the 2 × 2 in. heater with glass plates so as to restrict liquid inflow from the edges.

For the clean enclosed heater the average value of the asymptotic critical heat flux was 50 W/cm², which is drastically lower than values predicted by hydrodynamic theories. The average asymptotic value increased to 111 W/cm^2 for a heater surface covered with a deposit. Finley[45] attempted to reproduce the 50 W/cm^2 value of Ref. 44 but was unsuccessful. His average value was 96 W/cm^2, which is consistent with predictions from Eqs. (1)–(11) in Table 6-1.

Sun and Lienhard,[46] following the previous dimensional analysis of Lienhard and Keeling[47] which identified and described an induced convection effect, gave the following expression for $q_{\max}$:

$$q_{\max}/q_{\max F} = f[R', I, (1 + \rho_g/\rho_f)^{1/2}, \theta_c] \tag{6-4}$$

where $q_{\max F}$ is given by

$$q_{\max F} = 0.131 h_{fg}(\sigma_g \rho_g(\rho_f - \rho_g)]^{1/4} \tag{6-5}$$

The parameter R' is a dimensionless radius

$$R' = R[g(\rho_f - \rho_g)/\sigma^{1/2} \tag{6-6}$$

When R' is large, buoyant forces are great in comparison with capillary forces and vice versa. The parameter I is an induced convection scale parameter given by

$$I = (\rho_f R\sigma/\mu_f{}^2)^{1/2} \tag{6-7}$$

In (6-4), θ_c is the contact angle, whose effect is neglected.

Figure 6-18 shows the variation of $q_{\max}/q_{\max F}$ with I for a range of characteristic widths; W' replaces R' since these data are for nichrome ribbon. Figure 6-19 shows the variation of $q_{\max}/q_{\max F}$ with R'. Based on a

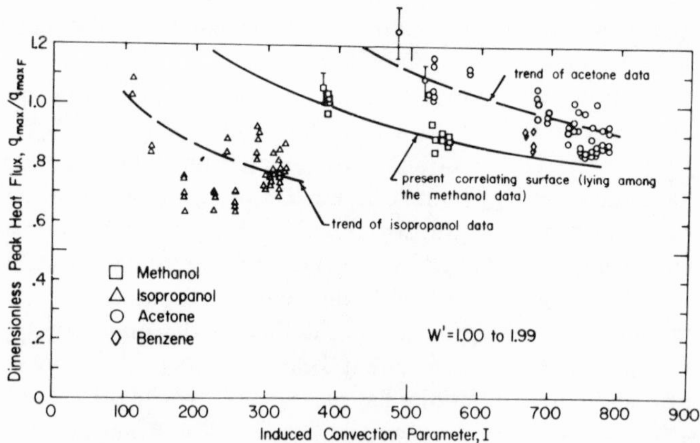

FIG. 6-18. Influence of induced convection on maximum heat flux[47]

(a)

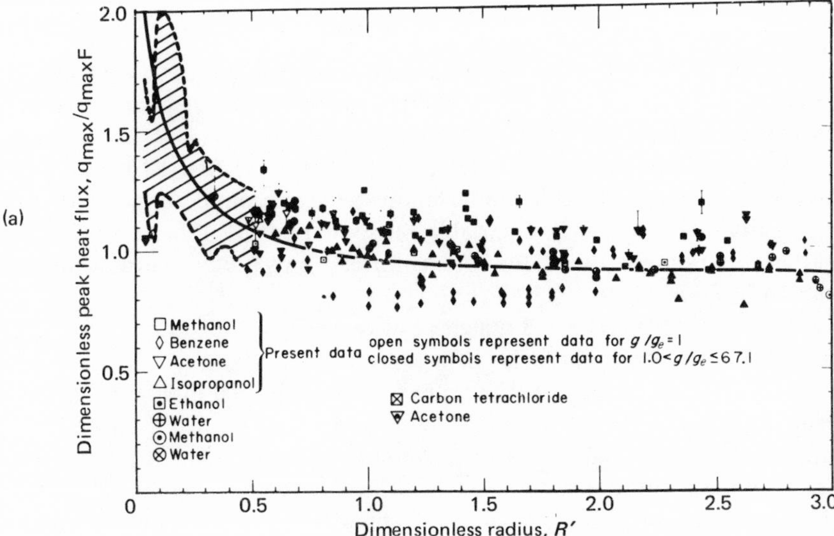

(b)

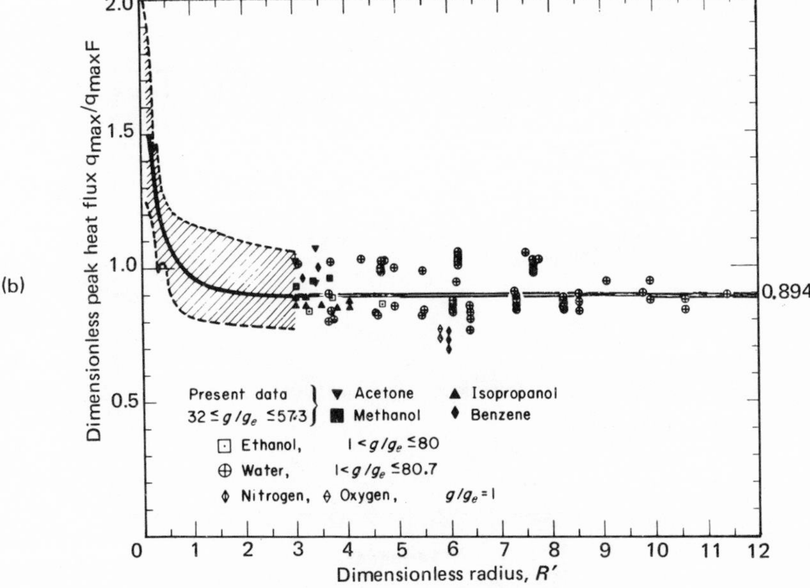

FIG. 6-19. Influence of dimensionless heater radius on maximum heat flux.[46]

stability analysis of vapor jets coming from the wire and experimental data, Sun and Lienhard showed that for horizontal cylinders

$$q_{\max}/q_{\max F} \simeq 0.890 + 2.27 \exp(-3.44\sqrt{R'}) \tag{6-8}$$

For values of $R' < 0.15$ the authors concluded that the peak heat flux is governed by another type of instability, which tends to corroborate the conclusions of Ivey and Morris.[41]

The correlations of Refs. 41, 46, and 47 are the first efforts toward developing a burnout prediction technique which accounts for some of the apparently nonhydrodynamic variables. It should be noted, however, that in all these studies only two datum points for cryogens are compared with the correlations (see Fig. 6-19).

Lyon[39] has investigated the effect of composition for nitrogen–oxygen mixtures on the critical heat flux. The critical flux tends to remain near that of 100% N_2 until values of 80% O_2 are used, as indicated by Fig. 6-20. This result is different from the results reported by van Stralen[50] for binary mixtures of water and various alcohols or ketones. Van Stralen found maxima in the critical heat fluxes for mixtures. Huber and Hoehne[51] added small quantities of benzene to diphenyl and also found a maximum critical heat flux for the mixture. Owens,[37] on the other hand, reported maxima and minima in the ratio of peak heat flux for binary mixtures to that for

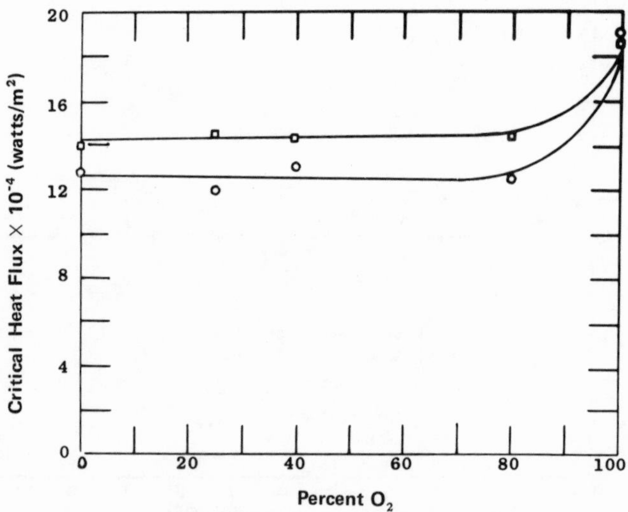

FIG. 6-20. Influence of composition on the maximum heat flux.[39] ○ Polished gold; □ polished copper.

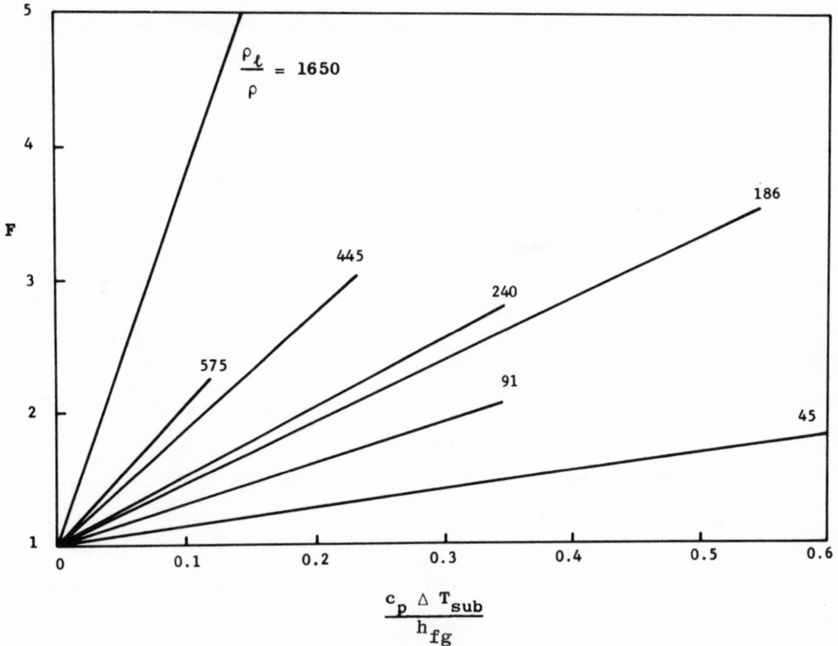

FIG. 6-21. Increase in maximum heat flux with increasing subcooling and decreasing pressure.

pure water, but concluded that the peak heat flux in the mixtures is always less than that for pure water.

6.2.1.2 *Subcooled:* Subcooling is one of the primary variables influencing the critical heat flux. All data agree that the critical heat flux increases almost linearly with ΔT_{sub} at all pressures. Also established is the fact that the enhancement of increased subcooling is greater at low pressures (Fig. 6-21).

When vapor bubbles depart or collapse in the boiling boundary of a subcooled pool, additional heat is required to bring the mass of in-flowing liquid to saturation temperature. Kutateladze[2] used this argument to develop the semiempirical expression which corrects q_{max} predicted by Eq. (5) of Table 6-1 for saturated conditions to subcooled conditions:

$$F = q_{max,\,sub}/q_{max,sat} = 1.0 + (\rho_l/\rho_v)^{0.80}(C_p\,\Delta T_{sub}/15.38h_{fg}) \qquad (6\text{-}9)$$

Also Zuber gives

$$\frac{q_{max,sub}}{q_{max,sat}} = 1.0 + \frac{5.3}{h_{fg}\rho_v}(k_l\rho_l C_p)^{1/2} \times \left[\frac{g(\rho_l - \rho_v)}{g_c\sigma}\right]^{1/4}\left[\frac{\sigma g g_c(\rho_l - \rho_v)}{\rho_v^{\,2}}\right]^{1/8}(T_{sat} - T_b) \qquad (6\text{-}10)$$

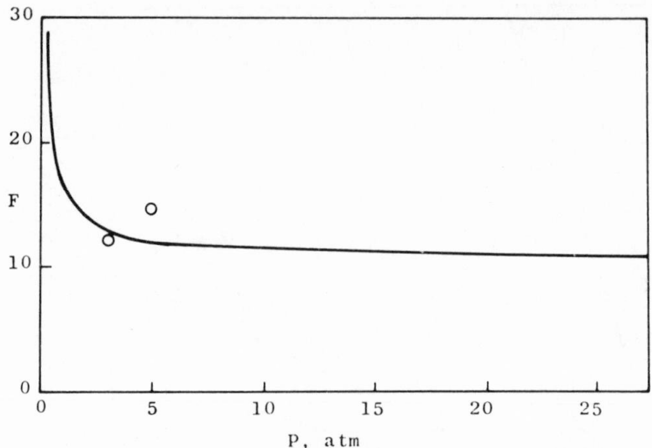

FIG. 6-22. Comparison of subcooled nitrogen data with Kutateladze's correction factor.

A more recent equation evaluated with all available data is that proposed by Ivey and Morris[48]:

$$F = 1 + (\rho_l/\rho_v)^{0.75} C_p \,\Delta T_{sub}/9.8 h_{fg} \tag{6-11}$$

All properties, with the exception of C_p, are evaluated at saturation conditions. C_p is evaluated at a reference temperature of

$$T_{ref} = T_{sat} - \tfrac{1}{2}\,\Delta T_{sub} \tag{6-12}$$

in all three equations.

The above expressions for subcooling have been compared with data for water and organic liquids. Figure 6-22 compares Eq. (6-9) with two points for subcooled nitrogen.[49] These were the only two points for subcooled cryogens the authors could find. The agreement is reasonably good.

Ivey and Morris[41] found the variation of critical heat flux with subcooling not to be strictly linear for horizontal wires of diameter 0.048–0.015 in. The variation of critical heat flux with wire diameter shown in Fig. 6-16 was not observed with subcooled liquid.

6.2.2 Confined Natural Convection

Critical heat flux in a natural circulation boiling loop is of design interest for boiling reactors, process heat exchanger equipment, and cryopumping systems of ground test facilities.[52] Gambill and Bundy[53] gave a physical description of three mechanisms by which burnout in a natural circulation loop can occur and proposed a semiempirical equation for predicting critical heat flux for low-pressure water.

Recently Bandy *et al.*[54] have applied optimal control theory to natural circulation boiling loops and produced significant increases in power by using a variable-area heating section and nonuniform heat flux. For a Freon-113 atmospheric pressure loop, the predicted power increase for a suboptimal test section was 16.7% over the conventional constant-area uniform heat flux channel, while the measured increase was 15%.

The authors are unaware of any studies of the critical heat flux in natural circulation loops specifically for cryogenic fluids.

6.2.3 Summary

The foregoing was intended to illustrate that the present methods of predicting pool boiling critical heat flux are based mainly on hydrodynamic models. Although a number of additional variables such as surface condition, heater geometry, etc. not included in the hydrodynamic model are known to influence the critical heat flux, the designer must use the existing correlation but remain cognizant of these secondary effects.

A graphical method of calculating critical heat fluxes which is convenient for design has been proposed by Frost and Dzakowic.[16]

6.3 FORCED CONVECTION BOILING, CRITICAL HEAT FLUX

As indicated in Chapter 4, two types of burnout in forced flow occur. (1) burnout in subcooled boiling, and (2) burnout in bulk boiling. The latter is probably the most commonly encountered with cryogenic fluids. However, very little experimental information on either burnout phenomenon is available for cryogenic fluids.

Burnout in forced flow is also very sensitive to flow oscillations and instabilities. MacBeth[55] has given an excellent up-to-date review and survey of the variables which influence burnout and a description of the types of systems recommended for stable two-phase flow conditions and, hence, maximum critical heat fluxes. This reference should be consulted for design information.

MacBeth[55] proposed two empirical burnout prediction equations for water in uniform heated tubes. One is for the low-velocity burnout regime, approximately defined by Fig. 6-23, where the appropriate burnout equation is

$$q_{\max} \times 10^{-6} = \frac{(G \times 10^{-6})(h_{fg} + \Delta h)}{158 d^{0.1}(G \times 10^{-6})^{0.49} + 4L/d} \tag{6-13}$$

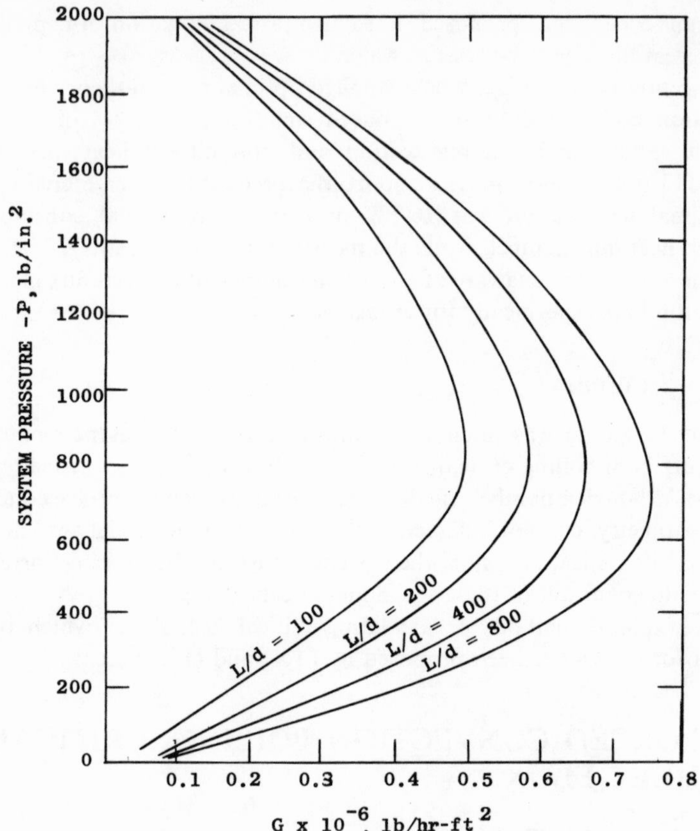

FIG. 6-23. Approximate boundary of the low-velocity burn-out regime for water flowing in round tubes (the low-velocity regime lies to the left of any given curve).[55]

This equation fits the published data for all system pressures with an rms error of 5.5%.

The other is a general form of burnout correlation optimized by computer to fit some 5000 data points divided into eight pressure groups. The form of the burnout equation is

$$q_{\max} \times 10^{-6} = [A' + \tfrac{1}{4}d(G \times 10^{-6})\,\Delta h]/(C' + L) \qquad (6\text{-}14)$$

where A' and C' are polynomial expressions in pressure, tube diameter, and mass flux.

Note that the foregoing equations are valid for axially, uniformly heated channels with flowing water. Scaling laws and scaling factors for adapting these equations to other fluids are also discussed by MacBeth.[55]

A procedure developed by Staniforth and Stevens[56] for obtaining scaling factors is as follows:

(1) Note is first made of the fact that h_{fg}, ρ_l, and ρ_v must be important properties influencing burnout.

(2) The ratio ρ_l/ρ_v is assumed to be the same for both the modeling fluid and the fluid of interest. This step establishes the corresponding pressures, but it must be noted that it is subject to the condition that there will not be another important dimensionless group of fluid properties.

(3) A scaling factor for Δh is assumed by making the ratio $\Delta h/h_{fg}$ the same for both the modeling fluid and the fluid of interest.

(4) The ratio L/d is taken to be the same, which is axiomatic in any modeling procedure.

(5) From the above, the burnout equation for, say, a round tube can be written as follows:

$$q_{\max} = f(L/d, d, G, \rho_l/\rho_v, \Delta h/h_{fg}) \tag{6-15}$$

which shows that scaling factors are still required for $q_{\max}$ d, and G.

(6) It is noted that $q_{\max}/h_{fg}G$ is a dimensionless group, and that, consequently, it is possible to write scaling factor for $q_{\max} = h_{fg}/h_{fg,\mathrm{m}} \times$ scaling factor for G where h_{fg} and $h_{fg,\mathrm{m}}$ refer, respectively, to the fluid of interest and to the modeling fluid. This equation means that only two scaling factors have to be found.

When this procedure is applied to the liquid hydrogen data of Lewis

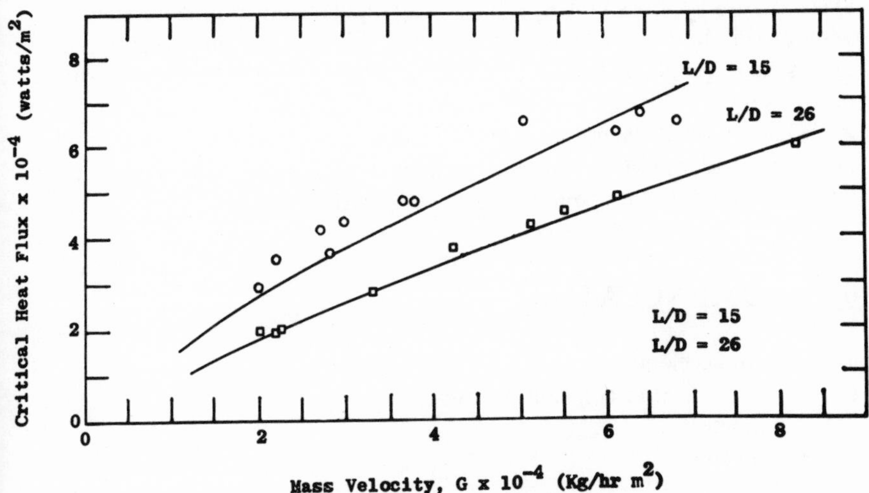

FIG. 6-24. Comparison of Staniforth and Stevens' scaling method[56] with liquid hydrogen data of Lewis *et al.*[57]

TABLE 6-5
Scaling Factors for the Liquid Hydrogen Data of Lewis et al.[57]

L/d	Scale factor $q_{max}/q_{max,m}$	Scale factor G/G_m
26	0.041	0.140
20	0.058	0.199
15	0.037	0.126
8.5	0.026	0.090

et al.[57] Fig. 6-24 results. Equation (6-13) for the low-velocity regime was used to correlate the data. Scaling factors. for L/d, d, ρ_l/ρ_v, $\Delta h/h_{fg}$ were taken as unity. The scaling factors for q_{max} and G are given in Table 6-5.

Although the use of Eqs. (6-13) and (6-14) does not provide a general method of predicting cryogenic burnout, the use of scaling factors appears promising once sufficient experimental data are available. Other correlations for burnout in bulk boiling which are purported to be applicable to all fluids do not show agreement with cryogenic data.[23]

For subcooled burnout, Gambill[58] has proposed a generalized prediction technique for flowing, subcooled wetting liquids. The method superimposes a forced convection heat transfer component on the burnout heat flux determined in the absence of forced flow. Comparison with 96% of 878 datum points shows a maximum deviation of 40% for seven fluids in axial, swirl, and cross flow in tubular, rectangular, and rod geometries over very broad flow conditions. The method of solution has recently been expressed in graphical form[16] permitting rapid design estimates of critical heat fluxes.

Gambill[58] does not compare his theory with any cryogenic fluids; however, Seader *et al.*[8] showed that the method is in reasonable agreement with the little data they found from their literature search.

6.4 NOMENCLATURE

a = local acceleration
C_p = specific heat at constant pressure
D = diameter
d = diameter
G = mass flux
g = acceleration of gravity

h = enthalpy
h_{fg} = latent heat of vaporization
I = parameter defined by Eq. (6-7)
k = thermal conductivity
L = length
q = heat transfer rate per unit area
R = radius
R' = parameter defined by Eq. (6-6)
T = temperature

Greek Letters

θ = contact angle
μ = viscosity
ρ = density
σ = surface tension

Subscripts

f = fluid
g = gas
l = liquid
m = model
r = thermodynamic reduced value
sat = saturated
sub = subcooled
v = vapor
0 = wire

6.5 REFERENCES

1. W. M. Rohsenow and P. Griffith, *Chem. Eng. Progr. Symp. Ser.* **52**(18), 47 (1956).
2. S. S. Kutateladze, *Fundamentals of Heat Transfer*, Academic Press, New York (1963).
3. N. Zuber, AECU-4439 (1959).
4. J. N. Addoms, D. Sc. Thesis, MIT, Cambridge, Massachusetts (1948).
5. Y. P. Chang and N. W. Snyder, *Chem. Eng. Progr., Symp. Ser.* **56**(30), 25 (1960).
6. R. C. Noyes, *Trans. ASME, J. Heat Transfer* **85C**, 125 (1963).
7. R. Moissis and P. J. Berenson, *Trans. ASME, J. Heat Transfer* **85C**, 221 (1963).
8. J. D. Seader, W. S. Miller, and L. A. Kalvinskas, NASA CR-243 (1965).
9. P. Griffith, ASME-AIChE Heat Transfer Conference, paper 57-H-21, Philadelphia, Pennsylvania (August 1957).
10. N. Zuber and M. Tribus, UCLA Rept. 58-5 (1958).
11. Y. P. Chang, *Trans. ASME, J. Heat Transfer* **86C**, 89 (1963).

12. E. G. Brentari, P. J. Giarratano, and R. V. Smith, NBS-TN-317 (1965).
13. D. N. Lyon, M. C. Jones, G. L. Ritter, C. Chiladakis, and P. G. Kosky, *AIChE J.* **11**(5), 773 (1965).
14. M. T. Cichelli and C. F. Bonilla, *Trans. AIChE* **41**, 755 (1945).
15. S. S. Kutateladze and V. M. Borishanskii, *A Concise Encyclopedia of Heat Transfer*, Pergamon Press, New York (1966), p. 209.
16. W. Frost and G. S. Dzakowic, AEDC-TR-69-106 (1969).
17. C. T. Sciance, C. P. Colver, and C. M. Sliepcevich, in *Advances in Cryogenic Engineering*, Vol. 12, Plenum Press, New York (1967), p. 395.
18. V. M. Borishanskii, in *Symp. Problems of Heat Transfer and Hydraulics of Two-Phase Media* (S. S. Kutateladze, ed.), Pergamon Press, New York (1969).
19. C. B. Cobb and E. L. Park, Jr., *Chem. Eng. Progr. Symp. Series* **65**(92), 188 (1969).
20. C. M. Usiskin and R. Siegel, *Trans. ASME, J. Heat Transfer* **83C**, 243 (1961).
21. H. Merte and J. A. Clark, *Trans. ASME, J. Heat Transfer* **86C**, 351 (1964).
22. J. E. Sherley, in *Advances in Cryogenic Engineering*, Vol. 8, Plenum Press, New York (1963), p. 495.
23. J. A. Clark, in *Advances in Heat Transfer*, Vol. 5, Academic Press, New York (1968), p. 325.
24. Y. A. Kirchenko and M. L. Dolgoi, *High Temperature* **8**(1), 120 (1970).
25. C. P. Costello and J. Adams, *Intern. Developments in Heat Transfer*, Paper No. 30, ASME (1961), p. 255.
26. H. J. Ivey, AEEW-R99 (1961).
27. W. R. Beasant and H. W. Jones, UK Report AEEW-R, Winfrith (1963), p. 275.
28. C. P. Costello and J. M. Adams, *AIChE J.* **9**(5), 663 (1963).
29. W. R. Gambill, *AIChE J.* **10**(4), 502 (1964).
30. C. P. Costello and W. J. Frea, *Chem. Eng. Progr. Symp. Ser.* **61**(57), 258 (1965).
31. E. A. Farber and R. L. Scorah, *Trans. ASME* **70**, 369 (1952).
32. H. J. Ivey and D. J. Morris, AIChE Preprint 160, Chicago, Illinois (1960).
33. L. Bernath, *Chem. Eng. Progr. Symp. Ser.* **56**(30), 95 (1960).
34. C. R. Class *et al.*, Beechcraft Research and Development, Inc., Engineering Report 6154, WADC TR 58-528, ad 214256 (1958).
35. R. F. Gaertner, 63-RL-3449C, General Electric Report, Schenectady, New York (1963).
36. G. E. Walker, Jr., M. S. Thesis, Mech. Engr. Dept., Univ. of Washington, Seattle, Washington (1965).
37. W. L. Owens, AEW-R 180 (1964).
38. P. J. Berenson, Tech. Rept. No. 17, Heat Transfer Lab., MIT, Cambridge, Massachusetts (March 1960).
39. D. N. Lyon, *Intern. J. Heat Mass Transfer* **7**, 1097 (1964).
40. S. Ishigai *et al.* in *Intern. Developments in Heat Transfer*, ASME, Boulder Colorado (1961), p. 224.
41. H. J. Ivey and D. J. Morris, in *Third Intern. Heat Transfer Conference*, AIChE, Chicago, Illinois (1966).
42. C. C. Pitts, Dept. of Mech. Engr. Rept., Stanford Univ. Stanford, California (1964).
43. W. J. Frea, Dept. of Mech. Eng., Heat Transfer Lab., Univ. of Washington, Seattle, Washington (1963).
44. C. P. Costello *et al.*, *Chem. Eng. Progr. Symp. Ser.* **61**(59), 271 (1965).
45. B. G. Finley, M.S. Thesis, Univ. of Tennessee Space Institute, Tullahoma, Tennessee (1968).
46. K. H. Sun and J. H. Lienhard, *Intern. J. Heat Mass Transfer* **13**, 425 (1970).

47. J. H. Lienhard and K. B. Keeling, Jr., ASME Paper No. 69-HT 48 (1969).
48. H. J. Ivey and D. J. Morris, UK Report AEEW-R-137, Winfrith (1962).
49. H. Merte and E. W. Lewis, Heat Transfer Lab., Dept. Mech. Eng., Univ. of Michigan, Ann Arbor, Michigan (1967).
50. S. J. D. van Stralen, *Chem. Eng. Sci.* **5**, 290 (1956).
51. D. A. Huber and J. C. Hoehne, *Trans. ASME, J. Heat Transfer* **85C**, 215 (1963).
52. M. J. Triplett, in *Proc. Space Simulation Conference, September 1970*, NBS special publication 336.
53. W. R. Gambill and R. D. Bundy, *Nucl. Sci. Eng.* **18**, 80 (1964).
54. D. B. Bandy *et al.*, *Chem. Eng. Progr. Symp. Ser.* **65**(92), 231 (1969).
55. R. V. MacBeth, in *Advances in Chemical Engineering* (T. B. Drew *et al.*, eds.), Vol. 7, Academic Press, New York (1968).
56. R. Staniforth and G. F. Stevens, *Proc. Inst. Mech. Eng.* (*London*) **180**(Pt. 36) (1965–1966).
57. J. P. Lewis, J. H. Goodykoontz, and J. F. Kline, NASA TN-4382 (1958).
58. W. R. Gambill, *Chem. Eng. Progr. Symp. Ser.* **59**(47), 71 (1963).

FILM BOILING 7

D. CLEMENTS

7.1 INTRODUCTION

Film boiling is a form of change-of-phase heat transfer which is characterized by a solid body transferring heat to a surrounding liquid through an intermediate vapor film. The vapor film forms a continuous, relatively regular blanket over the solid surface. This liquid-over-vapor arrangement is inherently unstable. However, the solid–liquid temperature differences characteristic of film boiling are so large that no liquid can remain in contact with the solid, thus maintaining the stable film.

A knowledge of film boiling is particularly important when one is dealing with the handling and storage of cryogenic liquids. The extremely large temperature difference between a solid at room temperature and a cryogenic liquid often gives rise to film boiling during cooldown. The frequency with which film boiling can occur in cryogenic work makes an understanding of this mode of heat transfer essential in the larger problem of heat transfer at cryogenic temperatures.

The discussion here will be limited specifically to the area of stable film boiling in a quiescent liquid pool. Forced convection film boiling and Leidenfrost boiling will not be considered. First, our attention will be focused on a description of the physical factors which affect film boiling behavior and the experimental data available in the literature. The data are discussed by substance, with figures and tables to aid in locating data for use in design calculations. The second area of interest is a general presentation

---

D. CLEMENTS U.S. Army Ordnance Center and School, Aberdeen Proving Ground, Maryland.

of the predictive correlations for design calculations in film boiling. The correlations are also compared with some representative sets of data as a means of gaining some insight into the accuracy of the correlations. For a much more comprehensive review of the entire field of film boiling, the reader is directed to the review by Jordan[1] or that by Clements and Colver.[2] Detailed reviews of cryogenic interest include those of Brentari *et al.*[3,4] and Richards *et al.*[5]

7.2 CRYOGENIC FILM BOILING DATA

It is convenient at the beginning of the discussion of cryogenic film boiling data to make some generalizations concerning factors which influence film boiling behavior. The generalizations will be in terms of the effect a particular factor has upon the film boiling heat transfer coefficient at a given temperature difference ΔT. There is an inverse relationship between the characteristic heater dimension and the heat transfer coefficient. That is, an increase in characteristic heater dimension results in a decrease in heat transfer coefficient. For a horizontal, cylindrical heater the characteristic dimension is the diameter, for a strip heater it is the width, and for a vertical tube heater the characteristic dimension is the heater length. An increase in either system pressure or gravitational field tends to increase the film boiling heat transfer coefficient. Effects of heater orientation, surface condition and material of construction, liquid composition, and external electrical fields have also been studied.

7.2.1 Hydrogen Data

Hydrogen film boiling data from horizontal wires have been reported by Weil,[6] Weil and Lacaze,[7] and Mulford and Nigon.[8] These data all fall within the band of scatter shown in Fig. 7-1. The data of Class *et al.*,[9] using a horizontal strip heater facing upward, clearly show the expected increase of heat transfer coefficient with increasing pressure. Other heater orientations (45° and vertical) indicate a small decrease in heat transfer coefficient as the angle with the horizontal is increased. Also given in Fig. 7-1 are data of Graham *et al.*[10] The data at 3.6 atm for 1 g and 7 g show the characteristic increase in heat transfer coefficient with an increase in acceleration, but it is apparent that the effect of pressure at 1 g is contrary to what is considered normal. The explanation for this is not readily apparent. Experimental data for film boiling of hydrogen are summarized in Table 7-1.

TABLE 7-1
Hydrogen Film Boiling Data[a]

Author	Heater	Pressure	Notes
Weil[6]	1-mm lead wire	750 Torr	Boiling curve study
Weil and Lacaze[7]	0.1-mm copper wire	750 Torr	Boiling curve study
Mulford and Nigon[8]	12-mm-diameter copper tube	590 Torr	Boiling curve study
Class *et al.*[9]	22 in. by 1 in. nickel alloy strip	1.0, 5.1, and 8.7 atm	Pressure and orientation effects and effect of grease on surface studied
Graham *et al.*[10]	Chromel A ribbon	3.6 and 6.6 atm	Pressure and gravity effects studied ($g = 1$ and $g = 7$)

[a] Unless otherwise indicated, all heater orientations are horizontal.

TABLE 7-2
Helium I Film Boiling Data

Author	Heater	Pressure	Notes
Frederking *et al.*[11–14]	5.5–51-μm platinum wires	6–720 Torr	Pressure[12] and dimensional effects studied
Frederking *et al.*[15]	0.25- and 0.375-in.-diameter copper spheres	—	Cooldown study
Frederking *et al.*[16]	0.187-in.-diameter copper plate	1 atm	—
Lyon[17]	0.991-cm-diameter platinum cylinder	—	Bath temperatures of 2.27, 4.21, and 5.13 K
Holdredge and McFadden[18]	0.145- and 0.245-cm diameter nickel–iron film on glass cylinder	—	Bath temperatures of 2.6, 3.2, 3.8 K
Grassmann and Karagounis[19]	10- and 20-μm Wollaston wires	—	Bath temperatures of 4.2 and 2.83 K

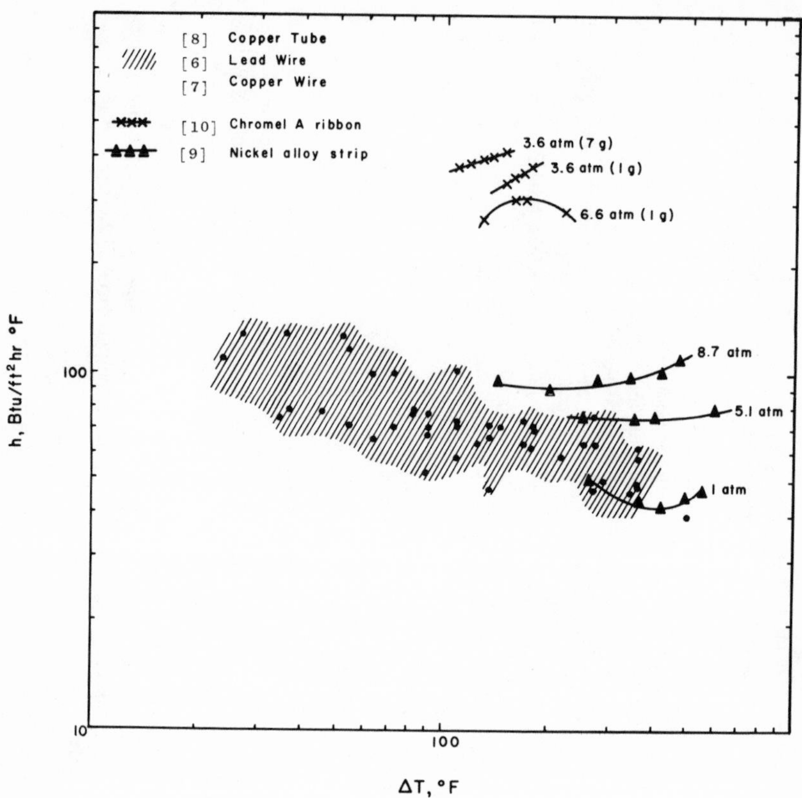

FIG. 7-1. Hydrogen film boiling data.

7.2.2 Helium I Data

Much of the helium I film boiling work, summarized in Table 7-2, has been conducted by Frederking and co-workers. The effects of heater dimension on helium I film boiling are shown quite clearly by selected data of Frederking and others in Fig. 7-2. Results for several platinum wire heaters[11–14] as well as two copper spheres[15] and a flat copper plate[16] are shown. Other data using a platinum wire heater [12] indicate a consistent effect of pressure on film boiling in the pressure range 6–720 Torr.

Film boiling was noted at very low heat fluxes in a study reported by Lyon.[17] The data were taken at three bath saturation temperatures, 5.13, 4.21, and 2.27 K. It is interesting to note that the data at 2.27 K fall in between the other two curves and that the 4.21 K curve was the highest of

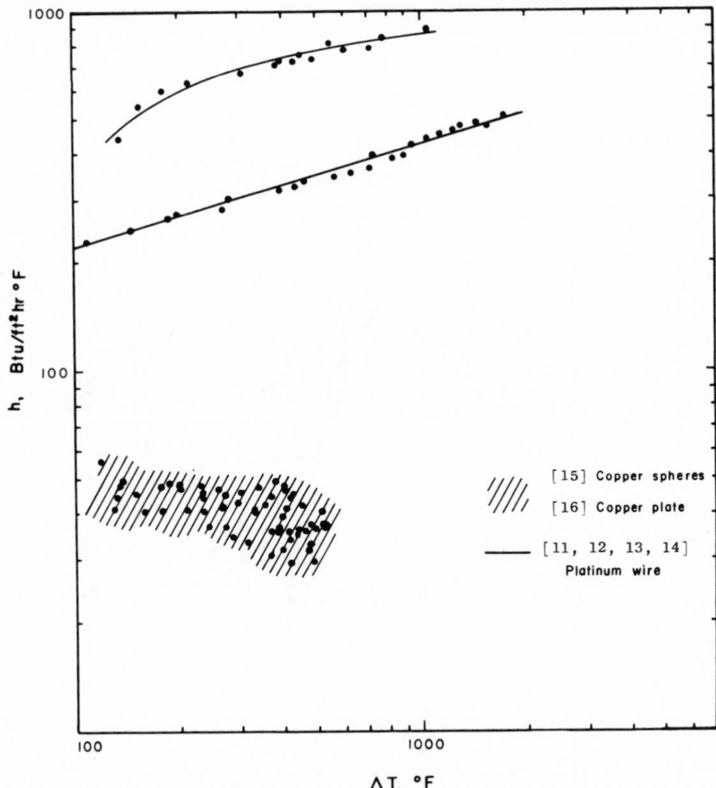

FIG. 7-2. Effect of heater diameter on helium I film boiling.

the three. When extrapolated to higher ΔT's, the data at 4.21 K agree well with the large-diameter data of Frederking *et al.*[16]

Recently Holdredge and McFadden[18] have presented data at bath temperatures of 2.6, 3.2, and 3.8 K for two different sizes of heater. The data show a regular increase in heat transfer coefficient with increasing bath temperature and the expected diameter effect. Data from two different Wollaston wire heaters at two different bulk temperatures have been published by Grassmann and Karagounis.[19]

7.2.3 Nitrogen Data

It is evident from Table 7-3 that film boiling in nitrogen has been more widely studied than for any other cryogenic substance. The bulk of the data, however, have been taken in single studies using large-diameter heaters. The large-diameter data of Flynn *et al.*,[20] Harman and Gordy,[21] Bromley,[22,23]

TABLE 7-3
Nitrogen Film Boiling Data

Author	Heater	Pressure	Notes
Flynn *et al.*[20]	0.625-in.-diameter copper tube	1 atm	Boiling curve study
Harman and Gordy[21]	10 mm by 2 mm semiconductor plates	1 atm	Used *n*-type germanium and *p*-type silicon heaters; cooldown study
Bromley[22,23]	0.35-in.-diameter carbon tube	1 atm	—
Cowley *et al.*[24]	4-cm-diameter by 10-cm copper cylinder	1 atm	Cooldown study; effect of surface coatings studied
Hsu and Westwater[25,26]	0.5-in.-diameter vertical stainless steel cylinder	1 atm	Heaters 0.219, 0.240, and 0.367 ft long were used
Ruzička[27]	10-, 15-, and 20-mm-diameter copper tubes	1 atm	Boiling curve study
Frederking *et al.*[16	0.25- and 0.375-in.-diameter copper spheres	1 atm	Cooldown study
Rhea and Nevins[28]	1-, 0.75-, 0.5-in.-diameter copper spheres	1 atm	Geometry, surface condition, and effects of imposed oscillations studied
Park *et al.*[29]	0.65-in.-diameter gold or copper tubes	1–30 atm	Pressure and heater material effects studied
Lyon *et al.*[30]	3.325-cm-diameter by 0.2-cm-wide platinum ring	31.07 atm	Present only three data points
Frederking[11,12]	10-, 16.2-, 31-, 51-, 100-, and 200-μm platinum wires	140–720 Torr	Geometry and pressure[12] effects studied
Weil[6]	1-mm lead wire	740 Torr	—
Weil and Lacaze[31]	1.5-mm copper wire	740 Torr	—
Sauer and Ragsdell[32]	0.5-, 1-, 2-in.-wide Kanthal-A-1 or Inconel-600 strips	1 atm	Dimensional and surface finish effects studied
Flanigan and Park[33]	0.55-, 0.75-, and 0.95-in.-diameter cylinder	1–30.2 atm	Pressure and dimensional effects studied
Merte and Clark[34–36]	1-in.-diameter copper sphere	1 atm	Reduced gravity effect on boiling curve studied
Price and Sauer[37]	2-in.-wide Inconel-600 strip	1 atm	Heater orientation effect studied
Pai and Bankoff[39,40]	Sintered stainless steel plates (three grades)	1 atm	Surface texture and vapor removal effects studied
Wayner and Bankoff[41]	1.75- by 1.75-in. stainless steel plate	1 atm	Vapor suction studied

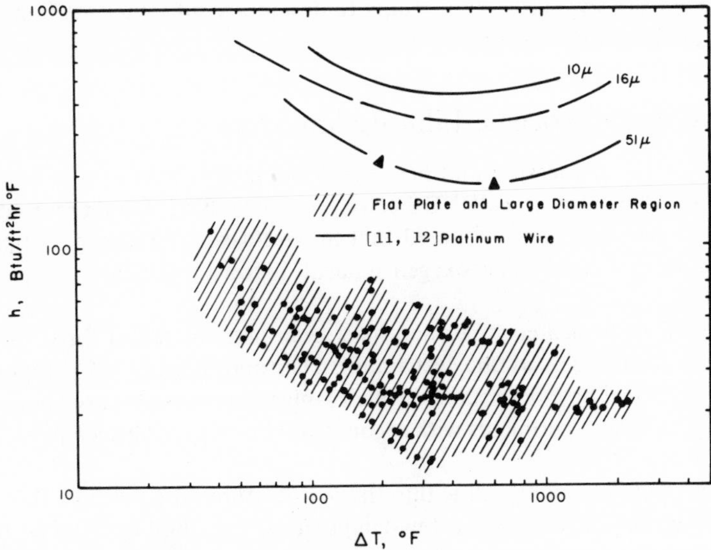

FIG. 7-3. Effect of heater diameter on nitrogen film boiling data.

Cowley *et al.*,[24] Hsu and Westwater,[25,26] Ruzička,[27] Frederking *et al.*,[16] Rhea and Nevins,[28] and Park *et al.*[29] all fall within the shaded region in Fig. 7-3. The data of Lyon *et al.*[30] line up with those in the shaded region, but at much lower ΔT's. Also shown are small-diameter data of Frederking,[11,12] which show the effect of characteristic heater dimension. The small wire heater data of Weil[6] and Weil and Lacaze[31] also fall within the large-diameter region. Recently Sauer and Ragsdell[32] have presented data taken using horizontal strip heaters which show that there is a consistent dimensional effect even with fairly large ($\frac{1}{2}$–2 in. wide) heaters. The dimensional effect for horizontal, large-diameter tube heaters has also been studied by Flanigan and Park.[33]

Merte and Clark[34–36] noted that reduced gravitational acceleration reduced the heat transfer coefficient for film boiling from copper spheres. A very regular (about 3 Btu/ft²-hr-°F per 30° of angle) increase in heat transfer coefficient resulted from increasing the angle with the horizontal for a strip heater. This was reported by Price and Sauer.[37]

Continuous removal of the vapor layer can considerably improve the heat transfer rate in film boiling. This method was first proposed by Bankoff[38] and then was verified by Bankoff and co-workers.[39–41] Increasing surface smoothness resulted in increased values of the heat transfer coefficient from data reported by Pai and Bankoff[39,40] without vapor suction. An imposed electric field causing stable film boiling to destabilize into nucleate boiling is a study published by Bochirol *et al.*[42]

7.2.4 Other Cryogenic Film Boiling Data

The effects of pressure and heat diameter on film boiling oxygen were verified by Banchero *et al.*[43] Other oxygen data have been reported by Giaque.[44] Bochirol *et al.*[45] noted that an electrical field also destabilized film boiling in oxygen. The oxygen data at 1 atm of Banchero *et al.* are shown in Fig. 7-4.

The effects of heater dimension have been investigated in three studies using neon and argon as the test liquid. Hsu and Westwater[25] reported data from vertical bayonet heaters, while Flanigan and Park[33] used horizontal heaters in their experiments with argon. Data for neon have been published by Astruc *et al.*[46]

While they do not all fall within the temperature range usually thought of as cryogenic, several other low-temperature film boiling studies bear mentioning. The light hydrocarbon series of methane, ethane, propane, and *n*-butane gave a consistent pressure effect in the work of Sciance *et al.*[47,48] Capone and Park[49] noted a similar result when dealing with carbon monoxide. The film boiling behavior of carbon dioxide very near its critical

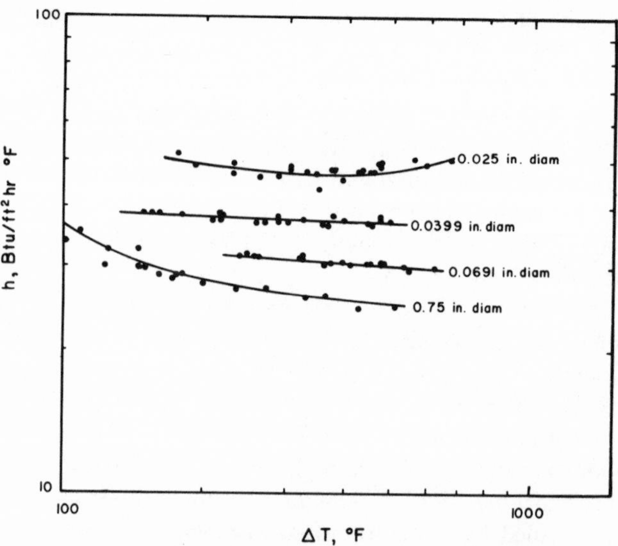

FIG. 7-4. Effect of heater diameter on oxygen film boiling data of Banchero *et al.*[43] All heaters are stainless steel.

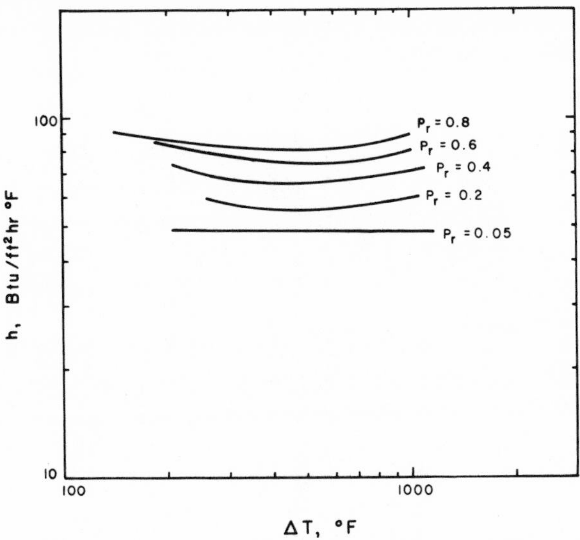

FIG. 7-5. Effect of pressure on methane film boiling data.[47,48]

point has been investigated by Grigull and Abadzic[50] and Abadzic and Goldstein.[51]

Two investigations of mixture effects in low-temperature film boiling heat transfer have been published. The liquefied natural gas (LNG) and the liquefied petroleum gas (LPG) results of Brown and Colver[52,53] verified a consistent pressure effect in film boiling, but because of the complexity of the mixtures no conclusions could be drawn regarding compositional effects. The data of Wright *et al.*[54] for ethane, ethylene, and three ethane–ethylene mixtures showed that mixtures in film boiling tend to act as pseudosubstances, with the results intermediate between the two pure components. Pressure effects on film boiling methane are shown in Fig. 7-5. The oxygen, neon, argon, and carbon compound data are summarized in Table 7-4.

7.3 CORRELATING EXPRESSIONS IN FILM BOILING

The phenomenon of film boiling is inherently an unstable situation because of the density inversion present. A density inversion of the type seen in film boiling is a physical realization of the sort of inversion which may be described mathematically on the basis of Taylor instability theory.[55,56]

TABLE 7-4
Other Low-Temperature Film Boiling Data

Fluid	Author	Heater	Pressure	Notes
Oxygen	Banchero *et al.*[43]	0.75-, 0.025-, 0.0399-, 0.0691-, and 0.122-in.-diameter stainless steel tubes	0.49–33.2 atm	Diameter, pressure, heater material, immersion depth effects studied
	Giaque[44]	0.2-mm platinum wire	1 atm	Cited from Ref. 43
	Bochirol *et al.*[45]	0.2-mm platinum wire	1 atm	Electric field effect studied
Argon	Hsu and Westwater[25]	0.5-in.-diameter vertical stainless steel cylinder	1 atm	Heaters 2 and 4 in. long were used
	Flanigan and Park[33]	0.55-, 0.75-, and 0.95-in.-diameter cylinder	3.8–30.8 atm	Pressure and dimension effects studied
Neon	Astruc *et al.*[46]	0.02-cm-diameter platinum, 0.2- and 0.05-cm lead wires	1 atm	Dimension effect studied
Methane	Sciance *et al.*[47,48]	0.81-in.-diameter gold-plated tube	1–45.7 atm	Pressure effect studied
	Park *et al.*[29]	0.65-in.-diameter gold-plated tube	1.6–41.7 atm	Pressure effect studied

Ethane	Sciance *et al.*[47,48]	0.81-in.-diameter gold-plated tube	1–48.2 atm	Pressure effect studied
	Wright *et al.*[54]	0.81-in.-diameter gold-plated tube	2.4–30 atm	Pressure effect studied
Ethylene	Wright *et al.*[54]	0.81-in.-diameter gold-plated tube	2.5–31.3 atm	Pressure effect studied
Ethane–ethylene mixtures	Wright *et al.*[54]	0.81-in.-diameter gold-plated tube	2.4–31 atm	Pressure and composition (25, 50, 75%) effects studied
Propane	Sciance *et al.*[47,48]	0.81-in.-diameter gold-plated tube	1–37.9 atm	Pressure effect studied
n-Butane	Sciance *et al.*[47,48]	0.81-in.-diameter gold-plated tube	1–33.7 atm	Pressure effect studied
LNG	Brown and Colver[52]	0.81-in.-diameter gold-plated tube	2.5–33 atm	Pressure effect studied
LPG	Colver and Brown[53]	0.81-in.-diameter gold-plated tube	1–34.3 atm	Pressure effect studied
Carbon monoxide	Capone and Park[49]	0.75-in.-diameter heater	3.4–328 atm	Pressure effect studied
Carbon dioxide	Grigull and Abadzic[50]	0.1-mm platinum wire	60.5, 73.4 bar	Photographic study of bubble spacing and wavelength
	Abadzic and Goldstein[51]	0.076- and 0.38-mm-diameter platinum wires	61.4–73.0 bar	Effect of pressures near critical point studied

Film stability, according to the theory, is based upon the characteristic wavelengths of disturbances propagated in the vapor–liquid interface. The existence and the properties of these interfacial waves have been studied photographically by Siegel and Keshock,[57] Lienhard and Wong,[58] and Grigull and Abadzic.[50] Visually the wave nature of film boiling is indicated by the rather regular spacing of the nodes from which bubbles are released and the regular period of bubble release. Lienhard and Sun[59] have recently extended the study of the wave nature of film boiling.

Two quantities which reflect the essential wave nature of film boiling have been widely used as characterizing dimensions for correlational purposes. These are the Laplace reference length

$$B = [g_c\sigma/g(\rho_l - \rho_v)]^{1/2} \tag{7-1}$$

and the critical film boiling wavelength

$$\lambda_c = 2\pi[g_c\sigma/g(\rho_l - \rho_v)]^{1/2} \tag{7-2}$$

Also used, particularly when determining the minimum film boiling heat flux, is the "most dangerous" wavelength

$$\lambda_d = 2\pi[3g_c\sigma/g(\rho_l - \rho_v)]^{1/2} \tag{7-3}$$

Two principal approaches, direct analysis of the heat transfer through a vapor film and analysis on the basis of instability theory, have formed the beginning points for deriving the bulk of the film boiling correlations in the literature. The earliest work is that of Bromley,[22,23] who used purely hydrodynamic considerations, with no reference to instability theory. Indeed, Taylor's instability theory was presented at the same time that Bromley presented his film boiling correlation. Bromley's model for stable film boiling from a horizontal, cylindrical heater in a quiescent liquid pool was based upon the following assumptions:

1. Heat transfer is by conduction through the thin vapor layer.
2. The vapor kinetic energy is negligible and latent heat absorption at the vapor–liquid interface is the primary contribution to heat transfer.
3. The heater wall temperature is constant.
4. Vapor film physical properties can be evaluated at the average film temperature $T_f = T_w + \frac{1}{2}(T_w - T_{sat})$.

These assumptions lead to the correlational expression

$$h_c = c_1[k_v{}^3\rho_v(\rho_l - \rho_v)gh'_{fg}/(D\Delta T\mu_v)]^{1/4} \tag{7-4}$$

Bromley found $c_1 = 0.62$ best represented all of his data. A somewhat more convenient form of Eq. (7-4) is

$$\mathrm{Nu}_D = 0.62[\mathrm{Ra}\theta']_f^{1/4} \tag{7-5}$$

where

$$\text{Ra} = [\rho_v(\rho_l - \rho_v)gC_{pv}D^3/k_v\mu_v] \tag{7-6}$$

and

$$\theta' = h'_{fg}/(C_{pv}\,\Delta T) \tag{7-7}$$

is the dimensionless heat parameter. Here the subscript D means that the Nusselt and Rayleigh numbers are based on the characteristic heater dimension. The subscript B would indicate a dimensionless group based on the Laplace reference length. Bromley also suggested an expression which included the radiative contribution to heat transfer, but this contribution is not normally of great importance in cryogenic work.

The dimensionless heat parameter may be modified, as it is in Eq. (7-4), by using a modified latent heat defined either by

$$h'_{fg} = h_{fg} + 0.5C_{pv}\,\Delta T \tag{7-8}$$

which was suggested by Bromley[22,23] to account for the sensible heat required to vaporize and raise one unit mass of liquid to the average film temperature, or by

$$h''_{fg} = h_{fg}[1 + 0.34(C_{pv}\,\Delta T/h_{fg})]^2 \tag{7-9}$$

Equation (7-9) was proposed for use in film boiling by Bromley[60,61] and was based on an analogy with results on the effect of sensible heat in laminar film condensation.

Banchero *et al.*[43] had to modify Eq. (7-5) somewhat to adequately express the diameter dependence of their oxygen data. The final result was

$$\text{Nu}_D = c_2[(1/\text{D}) + 36.5][\text{Ra}\theta']^{1/4} \tag{7-10}$$

where $0.441 < c_2 < 0.302$ at $100 < \Delta T < 500°\text{F}$. Another attempt to account for dimensional effects was made by Breen and Westwater,[62] who proposed

$$\text{Nu}_{\lambda_c} = [0.59 - (0.069\lambda_c/D)][\text{Ra}\theta'']^{1/4} \tag{7-11}$$

or, alternatively,

$$\text{Nu}_{\lambda_c} = 0.60[\text{Ra}\theta'']^{1/4}, \qquad \lambda_c/D < 0.8 \tag{7-12a}$$

$$\text{Nu}_D = 0.62[\text{Ra}\theta'']^{1/4}, \qquad 0.8 < \lambda_c/D < 8.0 \tag{7-12b}$$

$$\text{Nu}_{\lambda_c} = 0.16[\lambda_c/D]^{0.83}[\text{Ra}\theta'']^{1/4}, \qquad \lambda_c/D > 8.0 \tag{7-12c}$$

Recently Baumeister and Hamill[63,64] have presented the relations

$$\text{Nu}_B = 0.35[B/D]^{3/4}[\text{Ra}\theta'']^{1/4}, \qquad B/D > 10 \tag{7-13a}$$

$$\text{Nu}_B = 0.485[(B/D)^3 + 2.25(B/D)\Delta^2]^{1/4}[\text{Ra}\theta'']^{1/4}, \qquad B/D < 10 \tag{7-13b}$$

where

$$\Delta = \exp[4.35(k_v \mu_v \,\Delta T / h''_{fg} D \rho_l \sigma g_c)^{1/4}], \qquad 3 \leqslant B/D \leqslant 10 \tag{7-14a}$$

$$\Delta = 1, \qquad B/D < 3 \tag{7-14b}$$

Other relations based on Bromley's equation have been introduced to account for effects other than only the effect of heater dimension. Pomerantz[65] noted the analogy between dimensional effects and acceleration effects to arrive at the expression

$$\mathrm{Nu}_D = 0.62(D/\lambda_c)^{0.172}[\mathrm{Ra}\theta'']^{1/4} \tag{7-15}$$

Apparently the only correlation derived expressly for a vertical, bayonet heater is that of Hsu and Westwater.[25] They suggested the overall average heat transfer coefficient for a vertical heater be given by

$$\mathrm{Nu}_L = \frac{4}{3}\frac{L_0}{y^*} + \frac{A_2 + \frac{1}{3}}{A_1}\left\{\left[\frac{2}{3}\left(\frac{A_1}{A_2 + \frac{1}{3}}\right)(L - L_0) + \left(\frac{1}{y^*}\right)^2\right]^{3/2} - \left(\frac{1}{y^*}\right)^3\right\} \tag{7-16}$$

where†

$$A_1 = [g(\rho_l - \rho_v)/\rho_v](\bar{\rho}_v/\mu_v \mathrm{Re}^*)^2 \tag{7-17a}$$

$$A_2 = \frac{\mu_v + (f\rho_v\mu_v \mathrm{Re}^*/2\bar{\rho}_v) + (k_v \,\Delta T/h_{fg})}{k_v \,\Delta T/h_{fg}} \tag{7-17b}$$

$$y^* = [2\mu_v \mathrm{Re}^*/g\bar{\rho}_v(\rho_l - \bar{\rho}_v)]^{1/3} \tag{7-17c}$$

$$L_0 = \mu_v \mathrm{Re}^* h''_{fg} y^*/2k_v \,\Delta T \tag{7-17d}$$

$$\mathrm{Re}^* = (\bar{\rho}_v u^* y^*)/\mu_v \tag{7-17e}$$

Hsu and Westwater used the value $\mathrm{Re}^* = 100$ for the critical vapor Reynolds number.

Taylor instability theory has formed the second basis for modeling stable film boiling behavior. Frederking *et al.*[16,66] suggested that the instability model was dependent upon whether bubble release was assumed to be regular and whether vapor flow in the film layer was assumed to be laminar. Of the four possible combinations of these assumptions, the consensus of the correlations in the literature tacitly indicates that vapor flow is laminar. This is indicated by the universal use of the group [Raθ] which results from the assumption of laminar vapor flow rather than the group [RaθPr] which results from assuming turbulent vapor flow. Regular vapor release results in an exponent of $\frac{1}{4}$ on the Raθ group, while random release

† ρ_v is the density of the vapor in the turbulent core and ρ_v is the average density of the vapor in the laminar sublayer. [ed. note]

gives an exponent of $\frac{1}{3}$. Correlations resulting from an analysis based on instability theory all result in a form

$$\mathrm{Nu}_L = c_3[\mathrm{Ra}\theta']^{n_1}, \qquad \tfrac{1}{4} \leqslant n_1 \leqslant \tfrac{1}{3} \tag{7-18}$$

where c_3 is empirically determined.

Sciance *et al.*[47,48] used Eq. (7-18) as the starting point in determining an expression which would adequately account for the effect of pressure on film boiling data for a series of light hydrocarbons from a relatively large horizontal cylinder. They found that the addition of the reduced saturation temperature $T_r = T_{\mathrm{sat}}/T_c$ was adequate to represent pressure effects. The final, best fit, equation to correlate their data was

$$\mathrm{Nu}_B = 0.369[\mathrm{Ra}\theta' T_r^{-2}]^{0.267} \tag{7-19}$$

This equation has also proven quite accurate in correlating data for LNG, LPG, and ethane–ethylene mixtures taken on large-diameter heaters.

Recently, Clements and Colver[67] have extended Eqs. (7-18) and (7-19) to allow for both pressure and dimensional effects. Their final result was

$$\mathrm{Nu}_D = 0.94[\mathrm{Ra}\theta' T_r^{-2}]^{1/4} \tag{7-20}$$

This equation was shown to be satisfactory for a large variety of materials and a large range of heater diameters.

A unique approach to correlation of film boiling data has been taken by Park and co-workers recently. They have employed the results of the theory of corresponding states to generate correlations which do not necessitate the calculation of large numbers of physical property values. Capone and Park[49] assumed

$$h = f(T, D, P_r) \tag{7-21}$$

to arrive at the best-fit expression

$$\begin{aligned} h = {} & 255.83 + 94.69P_r - 86.79P_r^2 + 21.02P_r^3 - 0.3158\,\Delta T \\ & + 4.13 \times 10^{-4}(\Delta T)^2 - 438.02D + 286.09D^2 \end{aligned} \tag{7-22}$$

This idea was carried further by Flanigan and Park,[33] who proposed the relation

$$h = \alpha_2[(1/D) + 36.5]P_r^{1/4} \tag{7-23}$$

where

$$\alpha_2 = 13.38 - 15.53T_r + 6.14T_r^2 - 0.588T_r^3 \tag{7-24}$$

This expression correlated their data for nitrogen and argon plus Park's[29] nitrogen and methane data, the oxygen data of Banchero *et al.*,[43] and the methane data of Sciance *et al.*[47,48] with a deviation of about $\pm 20\%$.

TABLE 7-5
Film Boiling Correlations for Horizontal Heaters

Author	Correlation	Average deviation of test data set, %	Notes
Bromley[22,23]	$\mathrm{Nu}_D = 0.62[\mathrm{Ra}\,\theta']^{1/4}$	30.2	Coefficient based on data for nitrogen, *n*-pentane, water, ethanol, carbon tetrachloride, diphenyl ether
Breen and Westwater[62]	$\mathrm{Nu} = [0.59 - (0.069/D)][\mathrm{Ra}\,\theta'']^{1/4}$	58.5	Coefficient based on data for isopropanol, Freon 113, nitrogen, *n*-pentane, water, ethanol, benzene, carbon tetrachloride, diphenyl ether, oxygen, helium I
Chang[69]	$\mathrm{Nu}_D = 0.294[\mathrm{Ra}\,\theta]^{1/3}$	40.0	Coefficient based on data for benzene and water
Pomerantz[65]	$\mathrm{Nu}_D = 0.62(D/\lambda_c]^{0.172}[\mathrm{Ra}\,\theta'']^{1/4}$	36.3	Exponent on (D/λ_c) based on Freon 113 data
Frederking *et al.*[16]	$\mathrm{Nu}_B = 0.2[\mathrm{Ra}_B\theta']^{1/4}$	47.8	Coefficient based on data for nitrogen, helium I, water, Freon 11, carbon tetrachloride, *n*-pentane
Sciance *et al.*[47,48]	$\mathrm{Nu}_B = 0.369[\mathrm{Ra}_B\theta' T_r^{-2}]^{0.267}$	32.0	Coefficient and exponent based on data for methane, ethane, propane, *n*-butane
Clements and Colver[67]	$\mathrm{Nu}_D = 0.94[\mathrm{Ra}_D\theta' T_r^{-2}]^{1/4}$	21.8	Coefficient based on data for nitrogen, oxygen, water, methane, ethane, propane, *n*-butane, ethylene, ethane–ethylene mixtures
Berenson[68]	$\mathrm{Nu}_B = 0.425[\mathrm{Ra}\,\theta']^{1/4}$	38.3	Coefficient based on data for *n*-pentane, carbon tetrachloride

7.4 TESTING OF CORRELATIONS WITH EXPERIMENTAL DATA

Up to this point we have discussed what experimental data are available and what correlations have appeared in the literature. The only real basis for the designer to decide on what correlation to use if adequate data are not available is to directly compare the existing data with the correlations. This comparison is shown in Table 7-5, where the average absolute deviations between the experimental and the predicted values provide a means of measuring the accuracy of the several correlations. The data set used for the comparisons made in Table 7-5 includes data for nitrogen, oxygen, water, methane, ethane, ethylene, ethane–ethylene mixtures, propane, and *n*-butane. The complete set of data references is given in by Clements and Colver.[67]

7.5 NOMENCLATURE

A_1 = distance parameter defined in Eq. (7-17a), ft^{-3}
A_2 = dimensionless parameter defined in Eq. (7-17b)
B = Laplace reference length defined in Eq. (7-1), ft
C_p = heat capacity, Btu/lb_m-°F
$c_1, \ldots, c_d$ = empirical constants
D = diameter, ft
f = friction factor
g = gravitational acceleration, ft/sec^2
g_c = gravitational constant, lb_m-ft/lb_f-sec^2
h = heat transfer coefficient, Btu/ft^2-hr-°F
h_{fg} = latent heat of vaporization, Btu/lb_m
h'_{fg} = effective heat of vaporization defined in Eq. (7-8), Btu/lb_m
h''_{fg} = effective heat of vaporization defined in Eq. (7-9), Btu/lb_m
k = thermal conductivity, Btu/ft-hr-°F
L = characteristic dimension; heater length, ft
L_0 = critical height defined in Eq. (7-17d), ft
Nu = Nusselt number, $qL/k\,\Delta T$ or hL/k
$n_1, \ldots, n_k$ = empirical constants
P = pressure, atm
P_r = thermodynamic reduced pressure
Pr = Prandtl number, $C_p\mu/k$
q = heat flux, Btu/ft^2-hr
Ra = Rayleigh number, $[L^3 g\rho_f(\rho_l - \rho_f)\mathrm{Pr}]/\mu_v^2$

Re*	= critical Reynolds number, defined by Eq. (7-17e)
T	= temperature, °F
ΔT	= $T_w - T_{sat}$, °F
u^*	= maximum vapor velocity at y^*
y^*	= thickness of vapor layer defined in Eq. (7-17c), ft

Greek Letters

θ	= dimensionless heat parameter, $h_{fg}/(C_p\, \Delta T)$
θ'	= dimensionless heat parameter, $h'_{fg}/C_p\, \Delta T)$
θ''	= dimensionless heat parameter, $h''_{fg}/C_p\, \Delta T)$
λ	= wavelength, ft
λ_c	= critical wavelength defined in Eq. (7-2), ft
λ_d	= most dangerous wavelength defined in Eq. (7-3), ft
μ	= viscosity, lb_m/ft-hr
ρ	= density, lb_m/ft^3
σ	= surface tension, lb_f/ft
Δ	= dimensionless parameter defined in Eq. (7-14)

Subscripts

c	= critical
d	= most dangerous
f	= film
l	= liquid
r	= reduced
sat	= saturated
v	= vapor
w	= wall

7.5 REFERENCES

1. D. P. Jordan, *Advan. Heat Transfer* **5**, 55 (1968).
2. L. D. Clements and C. P. Colver, *Ind. Eng. Chem.* **62**, 26 (1970).
3. E. G. Brentari, P. T. Giarratano, and R. V. Smith, NBS Tech. Note 317 (1965).
4. E. G. Brentari, P. T. Giarratano, and R. V. Smith, in *International Advances in Cryogenic Engineering*, Plenum Press, New York (1965), p. 325.
5. R. J. Richards, W. G. Steward, and R. B. Jacobs, NBS Tech. Note 122 (1961).
6. L. Weil, in *Proc. 8th Intern. Congr. Refrig.* London (1951), p. 181.
7. L. Weil and A. Lacaze, *J. Phys. Radium* **12**(9), 890 (1951).
8. R. N. Mulford and J. P. Nigon, LA-1416 (1952).
9. C. R. Class, J. R. DeHaan, M. Piccone, and R. B. Cost, in *Advances in Cryogenic Engineering*, Vol. 5, Plenum Press, New York (1960), p. 254.

10. R. W. Graham, R. C. Hendricks, and R. C. Ehlers, in *International Advances in Cryogenic Engineering*, Plenum Press, New York (1965), p. 342.
11. T. H. K. Frederking, *AIChE J.* **5**(3), 403 (1959).
12. T. H. K. Frederking, *Forschung* **27**(1), 17 (1961).
13. T. H. K. Frederking and P. Grassmann, *Bull. Inst. Intern. Froid, Annexe 1958-1*, 317 (1958).
14. P. Grassmann, A. Karagounis, J. Kopp, and T. Frederking, *Kältetecknik* **10**(7), 206 (1958).
15. T. H. K. Frederking, R. C. Chapman, and S. Wang, in *International Advances in Cryogenic Engineering*, Plenum Press (1965), p. 353.
16. T. H. K. Frederking, Y. C. Wu, and B. W. Clement, *AIChE J.* **12**(2), 238 (1966).
17. D. N. Lyon, in *Advances in Cryogenic Engineering*, Plenum Press, New York (1965), p. 371.
18. R. M. Holdredge and P. W. McFadden, in *Advances in Cryogenic Engineering*, Vol. 16, Plenum Press, New York (1971), p. 352.
19. P. Grassmann and A. Karagounis, in *Proceedings 5th Intern. Conference, Low Temperature Physics and Chemistry*, Madison, Wisconsin (1958), p. 41.
20. T. M. Flynn, J. W. Draper, and J. J. Roos, in *Advances in Cryogenic Engineering*, Vol. 7, Plenum Press, New York (1962), p. 539.
21. G. G. Harman and L. H. Gordy, *Cryogenics* **7**(2), 89 (1967).
22. L. A. Bromley, AEC D2295 (1948).
23. L. A. Bromley, *Chem. Eng. Progr.* **46**(5), 221 (1950).
24. C. W. Cowley, W. J. Timson, and J. A. Sawdye, *Ind. Eng. Chem., Process Des. Develop.* **1**(2), 81 (1962); also in *Advances in Cryogenic Engineering*, Vol. 7, Plenum Press, New York (1962), p. 385.
25. Y. Y. Hsu and J. W. Westwater, *Chem. Eng. Progr., Symp. Ser.* **56**(30), 15 (1960).
26. Y. Y. Hsu and J. W. Westwater, *AIChE J.* **4**(1), 58 (1958).
27. J. Ruzička, in *Problems of Low Temperature Physics*, Vol. 1, Pergamon Press, New York (1959), p. 323.
28. L. G. Rhea and R. G. Nevins, *Trans. ASME, J. Heat Transfer* **91C**(2), 267 (1969).
29. E. L. Park, C. P. Colver, and C. M. Sliepcevich, in *Advances in Cryogenic Engineering*, Vol. 11, Plenum Press, New York (1966), p. 516.
30. D. N. Lyon, P. G. Kosky, and B. N. Harman, in *Advances in Cryogenic Engineering*, Vol. 9, Plenum Press, New York (1964), p. 77.
31. L. Weil and A. Lacaze, *Compt. Rend.* **230**(1), 186 (1950).
32. H. J. Sauer and K. M. Ragsdell, in *Advances in Cryogenic Engineering*, Vol. 16, Plenum Press, New York (1971), p. 412.
33. V. J. Flanigan and E. L. Park, in *Advances in Cryogenic Engineering*, Vol. 16, Plenum Press, New York (1971), p. 402.
34. J. A. Clark and H. Merte, *Advan. Astronaut. Sci.* **14**, 177 (1963).
35. H. Merte and J. A. Clark, *Trans. ASME, J. Heat Transfer* **86C**(3), 351 (1964).
36. H. Merte and J. A. Clark, *Advances in Cryogenic Engineering*, Vol. 7, Plenum Press, New York (1962), p. 546.
37. C. E. Price and H. J. Sauer, *ASHRAE Trans.* **76**, 58 (1970).
38. S. G. Bankoff, *AIChE J.* **7**(3), 485 (1961).
39. V. K. Pai and S. G. Bankoff, *AIChE J.* **11**(1), 65 (1965).
40. V. K. Pai and S. G. Bankoff, *AIChE J.* **12**(4), 727 (1966).
41. P. C. Wayner and S. G. Bankoff, *AIChE J.* **11**(1), 59 (1965).
42. L. Bochirol, E. Bonjour, and L. Weil, *Bull. Inst. Intern. Froid., Annexe 1960–1*, 251 (1960).

43. J. T. Banchero, G. E. Barker, and R. H. Boll, *Chem. Eng., Progr. Symp. Ser.* **51**(17), 21 (1955).
44. W. F. Giaque, OSRD-491, Ser. 201 (1942).
45. L. Bochirol, E. Bonjour, and L. Weil, *Compt. Rend.* **250**(1), 76 (1960).
46. J. M. Astruc, P. Perroud, A. Lacaze, L. Weil, *Advances in Cryogenic Engineering*, Vol. 12, Plenum Press, New York (1967), p. 387.
47. C. T. Sciance, C. P. Colver, and C. M. Sliepcevich, *Chem. Eng. Progr. Symp. Ser.* **63**(77), 115 (1967).
48. C. T. Sciance, C. P. Colver, and C. M. Sliepcevich, in *Advances in Cryogenic Engineering*, Vol. 12, Plenum Press, New York (1967), p. 395.
49. G. J. Capone and E. L. Park, "Comparison of the Experimental Film Boiling Behavior of Carbon Monoxide with Several Film Boiling Correlations," presented at 3rd AIChE–IMIQ Joint Meeting, Denver, Colorado (1970).
50. U. Grigull and E. Abadzic, *Forschung* **31**(1), 27 (1965).
51. E. Abadzic and R. J. Goldstein, *Intern. J. Heat Mass Transfer* **13**, 1163 (1970).
52. L. E. Brown and C. P. Colver, in *Advances in Cryogenic Engineering*, Vol. 13, Plenum Press, New York (1968), p. 647.
53. C. P. Colver and L. E. Brown in *Proc. 48th Annual Convention, Natural Gas Processors Association*, Dallas, Texas (1969), p. 85.
54. R. D. Wright, L. D. Clements, and C. P. Colver, to appear in *AIChE J.*
55. R. Bellman and R. H. Pennington, *Quart. Appl. Math.* **12**(2), 151 (1954).
56. G. Taylor, *Proc. Roy. Soc.* (*London*) **201A** (1065), 192 (1950).
57. R. Siegel and E. G. Keshock, NASA TR R-216 (1965).
58. J. H. Lienhard and P. T. Y. Wong, *Trans. ASME, J. Heat Transfer* **86C**, 220 (1964).
59. J. H. Lienhard and K.-H. Sun, *Trans. ASME, J. Heat Transfer* **92C**, 292 (1970).
60. L. A. Bromley, *Ind. Eng. Chem.* **44**(12), 2966 (1952).
61. L. A. Bromley, N. R. LeRoy, and J. A. Robbers, *Ind. Eng. Chem.* **45**(11), 2639 (1953).
62. B. P. Breen and J. W. Westwater, *Chem. Eng. Progr.* **58**(7), 67 (1962).
63. K. J. Baumeister and T. D. Hamill, AIChE–ASME Heat Transfer Conference, Paper 67-HT-2 Seattle, Washington (1967).
64. K. J. Baumeister and T. D. Hamill, NASA TN D-4035 (1967).
65. M. L. Pomerantz, *Trans. ASME, J. Heat Transfer* **86C**(2), 213 (1964).
66. T. H. K. Frederking, Paper 21b presented at the 52nd National AIChE Meeting, Memphis, Tennessee (1964).
67. L. D. Clements and C. P. Colver, submitted to *J. Heat Transfer*.
68. P. J. Berenson, *Trans. ASME, J. Heat Transfer* **83C**(3), 351 (1961).
69. Y. P. Chang, *Trans. ASME, J. Heat Transfer* **81C**, 1 (1959).

MINIMUM FILM BOILING HEAT FLUX 8

W. FROST

8.1 INTRODUCTION

The transition from film boiling to nucleate boiling occurs at a heat flux known as the minimum film boiling heat flux $q_{\min}$ (see Fig. 4-4). Most theories used to predict the magnitude of this heat flux are based on hydrodynamic instability theories (see Refs. 1–3 for good descriptions of the mathematical model). Agreement of the theory with experiment has been good. However, because of the fact that the phenomenon occurs as a result of an instability, there is an inherent uncertainty in the value of $q_{\min}$. Thus, only the range within which $q_{\min}$ lies can be determined.

Prediction of the minimum heat flux is normally of lesser importance than prediction of the maximum or peak heat flux in fluid with high boiling temperatures since the transition to nucleate boiling is generally accompanied by a marked reduction in surface temperature and consequently involves no severe thermal loading of the system as described in Chapter 6. One does, however, encounter a need to predict the minimum film boiling heat flux in cryogenic fluids where a change to nucleate boiling occurs at operating temperatures. The several orders of magnitude change in the heat transfer coefficient accompanying the boiling transition becomes a design consideration in many instances.

W. FROST University of Tennessee Space Institute, Tullahoma, Tennessee.

TABLE 8-1

Summary of Pool Film Boiling Minimum Heat Flux Theories[4],[a]

Reference	φ	Equation No.
Zuber and Tribus[11]	$K_1[\sigma g g_c(\rho_l - \rho_v)/(\rho_l + \rho_v)^2]^{1/4}$, $K_1 = 0.099$–0.131	(1)
Zuber and Tribus[11] (alternate approach)	$K_2[\sigma g g_c/(\rho_l - \rho_v)]^{1/4}$, $K_2 = 0.109$–0.144	(2)
Zuber[1]	$0.177[\sigma g g_c(\rho_l - \rho_v)/(\rho_l + \rho_v)^2]^{1/4}$	(3)
Berenson[12]	$0.09[\sigma g g_c(\rho_l - \rho_v)/(\rho_l + \rho_v)^2]^{1/4}$	(4)

[a] $q_{\min}/h_{fg}(\rho_v)_f = \varphi$ (consistent units).

8.2 POOL BOILING

Seader *et al.*[4] have summarized several of the available theories for pool film boiling minimum heat flux on horizontal surfaces. Table 8-1 is taken from this reference. The correlations are expressed in terms of the parameter

$$\varphi = q_{\min}/h_{fg}(\rho_v)_f \tag{8-1}$$

In all cases, the physical properties are evaluated at the saturation temperature, except $(\rho_v)_f$, which is evaluated at the average vapor film temperature. Seader and co-workers[4] have plotted the various correlations for the minimum heat flux versus pressure for H_2, N_2, and O_2 as shown in Fig. 8-1 for nitrogen. Frost and Dzakowic[5] have expressed Eq. (1) of Table 8-1 in a convenient graphical form for predicting minimum heat flux with engineering accuracy in most fluids.

Lienhard and Schrock[6] have investigated the minimum heat flux on horizontal cylinders. Their correlation with an empirical constant evaluated from experiments with data for isopropanol and benzene is

$$\frac{q_{\min}}{h_{fg}(\rho_v)_f} = \frac{0.16}{D}\left[\frac{\sigma}{g(\rho_l - \rho_v)}\right]^{1/2}\left[\frac{g\sigma(\rho_l - \rho_v)}{(\rho_l + \rho_v)^2}\right]^{1/4}\left[1 + \frac{2\sigma}{D^2 g(\rho_l - \rho_v)}\right]^{-1/4} \tag{8-2}$$

Brentari *et al.*[7] recommend the use of Eq. (8-2) in their review of boiling heat transfer for oxygen, nitrogen, hydrogen, and helium. Figure 5-19 illustrates typical values of the minimum flux as predicted by Eq. (8-2).

Jordan's[3] survey of film and transition boiling reports that minimum film boiling heat flux from vertical surfaces has not been investigated, either experimentally or analytically.

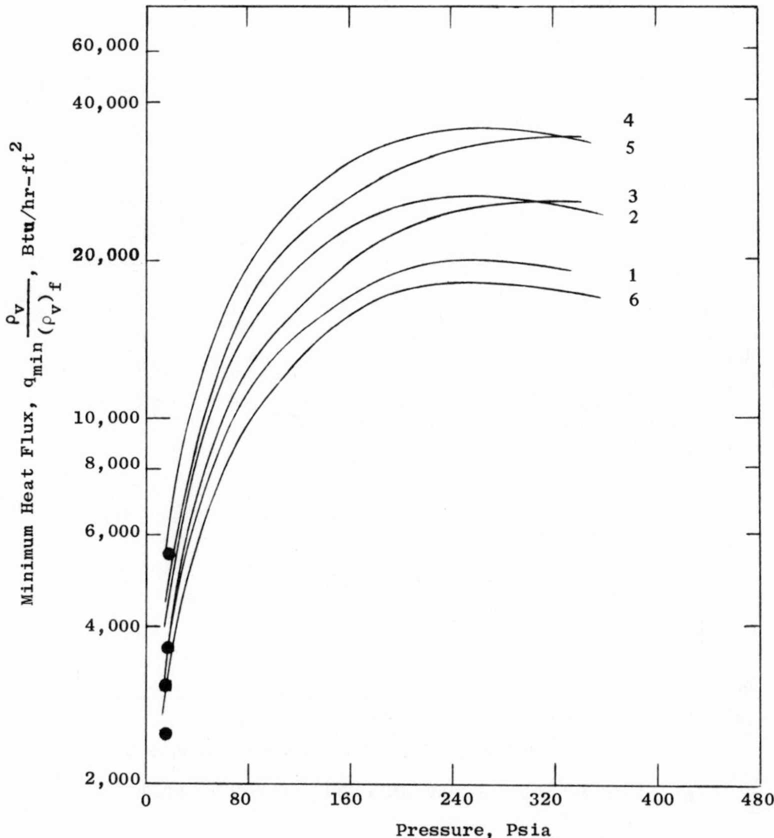

FIG. 8-1. Pool film boiling minimum heat flux for nitrogen. The circles represent the experimental data. (1) Zuber and Tribus[11], $K_1 = 0.099$; (2) Zuber and Tribus,[11] $K_1 = 0.131$; (3) Zuber and Tribus,[11] $K_2 = 0.109$; (4) Zuber and Tribus,[11] $K_2 = 0.144$; (5) Zuber [1]; (6) Berenson.[12]

8.3 FORCED CONVECTION BOILING

In forced convection boiling there is evidence[8] that the hysteresis effect associated with the minimum heat flux (Fig. 4-7) in pool boiling does not exist. Stevens, as reported in Ref. 8, has observed that on decreasing the heat flux from a stable film boiling state with Freon 12, the system returns to nucleate boiling along the same jump discontinuity as burnout (Fig. 8-2). McEwen and co-workers[9] also found the hysteresis effect to be absent, using water at 1500 psia; on the other hand, Sterman *et al.*[10] observed a hysteresis with isopropyl alcohol, ethyl alcohol, and distilled water at low

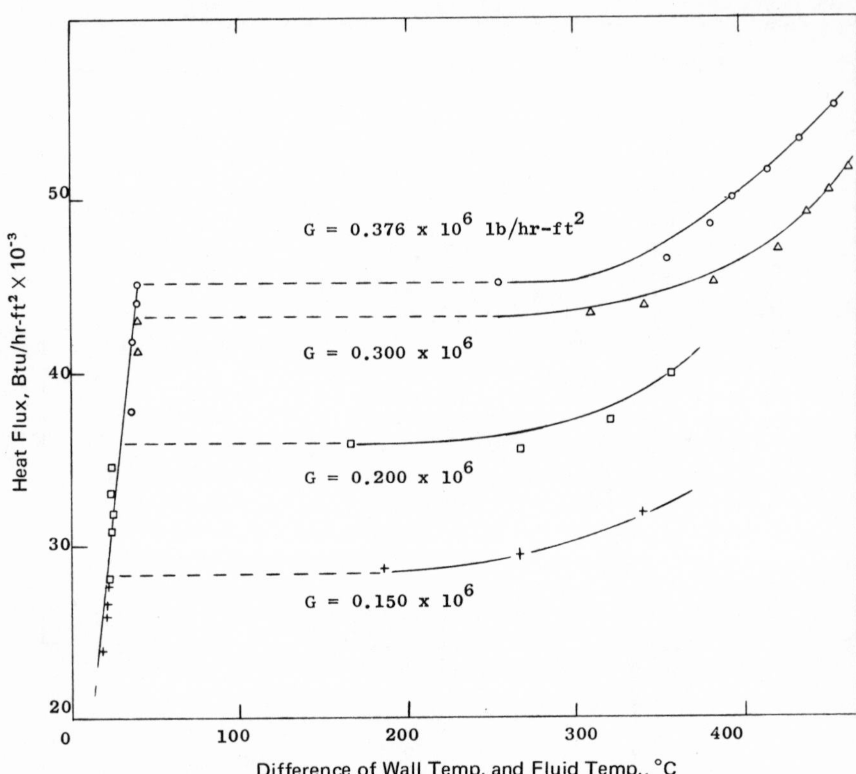

FIG. 8-2. Forced convection boiling curves for Freon 12.

pressures. Insufficient evidence is available to substantiate the correct nature of the forced convection boiling curve in the minimum boiling heat flux region.

8.4 NOMENCLATURE

D = diameter
g = acceleration of gravity
h_{fg} = latent heat of vaporization
$q_{\min}$ = minimum film boiling heat flux

Greek Letters

ρ = density
σ = surface tension

Subscripts

f = evaluated at film temperature
l = liquid
v = vapor

8.5 REFERENCES

1. N. Zuber, USAEC Report, AECU-4439 (1959).
2. G. Leppert and C. C. Pitts, in *Advances in Heat Transfer*, Academic Press, New York, (1964), p. 185.
3. D. P. Jordan, in *Advances in Heat Transfer*, Academic Press, New York, (1968), p. 55.
4. J. D. Seader, W. S. Miller, and L. A. Kalvinskaw, NASA CR-243 (1965).
5. W. Frost and G. S. Dzakowic, AEDC-TR-69-106 (1969).
6. J. H. Lienhard and V. E. Schrock, *Trans. ASME, J. Heat Transfer* **85**, 261 (1963).
7. E. G. Brentari, P. J. Giarratano, and R. V. Smith, NBS Tech. Note. No. 317 (1965).
8. R. V. MacBeth, in *Advances in Chemical Engineering*, Academic Press, New York, (1968), p. 207.
9. L. H. McEwen, J. M. Batch, D. J. Foley, and M. R. Kreiter, ASME Paper No. 57-SA-49 (1957).
10. L. S. Sterman, N. G. Stuishin, and V. G. Morozov, *J. Tech. Phys.* (*USSR*) **1**, 2250 (1956).
11. N. Zuber and M. Tribus, UCLA Report 58-5 (January 1958).
12. P. J. Berenson, MIT Heat Transfer Lab. Tech. Rept. No. 17, Cambridge, Massachusetts (March 1960).

VAPOR–LIQUID CONDENSATION ON CRYOGENIC SURFACES 9

K. D. TIMMERHAUS and M. A. LECHTENBERGER

9.1 INTRODUCTION

One of the major capital investments in any plant operated at cryogenic temperatures is heat transfer equipment. The optimum design of this equipment requires a detailed knowledge of the heat transfer characteristics of various cryogenic fluids and materials. Heat transfer to cryogenic fluids usually involves a change of phase, either boiling or condensation.

As noted in an earlier chapter, many investigations have been made into the characteristics of fluids boiling near room temperature or at cryogenic temperatures. One reason for this great interest is that with normal boiling liquids the resistance to heat transfer is generally higher than with condensing fluids. Thus, when a boiling film does provide one of the parallel resistances in a heat transfer problem, it is usually the controlling resistance. Conversely, condensing fluids generally exhibit relatively low resistance to heat transfer and consequently are not the controlling resistance in a parallel situation. On the other hand, studies with cryogenic fluids have shown that this is not necessarily true. With nitrogen, boiling and condensing resistances are approximately equal, while hydrogen exhibits a lower resistance to heat transfer for boiling than for condensation. Unfortunately, there is a dearth

K. D. TIMMERHAUS and M. A. LECHTENBERGER University of Colorado, Boulder, Colorado.

of experimental data covering the condensation of cryogenic fluids and the results of some of the studies available have been somewhat contradictory. For example, the studies with nitrogen and oxygen have been in reasonable agreement with condensation theory, while those with hydrogen and deuterium have varied significantly from theory over the same range of temperature difference.

9.2 FUNDAMENTAL CONSIDERATIONS

A vapor may condense on a cold surface in one of two ways—as a film or dropwise. Condensation of a pure vapor on a vertical surface that is clean and smooth is generally of the film type. In some cases, if the surface contains impurities or a promoter,[1] dropwise condensation may take place. Usually this type of condensation can be maintained only at low heat fluxes and the resulting heat transfer coefficient is of an order of magnitude higher than in film condensation where the heat transfer surface is covered by a continuous film of condensate. In the latter case, it is generally assumed that the condensate film exhibits the greatest resistance to heat transfer on the condensing side;[2] however, in special cases, such as with liquid metals having a high liquid thermal conductivity, the primary resistance to heat transfer is at the liquid–vapor interface rather than in the condensate film itself.

Since two fluid phases, vapor and condensate, are involved, condensation of a vapor on a colder surface is a complicated heat transfer process. The process of condensation on a vertical wall is shown in Fig. 9-1. Small pressure gradients near the vapor–liquid interface, resulting from the condensation of vapor and the downward flow of the liquid condensate, provide the driving force to move the vapor toward the condensing surface. Some of the molecules from the vapor phase strike the liquid surface and are reflected, while others penetrate and surrender their heat of condensation. The heat which is released is conducted through the condensate layer to the wall and then through the wall to the coolant located on the other side. The decreasing temperature gradient through the condensate and the wall provides the driving force for the heat transfer to the coolant. Simultaneously, the condensate drains from the surface by gravity.

The earliest successful theoretical study of laminar film condensation of a pure vapor on a vertical surface is attributed to Nusselt.[3] In this classical derivation, involving the solution of mass, momentum, and energy balance equations, Nusselt assumed that (1) the condensate behaved as a Newtonian fluid at steady state, (2) the vapor contained no noncondensibles and was saturated, (3) the condensate flow down the condensing surface was laminar, (4) the condensing surface was isothermal, (5) the physical properties of the

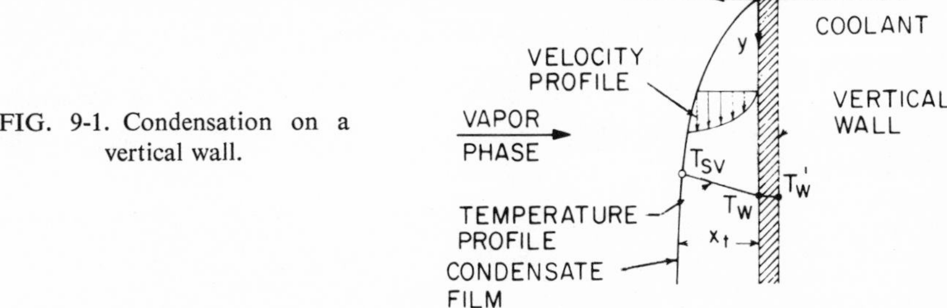

FIG. 9-1. Condensation on a vertical wall.

condensate film were constant through the film, (6) convective heat transfer in the condensate film was negligible in comparison to conductive heat transfer through the film, (7) there was negligible vapor drag and buoyancy effects on the condensate film, (8) the heat transfer due to subcooling of the condensate was negligible, and (9) the effect of curvature of the condensing surface on the velocity distribution in the condensing film was negligible. With these assumptions, Nusselt's relationships for the film thickness x_t as a function of the distance y down the condensing surface and the mean condensing film heat transfer coefficient are as follows:

$$x_t = [4k_f \mu_f y(T_{\text{sv}} - T_w)/g\rho_f h_{fg}]^{1/4} \tag{9-1}$$

$$h_c = 0.943[k_f{}^3 \rho_f{}^2 g h_{fg}/\mu_f L(T_{\text{sv}} - T_w)]^{1/4} \tag{9-2}$$

The condensate properties for these two equations are evaluated at a temperature given by

$$T_f = T_{\text{sv}} - 0.75\,\Delta T_c \tag{9-3}$$

where $\Delta T_c = T_{\text{sv}} - T_w$. Although the physical properties vary with temperature, the viscosity is the only property significantly affected when considering the temperature drop across a condensate film.

Since the first analysis by Nusselt, several researchers[4–12] have modified the original derivation to include those factors assumed negligible by Nusselt. Bromley *et al.*[4] considered the assumption of negligible subcooling and found that the heat of vaporization should be redefined as

$$h'_{fg} = h_{\text{sv}} + \tfrac{3}{8}C_p\,\Delta T_c \tag{9-4}$$

because the total heat flux is actually the sum of the latent heat plus the sensible heat given up by subcooling the liquid in the condensate film. This correction is significant at high pressures where the heat capacity is increasing as the heat of vaporization decreases. The error introduced by Nusselt's assumption of a constant tube temperature was also investigated by Bromley *et al.* with the conclusion that the conduction of heat around a tube has a negligible effect on the overall heat transfer coefficient.

Rohsenow[5] discarded the assumption of negligible convection in the condensate film and included the buoyancy force acting on the condensate film; however, the equations derived did not yield to an analytical solution and the final correlations were arrived at from a method of successive integrations. Rohsenow's final equations obtained from an approximate solution are

$$x_t = \left[\frac{4k_f\mu_f y(T_{sv} - T_w)}{g\rho_f(\rho_f - \rho_v)h'_{fg}[1 - \frac{1}{10}(C_p\Delta T_c/h'_{fg}) - 0.0328(C_p\Delta T_c/h'_{fg})^2]}\right]^{1/4} \tag{9-5}$$

$$h_c = 0.943\left[\frac{k_f{}^3\rho_f(\rho_f - \rho_v)gh'_{fg}[1 - \frac{1}{10}(C_p\Delta T_c/h'_{fg}) - 0.0328(C_p\Delta T_c/h'_{fg})^2]}{\mu_f\,\Delta t_c L[1 - \frac{1}{10}(C_p\Delta T_c/h'_{fg})]^4}\right]^{1/4} \tag{9-6}$$

The latter equation is very similar to the Nusselt equation for h_c with deviation occurring only when there are large temperature differences across the condensing film. Sparrow and Gregg[6] included the convection and buoyancy terms in the general boundary layer equations and by introduction of a similarity variable were able to obtain a numerical solution. This solution confirmed the approximate solution of Rohsenow.

In another study, Rohsenow *et al.*[7] again worked from Nusselt's original assumptions and examined the effect of vapor drag on the condensate film. This has resulted in the formulation of an expression which considers the existence of laminar flow of the condensate on the upper portion of a vertical plate and turbulent flow on the lower portion of the same vertical plate. Another researcher who has developed a model for condensation from high-velocity vapors and turbulent condensate flow is Dukler.[8] The equations involved were too complex for analytical solution but graphical relationships allowed determination of velocity distribution, the liquid film thickness, the local heat transfer coefficient, and the average coefficient over an entire condenser tube. Leonard and Estrin,[9] in an experimental study of steam condensation, have reported that Nusselt's equation was inadequate for flows having Reynolds numbers between 200 and 600 and indicated that Dukler's theory gave more accurate predictions in this range.

Laminar film condensation of pure saturated vapors on inclined circular cylinders has been treated analytically by Hassan and Jakob.[10] Their work indicated that for an $L/r < 6$, where L is the length and r is the radius of the cylinder, the cylinder has an optimum inclination which is something less than vertical; however, for $L/r > 6$ the optimum inclination is vertical. Attempts to verify the theory experimentally with steam showed wide disagreement and was attributed to possible rippling of the condensate film. Yang[11] has also analytically analyzed the effect on condensation of nonisothermal vertical surfaces but has presented no experimental confirmation of his theoretical predictions.

In the boundary layer treatment of laminar film condensation referred to earlier by Sparrow and Gregg,[6] the authors showed that the Prandtl number effect which arises from the retention of the acceleration terms is very small for Prandtl numbers greater than unity. In a somewhat similar study, Carpenter and Colburne[12] have obtained an empirical relation for the condensation of fluids having Prandtl numbers in the range of 2–5 of the form

$$h\mu/k\rho^{1/2} = 0.065(\mathrm{Pr})^{1/2}\tau_v^{1/2} \tag{9-7}$$

where Pr is the Prandtl number of the condensate and τ_v is the vapor shear stress.

The transition from laminar to turbulent flow occurs at a Reynolds number of approximately 1800.[13] Since Nusselt's equation for condensing film heat transfer coefficients was derived for laminar flow only, the Reynolds number, given by

$$\mathrm{Re} = 4\dot{m}_p/\mu \tag{9-8}$$

must be below 1800.

The subject of dropwise condensation of vapors has been studied extensively although no satisfactory correlations have yet been achieved.[14,15] The use of promoters in inducing dropwise condensation was originally disclosed in a U.S. patent granted to Nagle.[16]

For a more complete review of condensation of saturated vapors with particular emphasis on the modifications made to Nusselt's original equations, the reader is referred to the work by Wilhelm[17] and Sawochka.[18]

9.3 EXPERIMENTAL RESULTS

There have been few experimental investigations of condensation of cryogenic fluids. Those that have been made cover only the fluids of oxygen,[19] nitrogen,[19–22] hydrogen,[22,23] and deuterium.[23] These are shown in Figs. 9-2–9-5, respectively. Additional experimental results on condensing air and mixtures of nitrogen and oxygen have been performed by Haselden and Prosad[19] and Guther.[24]

An examination of the available experimental data for condensing pure vapors shows that the condensing film heat transfer coefficients for oxygen and nitrogen are represented quite well by the Nusselt correlation as evaluated by Ewald and Perroud.[23] However, for hydrogen and deuterium there is apparently a large deviation from the Nusselt correlation when the temperature drop across the condensate film is small, i.e., on the order of one

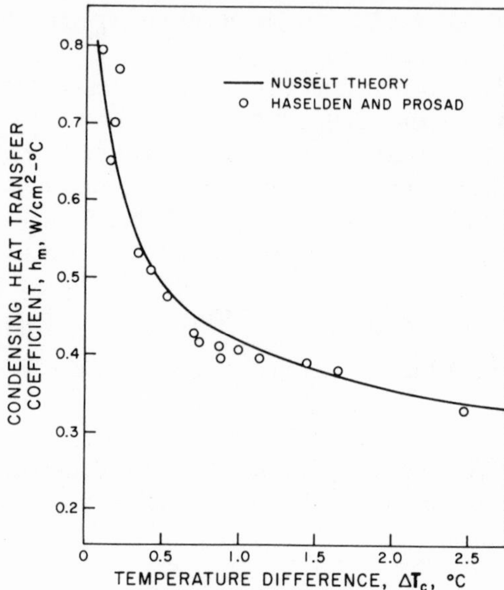

FIG. 9-2. Condensing film heat transfer coefficient for oxygen as a function of temperature difference across the condensing film.

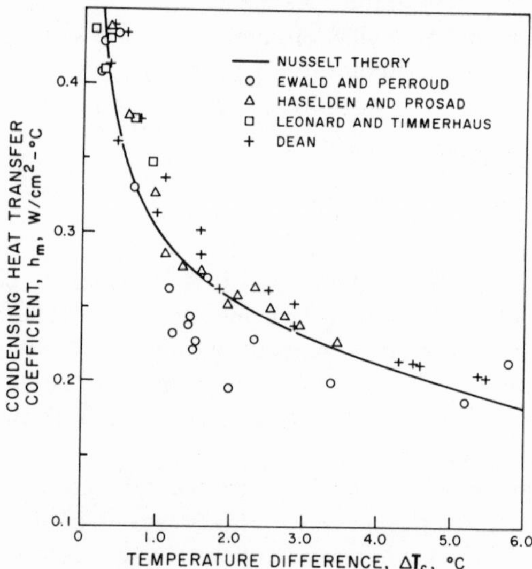

FIG. 9-3. Condensing film heat transfer coefficients for nitrogen as a function of temperature difference across the condensing film.

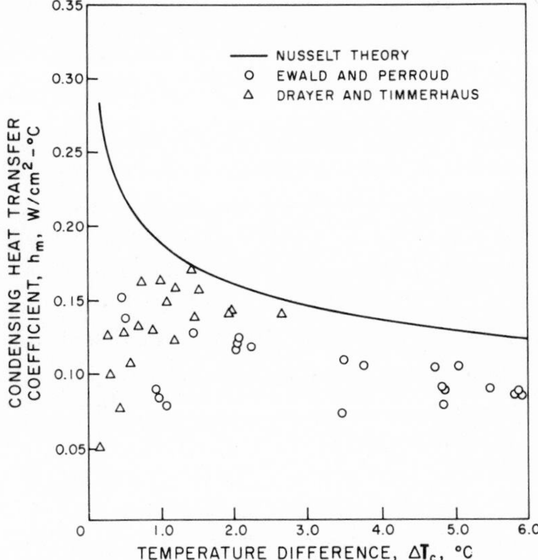

FIG. 9-4. Condensing film heat transfer coefficient for hydrogen as a function of temperature difference across the condensing film.

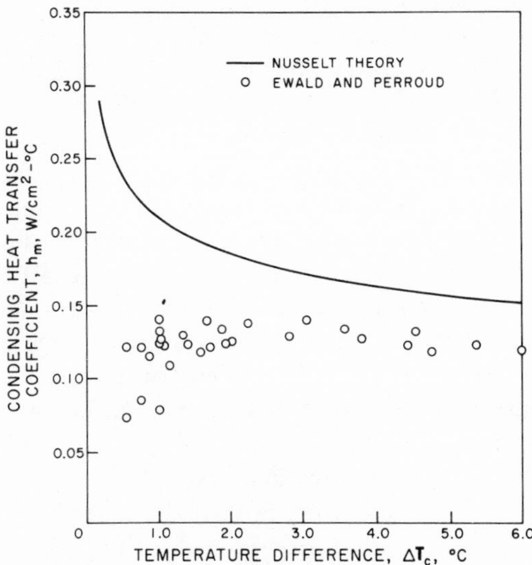

FIG. 9-5. Condensing film heat transfer coefficient for deuterium as a function of temperature difference across the condensing film.

or two degrees (centrigrade). It is worth noting that a similar situation was noted for nitrogen by Lechtenberger.[25] However, in the latter case, this occurred when the temperature drop across the condensate film was of the order of one-half degree or less. A careful inspection of the experimental system used indicates that with the condensation occurring on the outside of the cold tube surface, it is quite possible that there was sufficient heat leak into the system to superheat the vapor prior to contacting the condensing surface. Assuming the validity of this assumption, it would be normal to expect a lower condensing film heat transfer coefficient from inspection of Eq. (9-2). The experimental apparatus used by Ewald and Perroud[23] to evaluate the condensing film heat transfer coefficient for hydrogen and deuterium used the same principle as that used by Lechtenberger, i.e., the cryogenic vapor condensed on the outside of a cold tube surface and heat leak could have superheated the vapor prior to condensation. This could have accounted for the lower condensing film heat transfer coefficients that were obtained by these researchers when the temperature drop across the condensate film was small. This explanation, however, is not a very plausible one for the low condensing heat transfer coefficient values for hydrogen that were obtained by Drayer and Timmerhaus,[22] since in their apparatus the condensation was performed on the inside of a cold tube surface. With the colder fluid on the outside of the tube, there was little possibility for heat leak to the condensing cryogenic vapor. Thus, the results obtained for the above four cryogenic fluids, as far as condensing film heat transfer coefficients are concerned, are somewhat contradictory. Oxygen and nitrogen behave as predicted by the Nusselt correlation, while hydrogen and deuterium have shown noticeable deviations from this correlation.

The experimental data obtained for condensing film heat transfer coefficients for mixtures of cryogenic vapors show much more scatter than those obtained for pure cryogenic vapors. Part of this is due to the difficulty in measuring the temperature drop across the condensate film which, in the case of mixtures, is less than the temperature difference between the bulk of the vapor and the condensing surface. Nevertheless, Haselden and Prosad have indicated that the condensing film heat transfer coefficient for mixtures of nitrogen and oxygen would fall between the boundary curves for pure nitrogen and pure oxygen, and therefore until further work is done, the Nusselt relationship can be applied (using the temperature drop across only the condensate film) to give a preliminary estimate of the film coefficients.

In utilizing any of these coefficients for condensation, it should be borne in mind that the determination of each of the individual coefficients depends upon the measured values of heat flux and the temperature difference across the condensate film. Thus, the degree to which these variables are accurately measured determines the error involved in calculating the heat transfer coefficient. The most serious error in all of the measurements reported is in the accurate determination of the temperature difference across the condensate film. The smaller the temperature difference being measured across the condensate film, the greater will be the possible error. For example, in the study by Leonard,[26] the author determined that there was a possible 35% error that could be introduced into the coefficient from the measurement of temperature differences at the lowest Δt_c observed. At liquid hydrogen temperatures this could result in an even greater possible error in the coefficient due to additional complexities in the measurement of the temperature differences across the condensing film.

Finally, it is doubtful that all of the conditions imposed by Nusselt's assumptions are or can be approached in practice. These conditions are apt to be violated by commercial condenser–evaporator units using long vertical tubes and high vapor velocities. In these units the vapor viscous forces may combine with the turbulent condensate flow to give substantially larger condensing film coefficients than predicted by theory.

9.4 NOMENCLATURE

C_p = heat capacity
g = gravitational constant
h = individual film heat transfer coefficient
h_{fg} = latent heat
h'_{fg} = latent heat defined by Eq. (9-4)
h_c = condensing heat transfer coefficient
k = thermal conductivity
L = tube length
$\dot{m}_p$ = mass flow rate/unit tube perimeter
Pr = Prandtl number
Re = Reynolds number, Eq. (9-8)
t = temperature
Δt = temperature difference
x = independent length variable
y = independent length variable

Greek Letters

μ = viscosity
ρ = density
τ = shear stress

Subscripts

c = condensing, condensate
f = film
m = mean
sv = saturated vapor
t = thickness
v = vapor
w = wall

9.5 REFERENCES

1. H. Emmon, *Trans. AIChE* **35**, 109 (1939).
2. E. Baer and J. M. McKelvey, *AIChE J.* **4**, 218 (1958).
3. W. Nusselt, *Z. Ver. deut. Ing.* **60**, 541 (1916).
4. L. A. Bromley, R. S. Brodkey, and N. Fishman, *Ind. Eng. Chem.* **44**, 2962 (1952).
5. W. M. Rohsenow, *Trans. ASME* **78**, 1645 (1956).
6. E. M. Sparrow and J. L. Gregg, *J. Heat Transfer* **81**, 13 (1959).
7. W. M. Rohsenow, J. H. Webber, and A. T. Ling, *Trans. ASME* **78**, 1637 (1956).
8. A. E. Dukler, *Chem. Eng. Progr.* **55**(10), 62 (1959).
9. W. K. Leonard and J. Estrin, *AIChE J.* **13**(2), 401 (1967).
10. K. E. Hassan and M. Jakob, *Trans. ASME* **80**, 887 (1958).
11. K.-T. Yang, *J. Appl. Mech.* **33**, 203 (1966).
12. F. S. Carpenter and A. P. Colburne, in *Proc. General Discussion of Heat Transfer, Inst. Mech. Eng. and ASME* (July 1951), pp. 20–26.
13. W. H. McAdams, *Heat Transmission*, 3rd ed., McGraw-Hill Book Co., New York (1954), p. 6.
14. W. M. Nagle, G. S. Bays, L. M. Blenderman, and T. B. Drew, *Trans. AIChE* **31**, 593 (1934–35).
15. N. Fatica and D. L. Katz, *Chem. Eng. Progr.* **45**, 661 (1949).
16. W. M. Nagle, U.S. Patent 1,995,361, March 26, 1935.
17. D. H. Wilhelm, Ph.D. Dissertation, Ohio State Univ., Columbus, Ohio (1964).
18. S. G. Sawochka, Ph.D. Dissertation, Univ. of Cincinnati, Cincinnati, Ohio (1965).
19. G. G. Haselden and S. Prosad, *Trans. Inst. Chem. Eng.* (*London*) **27**, 195 (1949).
20. R. J. Leonard and K. D. Timmerhaus, in *Advances in Cryogenic Engineering*, Vol. 15, Plenum Press, New York (1970), p. 308.
21. J. W. Dean, M.S. Thesis, Univ. of Colorado, Boulder, Colorado (1958).
22. D. E. Drayer and K. D. Timmerhaus, in *Advances in Cryogenic Engineering*, Vol. 7, Plenum Press, New York (1962), p. 401.
23. R. Ewald and P. Perroud, in *Advances in Cryogenic Engineering*, Vol. 16, Plenum Press, New York, 1971, p. 475.
24. M. Guther, *Trans. Inst. Chem. Eng.* (*London*) **27**, 183 (1949).
25. M. A. Lechtenberger, M.S. Thesis, Univ. of Colorado, Boulder, Colorado (1972).
26. R. J. Leonard, M.S. Thesis, Univ. of Colorado, Boulder, Colorado (1965).

VAPOR–SOLID CONDENSATION 10

K. E. TEMPELMEYER

10.1 FROST FORMATION

When a gas molecule collides or interacts with a solid surface it may (1) reflect elastically, (2) reflect inelastically, in which case the energy exchange is characterized by the energy accommodation coefficient, or (3) lose sufficient energy that it remains on the surface, at least for a short time. Molecules which stick to the surface may remain there for long times by condensation or chemisorption mechanisms; they may be desorbed or reevaporated; or they may diffuse over the surface or under certain circumstances into the substrate material. The process of frost formation on cryogenically cooled surfaces involves nearly all of these mechanisms.

Because the detailed physical character of engineering cryosurfaces is unknown and the interaction potential is extremely complex, one can form, at best, only an idealized qualitative picture of the gas–surface interaction which leads to condensation. A schematic of the gas–surface interaction is shown in Fig. 10-1; for simplicity only the dimension normal to the surface is shown. A free molecule in the gas phase has some energy at point (*a*), and as it approaches the surface it experiences an attractive force. If in the interaction the molecule gives up some of its energy to the surface, it may not have sufficient energy to escape from the potential well. Once caught at some energy level in the well, say point (*b*), the molecule may undergo a relaxation

---

K. E. TEMPELMEYER The University of Tennessee Space Institute, Tullahoma, Tennessee.

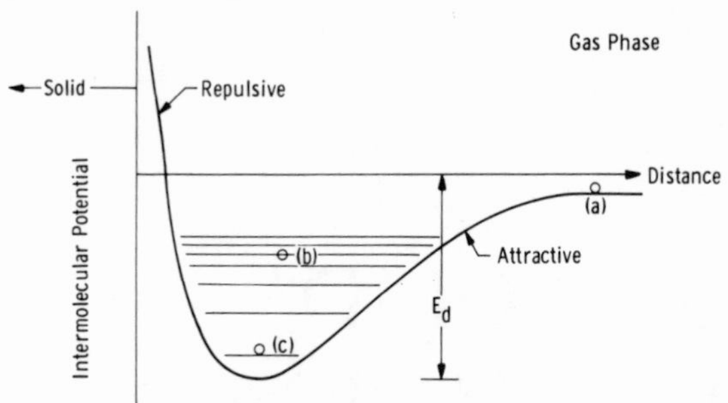

FIG. 10-1. Schematic of the gas–surface interaction.

process and attain thermal equilibrium with the solid. If the solid surface is cooled, this would result in the molecule residing near the bottom of the well (point *c*). In any event, once caught in the well a molecule will remain trapped there until it receives enough energy from the surface or from other incident molecules to escape. Frenkel[1] has deduced from statistical considerations that on the average an adsorbed particle will reside on the surface for the time

$$t_r = t_c e^{E_d/kT_s} \tag{10-1}$$

where E_d is the amount of energy needed to escape the surface, k is Boltzmann's constant, and t_c is a characteristic vibrational time of a particle adsorbed on the surface.

The residence time t_r on the surface can obviously be increased by decreasing the surface temperature T_s. In this case the adsorbed molecules would reside at the bottom of the well with a small statistical chance of receiving enough energy from the surface to be desorbed. This is the basis of frost formation. Molecules condensed on the surface have a low probability of being desorbed and remain in a solidified frost form. This will occur when the surface temperature is reduced to a few degrees less than the vapor pressure temperature. Moreover, gas can be continuously pumped as long as the frost surface is maintained at or below the required temperature.

Considerable information on vapor–solid condensation has been compiled in the cryopumping literature. Cryopumping is a process of forming a vacuum by freezing gases onto surfaces cooled to cryogenic temperatures. This vacuum technique is generally very attractive because the interior of a chamber can be completely lined with cryopanels and extremely large system pumping speeds obtained. Also, because the pumping surface is within the volume to be evacuated and, therefore, in direct contact with the gas to be pumped, the losses due to the resistance of interconnecting ducting to the pump is eliminated.

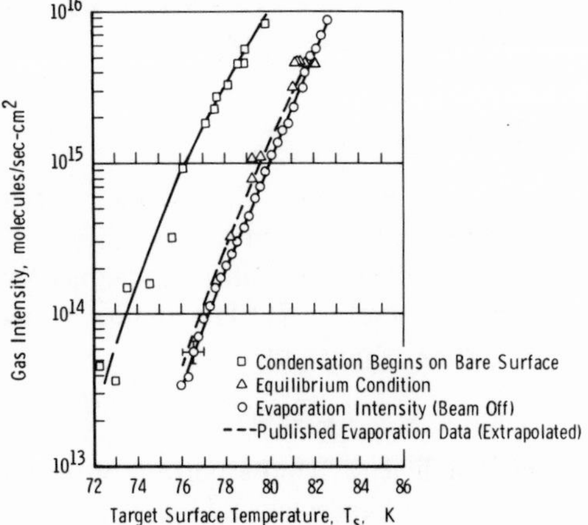

FIG. 10-2. Characteristic condensation and evaporation curves for CO_2.[2]

The effectiveness of a cryopump depends primarily upon the temperature of the cryosurface. Heald and Brown[2] have shown that gases will condense on a surface if the surface temperature is about 4 or 5 K lower than that specified by the vapor pressure curve (see Fig. 10-2). Figure 10-3 contains the vapor pressure curves for a number of gases which are commonly cryopumped and gives an indication of the surface temperatures required for cryopumped systems.

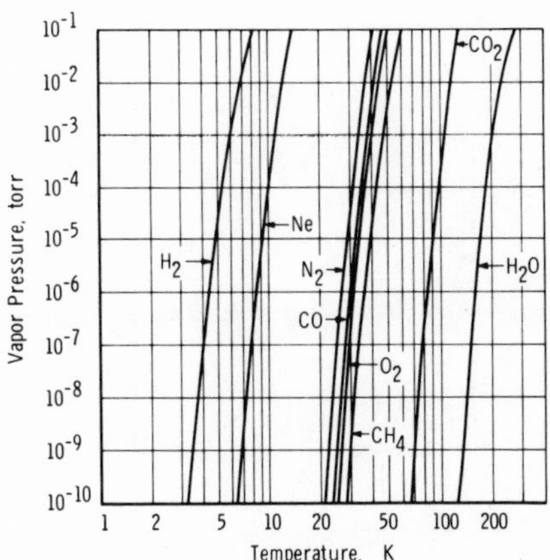

FIG. 10-3. Vapor pressure curves for several common gases.

10.2 MASS AND ENERGY CONSERVATION DURING CONDENSATION

10.2.1 Mass Balance

The system sketched in Fig. 10-4 illustrates a cryopanel cooled to a temperature T_s which views another surface at a higher temperature T_w. This latter surface may represent the wall of a chamber which encloses a gas at a pressure P_g and temperature T_g. The number of molecules per unit time $\dot{n}_i$ striking a cryosurface is given by

$$\dot{n}_i = P_g/(2\pi mkT_g)^{1/2} \tag{10-2}$$

When the surface is sufficiently cold a portion of the incident molecules stick to it and the remainder are reflected. It is convenient to define a *sticking probability* s as the ratio of the number of molecules that stick to the surface to the number that are incident, or

$$s \equiv (\dot{n}_i - \dot{n}_r)/\dot{n}_i \tag{10-3}$$

This ratio is, unfortunately, called the capture coefficient by many authors but it actually represents the probability of capture of the incident molecule, upon collision. The actual capture effectiveness of the surface is influenced by the fact that molecules that have been captured are also continuously reevaporated from the surface at a rate given by

$$\dot{n}_{\text{evp}} = \text{const} \times e^{-E_d/kT_f} \tag{10-4}$$

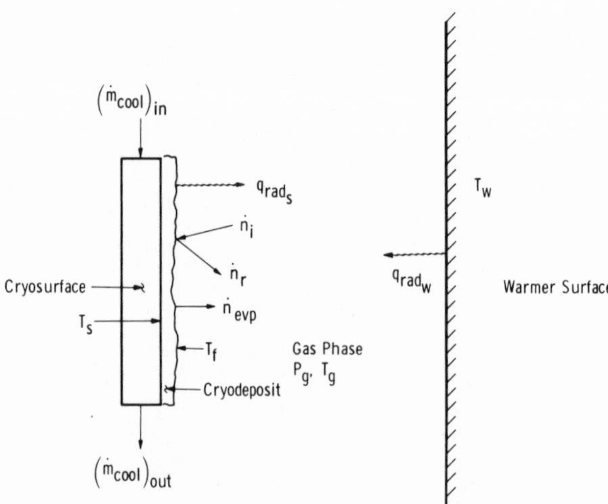

FIG. 10-4. Condensation pumping notation.

As a result, the rate at which molecules are effectively captured by the surface $\dot{n}_c$ is

$$\dot{n}_c = \dot{n}_i - \dot{n}_r - \dot{n}_{\mathrm{evp}} = s\dot{n}_i - \dot{n}_{\mathrm{evp}} \tag{10-5}$$

or, with Eqs. (10-2) and (10-4),

$$\dot{n}_c = [sP_g/(2\pi mkT_g)^{1/2}] - \mathrm{const} \times e^{-E_d/kT_f} \tag{10-6}$$

Consequently, it is more realistic to define the *capture coefficient* c as the ratio of the rate at which molecules are captured by the surface to the rate at which they strike

$$c \equiv (\dot{n}_i - \dot{n}_r - \dot{n}_{\mathrm{evp}})/\dot{n}_i \tag{10-7}$$

Because of the evaporation term, c is always less than s.

Equations (10-2)–(10-7) have been written in terms of molecular strike rates. They represent mass balances and can be expressed alternatively in terms of mass fluxes $\dot{m}$ since

$$\dot{m} = \dot{n}AM/A^{\circ} \tag{10-8}$$

where A is the area of the pumping surface, M is the molecular weight of the gas being pumped, and A° is Avogadro's number.

Sticking probabilities and capture coefficients, but particularly the latter, have been measured by a number of investigators in a variety of ways. The gas–surface interaction is a very complex phenomenon, and c depends upon many variables, including the gas species, pressure, gas temperature, cryosurface temperature, and cryosurface geometry. As a result, the earlier experimental data for c are conflicting. More recently, Heald and Brown[2–4] have made very careful capture coefficient measurements by directing a beam of molecules onto a cooled surface in a molecular beam system. The results of some of their measurements are given in Figs. 10-5 and 10-6. All of their results show that the value of c is very sensitive to the cryosurface temperature. It may increase from essentially zero to slightly less than unity as the cryosurface temperature is decreased 4 or 5 K. These figures also vividly illustrate the necessity of achieving a cryosurface temperature in the required range. An energy balance is useful for predicting cryopanel temperature.

Figure 10-5 also illustrates how the molecular strike rate or chamber pressure level influences the capture coefficient. From Eq. (10-2) it is seen, for example, that a room-temperature gas with a strike rate of 4×10^{15} molecules/cm^2-sec corresponds to a pressure level of about 10^{-5} Torr in the chamber. Figure 10-6 shows that the gas temperature only slightly influences the capture coefficient at temperatures up to 1400 K, as long as there is sufficient refrigeration capacity to maintain the cryosurface at the temperatures shown. Subsequent measurements by Brown *et al.*[5] and Arnold

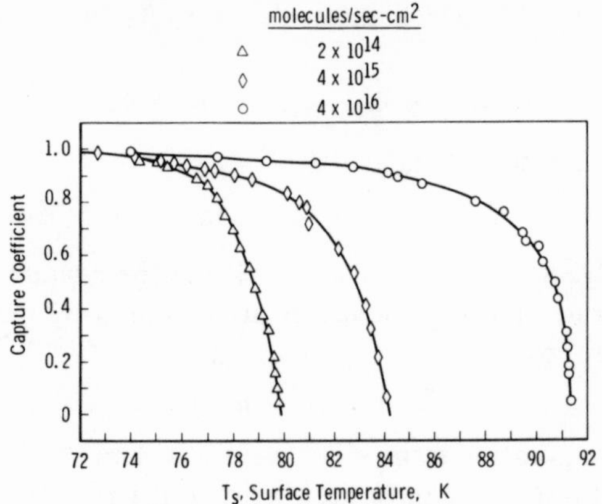

FIG. 10-5. Capture coefficient of CO_2 at various strike rates.[2]

et al.[6] have shown that the capture coefficient of argon and nitrogen at temperatures above about 1500 K decreases steadily with increasing gas temperature even when the cryosurface temperature is maintained below 20 K; and at a gas temperature of about 2500 K, c is less than 0.1. Table 10-1 contains a summary of measured capture coefficients which are useful in determining the performance of cryogenic pumping systems for atmospheric gases. References 7 and 8 also contain summaries of capture coefficient data

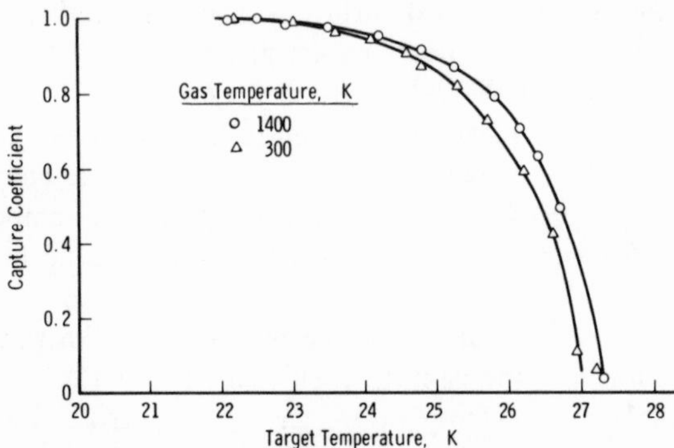

FIG. 10-6. Capture coefficient of N_2 at various temperatures. Nozzle diameter, 0.013 mm; skimmer diameter, 4 mm; collimator diameter, 4 mm; nozzle–skimmer separation, 85 nozzle diameters; total beam flux, 2×10^{15} mol/sec.[3]

TABLE 10-1

Summary of Capture Coefficients at Chamber Pressure Levels of 10^{-5}–10^{-6} Torr

Gas	Surface temp., K	c			
		77 K	300 K	1500 K	2500 K
N_2	15	1.0	1.0	1.0	0.10
	20	0.98	0.98	0.98	0.15
	25	0.9	0.85	0.80	0.10
	30	<0.10	<0.01	<0.01	<0.01
	77	0	0	0	0
Ar	15	1.0	0.98	0.85	0.50
	20	0.98	0.98	0.80	0.25
	25	0.98	0.98	0.75	<0.10
	30	0.20	0.10	<0.01	<0.01
	77	0	0	0	0
CO	15	1.0	0.85		
	20	1.0	0.85		
	25	1.0	0.80		
	30	<0.1	<0.1		
	77	0	0		
CO_2	20		1.0		
	70		1.0		
	77		0.90	Decomposes	
	80		0.75		
	85		0.10		

obtained from earlier experiments. Although they are believed to be less accurate, they may serve as a guide and together with the trends exhibited in Figs. 10-5 and 10-6 may be useful in estimating values of c for other gases.

Data shown in Figs. 10-5 and 10-6 were obtained by directing a beam of molecules onto the surface at a given angle. However, in most practical applications the molecules will strike the pumping surface with a random distribution of directions. Busby *et al.*[9] have shown that the capture coefficient tends to decrease as molecules strike the surface at a more glancing angle. The value of c decreases roughly as the cosine of the angle of incidence. Consequently, the overall capture coefficient of cryogenic systems tends to be lower than the values given in Table 10-1 and Figs. 10-5 and 10-6. Fewer capture coefficient data are available for complex geometric configurations; however, estimated overall capture coefficients for air are listed in Ref. 10. These values are highly approximate but may suffice for engineering estimates.

10.2.2 Energy Balance

Referring again to Fig. 10-4, we will assume that the pressure P_g in the system is sufficiently low so that heat transfer by conduction or convection in the gas is negligible and the only active heat transfer mode is radiation exchange between surfaces. The energy inputs to the cryogenic panel are due to (1) the incident gas which is condensed and captured and (2) the net radiant energy received from the warmer surrounding surfaces.

Gas molecules which are reflected and reevaporated from the surface represent an energy component leaving the surface, as does the flow of the cryogenic fluid $\dot{m}_{\text{cool}}$.

Summing these components, the energy delivered to a cryosurface by the impinging gas E_i in general can be written

$$E_i = s\dot{m}_i\left[\int_{T_s}^{T_v} C_{pf}\,dT + \int_{T_r}^{T_g} C_{pg}\,dT\right] + (1-s)\dot{m}_i\int_{T_e}^{T_g} C_{pg}\,dT + s\dot{m}_i E_a \tag{10-9}$$

The first term on the right-hand side assumes that the energy of the molecules that stick is accommodated to the cryosurface temperature T_s. It is divided into two terms because the condensing gas may be cooled through both the gas and solid frost phases where C_{pf} and C_{pg} are the specific heats of the cryodeposited frost and gas, respectively. The second term states that the energy of the fraction of molecules $1 - s$ that reflect from the surface is accommodated to a temperature T_e which is greater than T_s. The last term accounts for the fact that the molecules that stick and condense give up their heats of vaporization and fusion to the surface. The mass flow $\dot{m}_i$ of the incident gas is related to the incident molecular flux $\dot{n}_i$ used in some of the preceding equations by Eq. (10-8).

Molecules are both reflected and reevaporated from the surface (see Fig. 10-4). However, to be consistent with Eq. (10-9), the reflected molecules leave the surface at a temperature T_e and do not represent an energy rejection. The reevaporated molecules are assumed to leave at the temperature of the surface and carry with them their heats of fusion and vaporization (or sublimation); thus

$$E_r = \dot{m}_{\text{evp}} E_d \tag{10-10}$$

where

$$\dot{m}_{\text{evp}} = (AM/A^\circ)\text{const} \times e^{-E_d/kT_s}$$

from Eqs. (10-4) and (10-8).

In addition, the cryosurface receives a net radiant energy q_{rad} by means of radiation exchange with the surrounding surfaces. The radiation interchange depends upon the geometry of the surfaces and is covered in Chapter 14. For the simple two-zone schematic configuration shown in Fig. 10-4 the net radiation exchange can be written

$$q_{\text{rad}} = (q_{\text{rad},w} - q_{\text{rad},s}) = F_{ws}\sigma(T_w^4 - T_s^4) \tag{10-11}$$

where σ is the Stefan–Boltzmann constant; F_{ws} is the view factor between the cryosurface at T_s and the warmer surrounding wall at T_w, and depends upon the emissivities ϵ and reflectivities ρ of both surfaces. If the warmer wall completely encloses the cryosurface, which is often the case in closed systems, then

$$F_{ws} = [1 + (\rho_s/\epsilon_s) + (A_s\rho_w/A_w\rho_s)]^{-1} \tag{10-12}$$

where A_s and A_w are the areas of the cryosurface and surrounding wall, respectively. Thus, it is necessary to know the optical properties of the cryodeposited gas to carry out the radiation balance.

The energy removed from the cryosurface q_{cool} by the cryogenic coolant is determined by the methods described in preceding chapters and earlier in this chapter. Cryosystems may operate as (1) single-phase flow systems of liquid (liquid nitrogen) or gas (cooled helium) or (2) as two-phase flow systems. A heat transfer analysis appropriate to the type of flow desired or anticipated would be employed.

If the cryopanel has a mass M and a specific heat C_s, its temperature variation with time can be written

$$MC_s\, dT_s/dt = E_i - E_r + q_{\text{rad}} - q_{\text{cool}} \tag{10-13}$$

Equations (10-9) and (10-12) can be inserted into Eq. (10-13) to obtain a general expression for the energy balance. However, for most practical purposes some simplification is possible.

The spatial distributions of molecules which are reflected from a cold surface at various temperatures are shown in Fig. 10-7.[4] As the surface

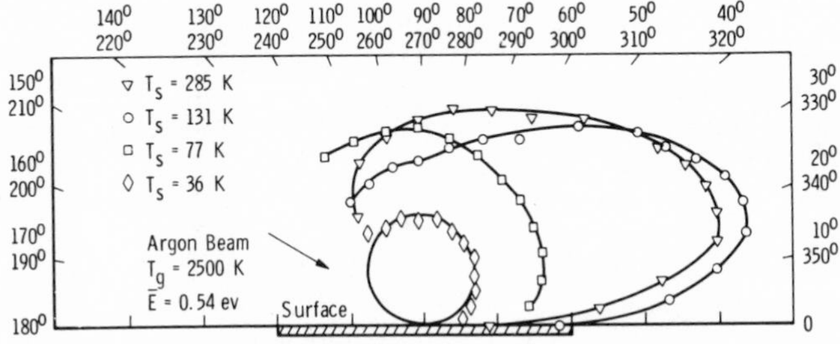

FIG. 10-7. Spatial distribution of argon molecules reflected from cooled copper surfaces.[4]

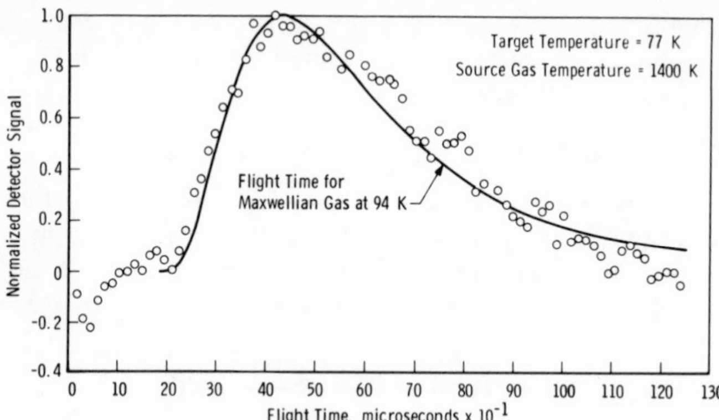

FIG. 10-8. Velocity distribution of argon molecules reflected from an engineering surface for a target temperature of 77 K and a source gas temperature of 1400 K.[12]

temperature approaches that at which the gas will condense, molecules tend to reflect in a diffuse manner (i.e., in random directions). This suggests that the molecules reside on the surface long enough for them to assume a temperature close to that of the surface. That is, their energy becomes nearly fully accommodated to the surface temperature. Powell[11] has developed a mass spectrometer probe to measure the velocity distribution of molecules. Figure 10-8, taken from Ref. 12, shows the velocity distribution of reflected

TABLE 10-2
Accommodation Coefficients for Argon Beams

Surface description	Surface temperature, K	Incidence angle θ_i, deg	Detector angle θ_r, deg	Partial accommodation coefficient, $\alpha(\theta_i\theta_r)$
Hand-polished copper	77	45	0	0.99
Hand-polished copper	276	45	0	0.97
CO_2 frost on copper	77	45	0	0.99
Hand-polished copper	77	45	45	0.99
Hand-polished copper	280	45	45	0.98
CO_2 frost on copper	77	45	45	0.99

argon molecules. It demonstrates that 1400 K argon is reflected from a 77 K surface at a mean temperature of about 94 K, which corresponds to an energy accommodation coefficient of about 0.98. Other measured energy accommodation coefficients for argon reflecting from copper and frost surfaces are given in Table 10-2.

Similar measurements have also shown that N_2 and CO_2 molecules striking a cryogenic cooled surface undergo a very high degree of energy accommodation. Consequently, one can assume a unity energy accommodation coefficient for molecules that are not captured and reflected. In addition, it is reasonable to assume that the adsorption and desorption energies are equal ($E_a \simeq E_d$)[13] and Eq. (10-13), together with Eqs. (10-9) and (10-10), simplifies to

$$MC_s\, dT_s/dt = \dot{m}_i\left[\int_{T_s}^{T_v} C_{pf}\, dT + \int_{T_v}^{T_g} C_{pg}\, dT\right] + (s\dot{m}_i - \dot{m}_{\text{evp}})E_a + q_{\text{rad}} - q_{\text{cool}} \tag{10-14}$$

From Eqs. (10-3) and (10-7) it can be shown that

$$s = c + (\dot{m}_{\text{evp}}/\dot{m}_i) \tag{10-15}$$

Because capture coefficient data are available, Eq. (10-14) is rewritten in terms of c rather than s,

$$MC_s\, dT_s/dt = \dot{m}_i\left[\int_{T_s}^{T_v} C_{pf}\, dT + \int_{T_v}^{T_g} C_{pg}\, dT + cE_a\right] + q_{\text{rad}} - q_{\text{cool}} \tag{10-16}$$

Equation (10-16) can be used to estimate the temperature history of the cryosurface, if various thermodynamic and transport properties of cryodeposited frost are known. Under steady-state conditions the cooling capacity required to maintain the surface at a given value of T_s is

$$q_{\text{cool}} = \dot{m}_i\left[\int_{T_s}^{T_v} C_{pf}\, dT + \int_{T_v}^{T_g} C_{pg}\, dT + cE_a\right] + F_{ws}\sigma(T_w^4 - T_s^4) \tag{10-17}$$

It should be emphasized that cryogenic pumping effectiveness of the surface depends upon the temperature at the surface of the frost T_f rather than the actual cryosurface temperature. Because a cryodeposit has a finite thermal conductivity, it also will experience a temperature gradient. Consequently, Eq. (10-17) is valid for thin frost layers where $T_s \simeq T_f$. For thick frosts heat conduction across the frost must also be taken into account. In this case

$$\begin{aligned} q_{\text{cool}} &= (K/A_f)(T_f - T_s) \\ &= \dot{m}_i\left[\int_{T_f}^{T_v} C_{pf}\, dT + \int_{T_v}^{T_g} C_{pg}\, dT + cE_a\right] + F_{ws}\sigma[(T_w^4 - T_s^4) \end{aligned} \tag{10-18}$$

Equations (10-16)–(10-18) can be used to determine the amount of cryogen required, to determine the cryosurface temperature, or to determine the amount of gas which may be removed from the system by cryopumping. They are applicable to either closed or flowing systems. Stephenson,[14] for example, showed their application in predicting the amount of a supersonic flow which would be captured by a flat plate normal to the flow. His calculations agreed quite well with experimental results.

The application of these equations requires the knowledge of a number of cryodeposit properties. Many of the needed properties are summarized in the following section.

10.3 PROPERTIES OF CRYODEPOSITS

A heat transfer analysis of a cryogenic system requires knowledge of several properties of the cryodeposit. These properties for the common atmospheric gases gathered from a number of sources are collected here. They are sufficiently accurate for engineering calculations.

10.3.1 Latent Heats

The heats of fusion, vaporization, and sublimation were taken from Refs. 15–17 and are used to determine E_a. They all are somewhat dependent on temperature; the values given in Table 10-3 correspond to the cryogenic temperature range.

Values of E_a can be obtained from the sum of the heats of fusion and vaporization, or from the heat of sublimation, if sublimation occurs at the pressure and temperature under consideration.

TABLE 10-3
Latent Heats of Various Gases

	Fusion		Vaporization		Sublimation	
Gas	cal/g	K	cal/g	K	cal/g	K
Ar	6.7	70	3838	70	48	45
O_2	3.3	60	51	70	—	—
N_2	6.1	60	48	50	59	40
CO_2	45	100	87	100	145	80
CO	8.0	60	50	50	75	40
H_2O	80.0	270	600	270	—	—

TABLE 10-4
Specific Heats of Solidified Gases

Gas	C_{pf}, cal/g-K	Temp., K
Ar	0.155	50
O_2	0.336	50
N_2	0.39	60
CO	0.417	50
CO_2	0.124	45
H_2O	0.156	75

10.3.2 Specific Heats

The specific heats of the gas phase C_{pg} and that of the cryopanel material C_s can be readily found in the literature. The specific heats of solidified gases were taken from Ref. 15 and are summarized in Table 10-4.

10.3.3 Thermal Conductivity

The thermal conductivities K of solidified gases are highly temperature dependent. Dean and Timmerhaus[18] have measured the thermal conductivity of solid H_2O at cryogenic temperatures. Their results are shown in Fig. 10-9. Similar results have been obtained by Rogers[19] for solid N_2 and by White and Woods[20] for a number of inert monatomic gases. These measured values are also given in Fig. 10-9.

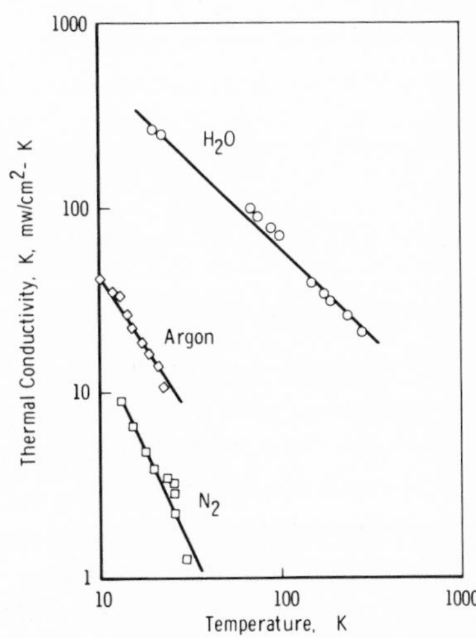

FIG. 10-9. Thermal conductivities of several solidified gases. The data for H_2O are from Ref. 18, for argon from Ref. 20, for N_2 from Ref. 19.

TABLE 10-5
Density of Cryodeposits

Frost	ρ_f, g/cm^3	Temp., K
Ar	1.70	30
N_2	0.90	20
O_2	1.36	25
CO_2	1.60	77
CO	1.04	30
H_2O	0.81	77

10.3.4 Capture Coefficient

Table 10-1 contains measured values of the capture coefficients c for a number of gases. It should be emphasized that c depends principally upon the cryosurface temperature and chamber pressure. The trends depicted in Figs. 10-5 and 10-6 can be used to estimate capture coefficients beyond the range of Table 10-1.

10.3.5 Density

The density of cryodeposits tends to be somewhat less than that of a solidified gas formed from the liquid phase. Some frost densities have been measured[15,19,21] and are tabulated in Table 10-5. Others can be estimated from values of the liquid density.

Values of the frost densities are useful in estimating the thickness of the frost.

10.4 NOMENCLATURE

A = area
A° = Avogadro's number
C = specific heat
C_p = specific heat at constant pressure
c = capture coefficient
E_a = energy of absorption
E_d = energy required to escape surface
E_i = energy delivered by impinging gas

E_r = energy of reflected molecules
F_{ws} = view factor between cryosurface and wall
K = thermal conductivity
k = Boltzmann's constant
M = molecular weight
m = mass of molecule
$\dot{m}$ = mass flow rate
n_c = number of effectively captured molecules
$\dot{n}_i$ = number of incident molecules per unit time
$\dot{n}_r$ = number of reflected molecules per unit time
p = pressure
q = heat flux
T = temperature
t = time
t_r = average time an absorbed particle resides on the surface
s = sticking probability

Greek Letters

ϵ = emissivity
σ = Stefan–Boltzmann constant
ρ = density
τ = characteristic vibration time of a particle adsorbed on the surface

Subscripts

f = frost
g = gas
i = incident
s = surface
w = wall

10.5 REFERENCES

1. J. Frenkel, *Physik* **26** 117 (1924).
2. J. H. Heald, Jr. and R. F. Brown, AEDC-TR-68-110 (September 1968).
3. J. H. Heald, Jr. and R. F. Brown, paper at 14th National Vacuum Symposium of the American Vacuum Society (October 1967).
4. R. F. Brown, D. M. Trayer, and M. R. Busby, *J. Vac. Sci. Technol.* **7**, 241 (1970).
5. R. F. Brown, R. L. Caldwell, and M. R. Busby, *Appl. Phys. Lett.* **14**, 219 (1969).
6. F. Arnold, M. R. Busby, and R. Dawbarn, AEDC-TR-70-172 (September 1970).

7. J. P. Dawson and J. E Haygood, *Cryogenics* **5**(2), 57 (1965).
8. M. M. Essenstadt, *J. Vacuum Sci. Tech.* **7**(4), 479 (1970).
9. M. R. Busby, J. D. Haygood, and C. H. Link, AEDC-TR-70-131 (July 1970).
10. Arthur D. Little, Inc., AEDC-TR-60-18 (January 1960).
11. H. M. Powell and J. H. Heald, Jr., paper presented at the *14th Annual National Vacuum Symposium* (October 1967).
12. R. F. Brown, H. M. Powell, and D. M. Trayer, in *Proc. 6th Rarefied Gas Dynamics Symposium* (1968), p. 1187.
13. K. E. Tempelmeyer, Ph.D. Dissertation, Univ. of Tennessee Space Institute (March 1970).
14. W. B. Stephenson, AEDC-TR-67-201 (January 1968).
15. D. Hodgman, *Handbook of Chemistry and Physics*, Chemical Rubber Publishing Co., 39th Ed.
16. J. W. Stewart, *Physics of High Pressures and the Condensed Phase*, John Wiley and Sons, New York (1965), Chapt. V.
17. R. P. Caren, *Cryogenic Technology* **4**(1) 15 (1968).
18. J. W. Dean and K. D. Timmerhaus, in *Advances in Cryogenic Engineering*, Vol. 8, Plenum Press, New York (1963), p. 299.
19. K. W. Rogers, NASA-CR-553 (August 1966).
20. G. K. White and S. B. Woods, *Phil. Mag.* **1958**, 785.
21. A. M. Smith, K. E. Tempelmeyer, P. R. Muller, and B. E. Wood, *AIAA J.* **7**(12), 2274 (1969).

PRESSURE DROP AND COMPRESSIBLE FLOW OF CRYOGENIC LIQUID–VAPOR MIXTURES 11

R. E. HENRY, M. A. GROLMES, and H. K. FAUSKE

11.1 INTRODUCTION

This chapter reviews the one-dimensional methods available to predict the fluid dynamics of two-phase systems. Most of the methods are derived from air–water or steam–water studies. However, as will be shown, these models also appear to characterize cryogenic two-phase flows.

The chapter is divided into three parts. The first part discusses pressure drop predictions for two-phase mixtures. The second part discusses pressure wave propagation through two-phase equilibrium mixtures and illustrates the role of the flow regime on the mixture compressibility. Finally, the third part evaluates critical flow of one-component, two-phase mixtures with special emphasis on duct geometry and flow regime as they affect the compressibility.

R. E. HENRY, M. A. GROLMES, and H. K. FAUSKE Argonne National Laboratory, Argonne, Illinois.

11.2 PRESSURE DROP

Only a rather small amount of two-phase pressure drop data for cryogenic fluids are available compared to the vast amount of data available for other nonmetallic fluids. Moreover, the data for cryogenic fluids appear to indicate conflicting results when compared with the widely used Lockhart–Martinelli[1] or Martinelli and Nelson[2] pressure drop correlations. For example, the Freon 11 data of Hatch and Jacobs,[3] the hydrogen data of Richards *et al.*[4] and the nitrogen data of Shen and Jao[5] are comparable to the Lockhart–Martinelli correlation, while the helium data of de La Harpe *et al.*[6] tend to show closer agreement with the homogeneous pressure drop relation. The limited friction factor data of Sugden and Timmerhaus[7] for Freon 11 can also be interpreted to indicate agreement with the homogeneous model, while the nitrogen and methane data of Lapin and Bauer[8] are compared with a modified Chenowith–Martin pressure drop correlation. This would seem to leave the designer of cryogenic equipment with considerable uncertainty with regard to two-phase pressure drop. However, the need for flow regime considerations in assessing the two-phase pressure drop will be illustrated and some simple approximations to the Lockhart–Martinelli correlation will be discussed. It will also be shown that some available data can also be favorably compared to the homogeneous pressure drop model where appropriate.

11.2.1 Equations for Steady Two-Phase Flow

The one-dimensional momentum and continuity equations for steady, two-phase, compressible flow in a constant-area circular duct can be written

$$\frac{dP}{dz} + 4\frac{\tau}{D} + g\rho_{\mathrm{TP}}\cos\theta + \frac{d}{dz}\{\rho_g\alpha u_g^2 + \rho_l(1-\alpha)u_l^2\} = 0 \tag{11-1}$$

and

$$(d/dz)\{\rho_g\alpha u_g + \rho_l(1-\alpha)u_l\} = 0 \tag{11-2}$$

where the two-phase density is given by

$$\rho_{\mathrm{TP}} = \alpha\rho_g + (1-\alpha)\rho_l \tag{11-3}$$

Equation (11-1) represents the total pressure gradient, the wall shear force or friction pressure gradient, the hydrostatic pressure head, and the momentum change, respectively. The second equation is a statement of conservation of vapor plus liquid mass. The vapor and liquid velocities are related to the total mass flux according to

$$u_g = (x/\alpha\rho_g)G \tag{11-4}$$

$$u_l = (1-x)G/(1-\alpha)\rho_l \tag{11-5}$$

With Eqs. (11-4) and (11-5), the total pressure drop can be restated as

$$-\Delta P = 4\tau(\Delta z/D) + g\rho_{\mathrm{TP}}(\cos\theta)\Delta z + G^2\,\Delta v \tag{11-6}$$

where the specific volume of the mixture v is defined by

$$v = \frac{x^2}{\alpha\rho_g} + \frac{(1-x)^2}{(1-\alpha)\rho_l} \tag{11-7}$$

Equation (11-6) can also be written in differential form,

$$-\frac{dP}{dz} = \frac{(4\tau/D) + g\rho_{\mathrm{TP}}\cos\theta}{1 - G^2(-\,dv/dP)} \tag{11-8}$$

to illustrate that a maximum or critical flow rate can be defined as

$$G_c = [-dv/dP]^{-1/2} \tag{11-9}$$

The critical flow rate and sonic velocity in two-phase flows will be reviewed later.

The friction pressure loss is determined from a measured total pressure drop by subtracting the contribution due to acceleration and hydrostatic head according to Eq. (11-6). For the cryogenic two-phase pressure drop data discussed in this review, the local quality is calculated from an energy equation based on thermodynamic equilibrium, in evaluating the momentum change $G^2\,\Delta v$ associated with the total pressure drop,

$$x = \frac{h_0 - h_f + q - E_k}{h_{fg}} \tag{11-10}$$

In section 11-4 it will be shown that the assumption of thermal equilibrium is not always appropriate and depends largely on the flow geometry. The kinetic energy term E_k is usually negligible but may be included. With the quality given by Eq. (11-10) some independent relation must also be used for the local vapor volume fraction α, which is related to the local quality, density ratio, and vapor-to-liquid velocity ratio according to

$$\alpha = \{1 + [(1-x)k\rho_g/x\rho_l]\}^{-1} \tag{11-11}$$

For homogeneous flow ($k = 1$), the mixture specific volume is simply

$$v_H = (x/\rho_g) + [(1-x)/\rho_l] \tag{11-12}$$

11.2.2 Annular Flow Model

There are many two-phase void fraction correlations available from the literature, such as those proposed by Levy,[9] Bankoff,[10] Armand,[11] etc. The Lockhart–Martinelli void correlation is the most widely used for annular flow and is related to the mass fraction, density ratio, and viscosity ratio as

$$\alpha = f(\chi_{tt}) \tag{11-13}$$

where

$$\chi_{tt} = \left(\frac{1-x}{x}\right)^{0.9}\left(\frac{\rho_g}{\rho_l}\right)^{0.5}\left(\frac{\mu_l}{\mu_g}\right)^{0.1} \tag{11-14}$$

The form of Eq. (11-13) has been expressed empirically by Wallis[12] as

$$\alpha = (1 + \chi_{tt}^{0.8})^{-0.378} \tag{11-15}$$

Wallis[13] also has shown on the basis of a simplified annular flow model that Eq. (11-13) is also well represented by

$$\chi_{tt}^2 = (1-\alpha)^2[1 + 75(1-\alpha)]/\alpha^{5/2} \tag{11-16}$$

The agreement of the above void correlation with a wide variety of two-phase void fraction data is illustrated later in Fig. 11-2. It is also pointed out that a simple relation can be obtained if the velocity ratio is assumed to vary with the square root of the density ratio[14] such that

$$\alpha = \left[1 + \frac{1-x}{x}\left(\frac{\rho_g}{\rho_l}\right)^{1/2}\right]^{-1} \tag{11-17}$$

While Eq. (11-17) does not necessarily lead to a satisfactory estimate of the void fraction, the momentum change calculated on the basis of the above void fraction relation does not differ appreciably from the momentum change based on the Lockhart–Martinelli void correlation. This is illustrated

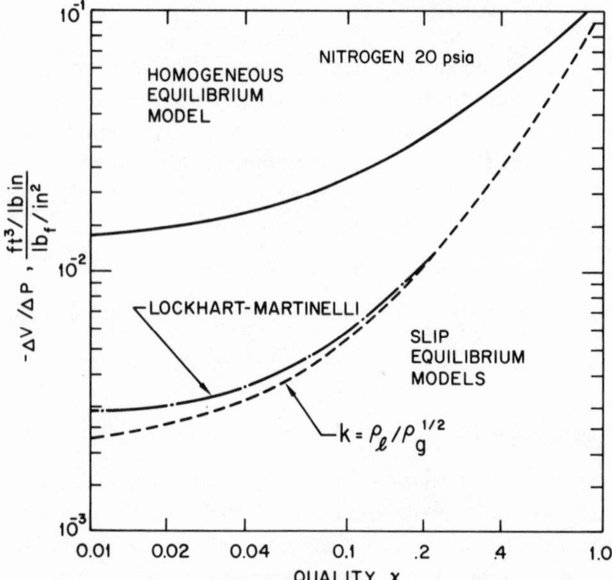

FIG. 11-1. Illustration of momentum change for homogeneous and slip models.

in Fig. 11-1. It should also be pointed out that for most of the pressure drop data reported in the cryogenic literature, the momentum change was evaluated according to the assumption of homogeneous flow. While some support for this approach is found in the two-phase hydrogen data of Graham *et al.*,[15] it is also seen in Fig. 11-1 that the momentum change for the homogeneous assumption differs substantially from that based on a slip void fraction. For the steam–water and sodium pressure drop and critical flow data discussed in Ref. 16 it is illustrated that the actual momentum change shows much better agreement with the slip models than the homogeneous assumption. It may well be that for some cryogenic pressure drop data, the homogeneous assumption is appropriate. However, evaluation of momentum change according to a homogeneous assumption may not be entirely consistent when the resulting friction pressure drop data are compared with a slip or annular flow correlation such as the Lockhart–Martinelli relation.

Two-phase friction pressure drop data are customarily represented by the parameter ϕ_l, which is defined in terms of the ratio of the two-phase to single-phase liquid pressure drop as

$$\phi_l = \left[\frac{(\Delta P/\Delta z)_{TP}}{(\Delta P/\Delta z)_l}\right]^{1/2} \tag{11-18}$$

The Lockhart–Martinelli correlation is often used for annular flow where the parameter ϕ_l is presented as an empirical function of the same parameter χ_{tt}, as illustrated in Fig. 11-2. However, a somewhat simpler approach originally proposed by Lottes and Flinn[17] can be used. The friction pressure drop in Eq. (11-6) is simply

$$\Delta P_F = 4\tau \, \Delta z/D \tag{11-19}$$

For turbulent single-phase flow the wall shear stress τ is related to the average velocity

$$\tau = f\rho_l u^2/2 \tag{11-20}$$

and the friction factor f is defined in terms of the Reynolds number

$$f = 0.07\text{Re}^{-0.25} \tag{11-21}$$

The single-phase liquid friction pressure loss is therefore

$$(\Delta P/\Delta z)_l = (4/D)(f\rho_l u_{l0}^2/2) \tag{11-22}$$

Lottes and Flinn proposed that the increased friction pressure drop for annular two-phase flow is attributed to the increased liquid velocity,

$$u_l = u_{l0}/(1-\alpha) \tag{11-23}$$

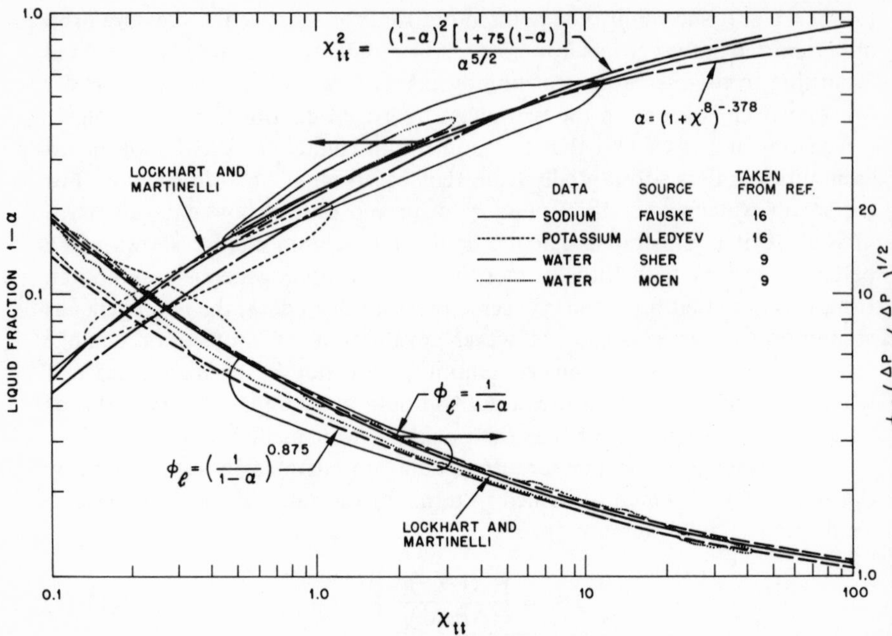

FIG. 11-2. Lockhart–Martinelli correlation and representative noncryogenic data.

The two-phase friction pressure gradient is therefore

$$(\Delta P/\Delta z)_{TP} = (4/D)[f\rho_l u_{l0}^2/2(1-\alpha)^2] \tag{11-24}$$

and the friction pressure loss ratio for the same friction factor f is then

$$\phi_l = 1/(1-\alpha) \tag{11-25}$$

Richardson[18] extended the above model by modifying the liquid friction factor for the increased liquid velocity in annular flow and arrived at the relation

$$\phi_l = [1/(1-\alpha)]^{0.875} \tag{11-26}$$

Both Eqs. (11-25) and (11-26) are shown in Fig. 11-2 along with the Lockhart–Martinelli correlation. Representative data from the literature for other than cryogenic fluids are shown for both pressure drop and void fraction. These data tend to show that for a wide variety of fluids where annular flow is dominant, good agreement is found with the suggested models and correlations.

11.2.3 Homogeneous Flow Model

If it is appropriate to assume negligible difference in vapor and liquid velocity ($k = 1$), a simplified homogeneous two-phase pressure drop model can be used.

For homogeneous two-phase flow, the two-phase pressure drop is again customarily defined in terms of the average mixture velocity and density as

$$(\Delta P/\Delta z)_{\mathrm{TP}} = 2fG^2/D\rho_{\mathrm{TP}} \tag{11-27}$$

At low void fractions where homogeneous flow is appropriate

$$\rho_{\mathrm{TP}} \approx (1 - \alpha)\rho_l \tag{11-28}$$

so that the pressure drop ratio at the same total mass flow rate for a two-phase mixture becomes

$$\phi_l = [1/(1 - \alpha)]^{1/2} \tag{11-29}$$

assuming the same friction factor is applicable for two-phase flow. Rose and Griffith[19] have shown that the friction factor is slightly augmented for homogeneous two-phase flow. One might recommend

$$f = 0.08\mathrm{Re}^{-0.25} \tag{11-30}$$

for homogeneous two-phase flow and

$$f = 0.07\mathrm{Re}^{-0.25} \tag{11-31}$$

for single-phase liquid flow in smooth tubes where the Reynolds number in Eq. (11-30) is defined in terms of the liquid viscosity

$$\mathrm{Re} = GD/\mu_l \tag{11-32}$$

The above development is by no means intended to be an exhaustive treatment of two-phase pressure drop. Only the Lockhart–Martinelli and some of the simpler analytical relations for two-phase pressure drop have been reviewed. However, it is suggested that Eqs. (11-25) and (11-26) for annular flow and Eq. (11-29) for homogeneous flow can be favorably compared with available cryogenic pressure drop data. To our knowledge no void fraction data are available for cryogenic fluids. In fact, because cryogenic pressure drop data are so limited, it is likely that additional data will be obtained in any given application. In these cases, the above relations can be suggested as reasonable guides.

11.2.4 Cryogenic Pressure Drop Data

Hatch *et al.*[20] and Hatch and Jacobs[3] have presented extensive two-phase pressure drop data for the adiabatic and diabatic horizontal flow of

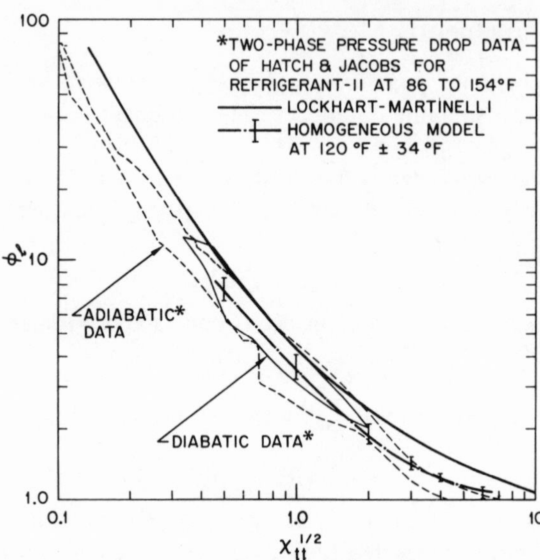

FIG. 11-3. Data of Hatch and Jacobs.[3]

Freon 11 at moderate pressures. Tubes of various diameters were employed. No diameter effect is noted, nor is any significant difference reported for adiabatic or diabatic conditions. Although the momentum change was calculated on the basis of a homogeneous assumption, all data are compared to the Lockhart–Martinelli correlation and show good agreement. Their data are summarized in Fig. 11-3. The good agreement with the annular flow models tends to diminish at higher values of the parameter χ_{tt} or as the liquid fraction increased above 70%. This may be an indication of a flow regime transition to bubbly or homogeneous flow. In fact, as is shown in Fig. 11-3, the homogeneous two-phase pressure drop model, Eq. (11-29) where α is taken as the homogeneous void fraction, tends to show good agreement with the Freon 11 data not only at low void fractions but also at void fractions up to 90%. This should not necessarily be interpreted as justification for the homogeneous model over the wide range of void fraction. It does, however, indicate that a consistent data reduction is needed with due consideration of the flow regime. For example, in Fig. 11-3 the good agreement with the homogeneous model at low void fraction may be valid. The same good agreement with the homogeneous model at high void fraction may in fact be a result of an underestimate of the actual two-phase pressure drop because of an overestimate of the momentum change based on the homogeneous assumption used in reducing the original data. If the momentum change at higher void fractions were evaluated on the

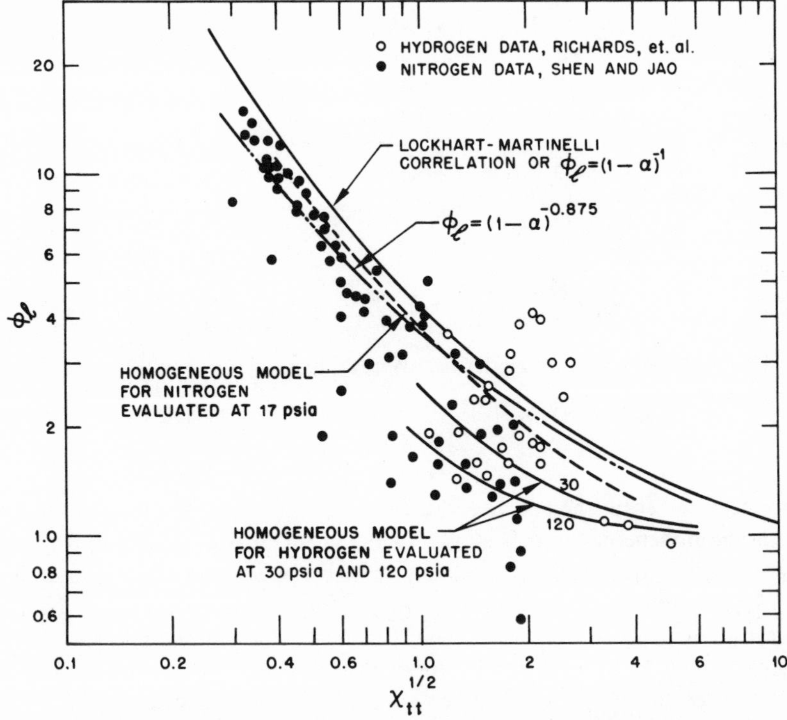

FIG. 11-4. Data of Richards *et al.* (H_2)[4] and Shen and Jao (N_2).[5]

basis of an annular flow assumption, the data may show even better agreement with the Lockhart–Martinelli pressure drop correlation. In Fig. 11-4, the two-phase pressure drop data of Richards[4] for hydrogen and the nitrogen data of Shen and Jao[5] are shown in comparison to the Martinelli correlation, as well as the simplified annular flow models of Eqs. (11-25) and (11-26). For these models the void fraction α is taken according to the Martinelli correlation (see Fig. 11-1). While some general agreement is shown with the annular flow models (particularly the data of Ref. 5 at lower values of χ_{tt}), again the comparison becomes less convincing at larger liquid fractions. Some data indicate a two-phase pressure drop less than the corresponding single-phase liquid pressure drop. For both Refs. 4 and 5 some of the data are reported to have been obtained under conditions where the momentum change is a major fraction of the measured total pressure drop. In this case the previous remarks that caution should be exercised in making a comparison with a slip pressure drop correlation when the momentum change is evaluated on the basis of a homogeneous assumption are even more appropriate. Although specific conditions for the data of Refs. 4 and 5 were not reported,

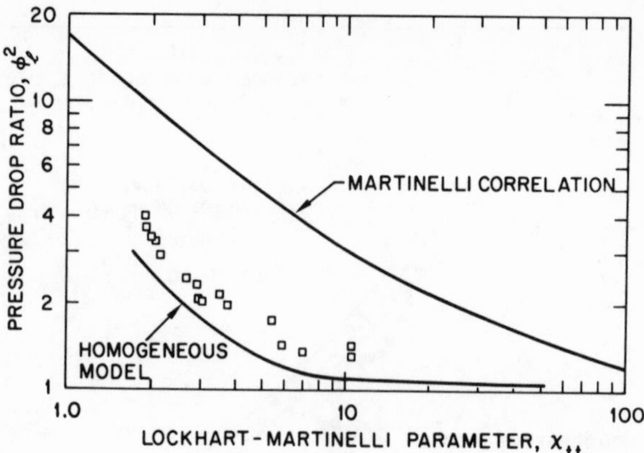

FIG. 11-5. Data of de La Harpe *et al.*[6] for helium I at 1.15 bar.

it was inferred from the text that the data must be within the range indicated for the homogeneous models shown in Fig. 11-4. It is evident that for some of the above data, a better comparison with the homogeneous model [Eq. (11-29)] is found at the large values of χ_{tt}. This can also be seen in the data of de La Harpe *et al.*[6] reproduced in Fig. 11-5. The limited data for He I at approximately 1 atm in 3-mm-ID tubes over a wide range in exit quality are shown to be in much closer agreement with the homogeneous formulation than the Martinelli correlation. Somewhat closer agreement with the homogeneous model is also indicated in Fig. 11-6, where the data of Sugden and Timmerhaus[7] are compared with Eq. (11-29) over the volume fraction range where homogeneous flow might be assumed.* Finally, Lapin and Bauer[8] present two-phase pressure drop data for high-pressure nitrogen and methane which deviate substantially from the Martinelli correlation. However, much of their data in the low-void-fraction range can also be compared favorably with the homogeneous model. This is shown in Fig. 11-7 for some representative methane data of Ref. 8. The data do in fact cover a wide range of liquid fraction and the abrupt changes indicated in the original figures may strongly suggest a flow regime transition.

Thus, there is a clear need for additional two-phase pressure drop and liquid fraction data for cryogenic fluids and more detailed reporting of experimental conditions as well as consistent data reduction. However, it can be shown with available data that some simple models for two-phase pressure loss can be recommended as long as flow regime considerations are taken into account.

* It is assumed that the vapor volume fractions of Ref. 7 are homogeneous values. This is probably unrealistic for these data for $\alpha > 0.6$.

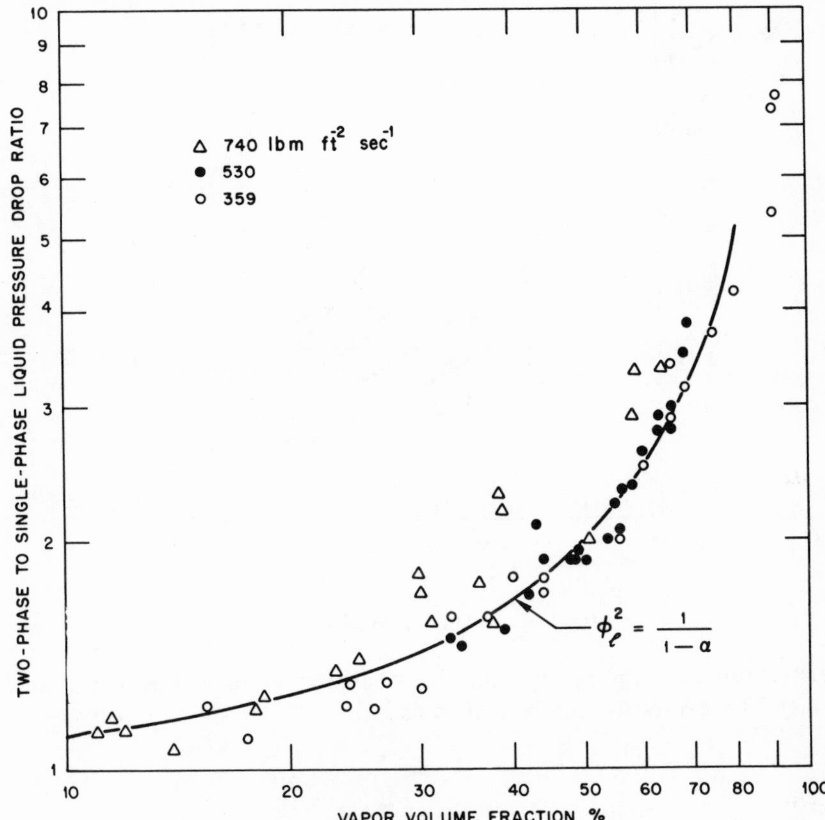

FIG. 11-6. Data of Sugden and Timmerhaus[7] for Freon 11. Note: The data represent the total two-phase pressure drop; the model represents two-phase friction pressure drop.

11.3 PRESSURE WAVE PROPAGATION

To the authors' knowledge, there is no experimental data describing pressure wave propagation through two-phase cryogenic mixtures except for two studies near the thermodynamic critical point.[21,22] Therefore, this section will discuss analytical methods which were verified by experiments with air–water and steam–water systems and are also believed to be applicable to cryogenic fluids.

As shown in Ref. 23, a one-dimensional expression for the pressure wave propagation velocity in a two-phase equilibrium mixture at rest can be

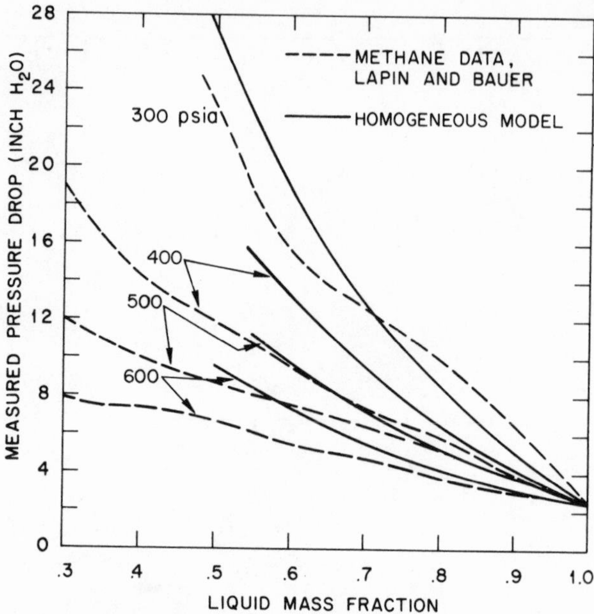

FIG. 11-7. Data of Lapin and Bauer.[8]

derived using the momentum and mass balances given in Eqs. (11-1) and (11-2). This expression can be written as

$$a_{\text{TP}} = [(1 - x)\rho_g + x\rho_l]^2 \times \left[x\rho_l^2 \frac{d\rho_g}{dP} + (1 - x)\rho_g^2 \frac{d\rho_l}{dP} - (\rho_l - \rho_g)\rho_l\rho_g \frac{dx}{dP} + x(1 - x)(\rho_l - \rho_g)\rho_g\rho_l \frac{dk}{dP}\right]^{-1} \quad (11\text{-}33)$$

The derivatives $d\rho_g/dP$, $d\rho_l/dP$, dx/dP, and dk/dP in Eq. (11-33) are determined by the rates of interphase heat, mass, and momentum transfer, respectively. The magnitude of these transfer rates will generally depend on the flow regime, the wave frequency, and the type of disturbance. It is also necessary to distinguish between the time associated with the first change in pressure experienced by the system (wavefront velocity) and the time for the system to experience the peak pressure (amplitude response). Generally the data in the literature are characteristic of the frontal velocity of a single pulse. The following models are also for such wavefront velocities.

For all but the very small void fractions, the liquid compressibility can be neglected, i.e., the liquid is incompressible,

$$d\rho_l/dP = 0 \quad (11\text{-}34)$$

As discussed in Refs. 24 and 25 and experimentally demonstrated in Ref. 26, for large gas or vapor volumes the interphase heat transfer rate is small, and as a first-order approximation, the gaseous phase can be assumed to behave in an isentropic manner,

$$d\rho_g/dP = \rho_g/\gamma P = 1/a_g^2 \tag{11-35}$$

Three separate investigations[26–28] have measured frontal velocities in low-quality steam–water equilibrium mixtures and both found that the mass transfer rates for such systems are negligible. For the low-quality region where the mass transfer term in Eq. (11-33) is of first-order importance, it is generally assumed that

$$dx/dP = 0 \tag{11-36}$$

Equation (11-36) indicates that steam–water (one-component) and air–water (two-component) systems behave in a comparable manner.

Substituting Eqs. (11-34)–(11-36) into Eq. (11-33) enables one to write the propagation velocity expression as

$$a_{\mathrm{TP}}^2 = \left\{\left[\alpha^2 + \alpha(1-\alpha)\frac{\rho_l}{\rho_g}\right]\frac{1}{a_g^2} + \alpha(1-\alpha)(\rho_l - \rho_g)\frac{dk}{dP}\right\}^{-1} \tag{11-37}$$

11.3.1 Bubbly Flow

As discussed in Refs. 25 and 30, the only important mechanism for momentum transfer within the wavefront is the virtual mass of the discrete phase. In bubbly flows the virtual mass of the bubble is large and as a first approximation the momentum transfer can be assumed to be complete,

$$dk/dP = 0 \tag{11-38}$$

Substituting Eq. (11-38) into Eq. (11-37) gives

$$a_{\mathrm{TP}}/a_g = [\alpha^2 + \alpha(1-\alpha)(\rho_l/\rho_g)]^{-1/2} \tag{11-39}$$

This result has been designated as the homogeneous frozen model in the literature.

11.3.2 Stratified, Annular, and Mist Flows

For mist flows, the virtual mass of the liquid droplet is small and can be neglected and for the separated smooth interface configurations of stratified, annular, and filament flows there are no virtual mass effects. For these flow regimes the separate equations of motion for the individual phases can be written as

$$dP + \rho_g u_g\, du_g = 0 \tag{11-40}$$

and

$$dP + \rho_l u_l \, du_l = 0 \tag{11-41}$$

By definition

$$dk/dP = (1/u_l)[(du_g/dP) - k(du_l/dP)] \tag{11-42}$$

In the Lagrangian reference frame $u_g \approx u_l \approx a_{\mathrm{TP}}$; hence

$$dk/dP \approx -(1/a_{\mathrm{TP}}^2)[(1/\rho_g) - (1/\rho_l)] \tag{11-43}$$

Substituting Eq. (11-43) into Eq. (11-37) gives

$$a_{\mathrm{TP}}/a_g = \{1 + [(1 - \alpha)/\alpha](\rho_g/\rho_l)\}^{1/2} \tag{11-44}$$

For a more detailed treatment of the virtual mass effect in bubbly, mist, and wavy annular flows, the reader is referred to Refs. 25 and 30.

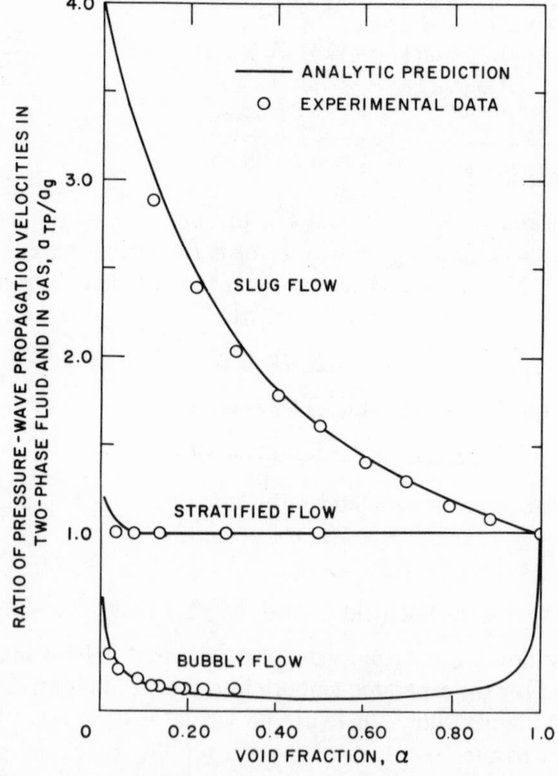

FIG. 11-8. Pressure wave propagation velocities for slug, stratified, and bubbly flows.

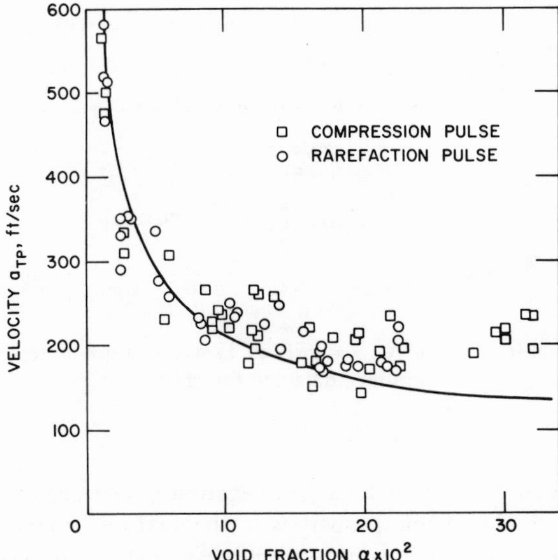

FIG. 11-9. A comparison between steam–water bubbly flow data and Eq. (11-39). The steam–water data are at the saturation temperature 265 ± 10°F.

11.3.3 Slug Flows

Slug flows were analyzed in Ref. 23. For a pulse propagating through an idealized gas–liquid configuration a series propagation mechanism was proposed:

$$a_{\text{TP}}/a_g = a_l/[\alpha a_l + (1 - \alpha)a_g] \tag{11-45}$$

Upon crossing the first liquid-to-gas interface the amplitude of such a pulse is reduced to essentially zero and any further propagation must be conducted by an inertial mechanism.

Equations (11-39), (11-44), and (11-45) are compared to the air–water data of Ref. 26 for corresponding flow regimes in Fig. 11-8. Figures 11-9 and 11-10 compare the bubbly and stratified steam–water data of Ref. 29 with Eqs. (11-39) and (11-44), respectively. Comparison of the mist flow steam–water data of England *et al.*[31] to Eqs. (11-39) and (11-44) shows the data to lie very close to Eq. (11-44), justifying the assumption in the analysis that the momentum transfer rate is generally negligible in this regime.

Comparing Figs. 11-8–11-10 shows that the analyses are capable of handling both one- and two-component mixtures. Since the interphase mass

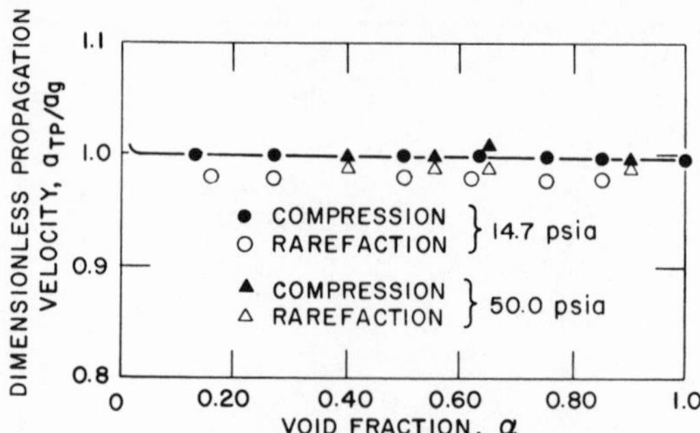

FIG. 11-10. A comparison between steam–water stratified data and Eq. (11-44).

transfer rate can be neglected, the propagation phenomenon is determined strictly by the fluid dynamic properties of the mixture. Therefore, for corresponding flow regimes one would expect these analyses to also be valid for cryogens. Figure 11-11 presents values for the propagation velocities in stratified and bubbly two-phase hydrogen mixtures at 10 psia.

11.4 CRITICAL FLOW

The geometric configuration of the discharge duct can significantly affect the flow regime of the two-phase mixture, which, as shown in the previous section, greatly affects the compressibility of the system. We shall first discuss flows in which the stagnation condition prior to the acceleration to the choking condition is either saturated vapor or a two-phase mixture. For these conditions the geometry has only a small effect on the flow pattern and the two-phase flows through various configurations can be related to an ideal nozzle flow in the same manner as single-phase flows. The latter part of the section will describe the two-phase critical discharge of initially saturated or subcooled liquids in which case the duct geometry plays a significant role.

It has been shown experimentally that the two-phase compressible flow of Freon 11 and 12 and carbon dioxide is dependent upon the geometry of the discharge duct.[32–34] This is in agreement with the experimental results of steam–water mixtures and there are no data which suggest that the two-phase critical flows of hydrogen, nitrogen, oxygen, etc. should be different. The models reviewed here are taken from the recent literature and emphasize the role of geometry.

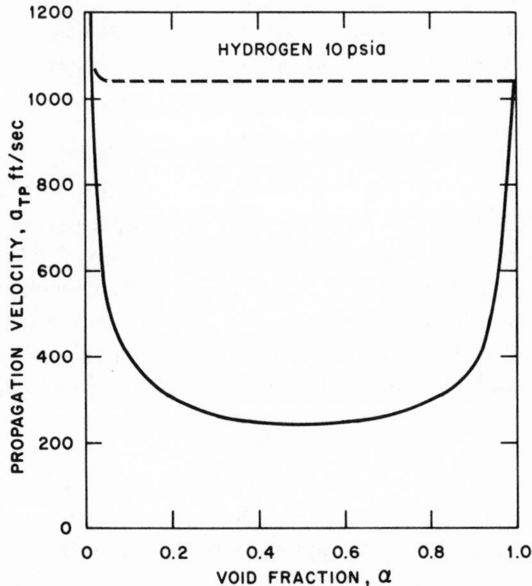

FIG. 11-11. Predictions for bubbly and stratified configurations of two-phase hydrogen mixtures: (– –) Eq. (11-44); (——) Eq. (11-39).

11.4.1 Two-Phase or Saturated Vapor Inlet Conditions

11.4.1.1 *Nozzles:* As derived in Ref. 35 and indicated in Section 11-2, the critical flow rate for a one-component, two-phase mixture can be expressed as

$$G_c^2 = -\left\{k\left[[1 + x(k-1)]x\frac{dv_g}{dP} + \{v_g[1 + 2x(k-1)] + kv_l[2(x-1) + k(1-2x)]\}\frac{dx}{dP} + k[1 + x(k-2) - x^2(k-1)]\frac{dv_l}{dP} + x(1-x)\left(kv_l - \frac{v_g}{k}\right)\frac{dk}{dP}\right]^{-1}\right\}_t \tag{11-46}$$

[This expression can be obtained by substituting Eqs. (11-7) and (11-11) into Eq. (11-9).] As discussed in Section 11-3, the derivatives dv_g/dP, dv_l/dP, dx/dP, and dk/dP are determined by the rates of interphase heat, mass, and momentum transfer.

In a converging nozzle, the acceleration and steep pressure gradients essentially occur between the upstream location, that has a diameter twice that of the throat and the throat itself. Therefore in normal nozzle configurations, there is little time for heat or mass transfer to occur and it is reasonable to assume that the *amounts* of heat and mass transferred in the expansion are negligible,

$$x_t \approx x_0 \tag{11-47}$$

$$T_{lt} \approx T_{l0} \tag{11-48}$$

The measured void fractions in Ref. 36 indicate that the steam–water velocity ratios in long, constant-area ducts for a throat pressure of 50 psia are between 1.0 and 1.5. These interphase velocity differences result from density differences which are suppressed by increased pressures. Since many of the cryogenic applications involve rather high levels of thermodynamic reduced pressure, it is assumed that the phase velocities are equal,

$$u_g = u_l = u \tag{11-49}$$

The validity of this approximation increases with increased pressure.

Since wall shear, heat exchange with the environment, and interfacial viscous terms are neglected, the system entropy can be assumed constant,

$$ds_0 = d[(1 - x)s_l + xs_g] = 0 \tag{11-50}$$

This result, along with the assumptions stating negligible amounts of interphase heat and mass transfer, implies that each phase expands isentropically,

$$s_{g0} = s_{gt} \quad \text{and} \quad s_{l0} = s_{lt} \tag{11-51}$$

$$P_0 v_{g0}^{\gamma} = P_t v_{gt}^{\gamma} \tag{11-52}$$

$$v_{l0} = v_{lt} \tag{11-53}$$

The temperature data reported by Smith *et al.*[37] for two-phase air–water critical flow in a venturi show that large heat transfer rates are in evidence at the throat. Hence, it is not reasonable to evaluate the derivative dv_g/dP in an adiabatic manner. A description of the actual heat transfer process requires a detailed knowledge of the flow configuration, which is unknown. As a compromise between simplicity and the real process, it is assumed that the vapor behavior at the throat can be described by the polytropic process

$$(dv_g/dP)_t = -(v_g/nP)_t \tag{11-54}$$

where n is the thermal equilibrium polytropic exponent derived by Tangren *et al.*[38]

$$n = \frac{(1 - x)(c_l/c_{pg}) + 1}{(1 - x)(c_l/c_{pg}) + (1/\gamma)} \tag{11-55}$$

This exponent reflects a significant heat transfer rate at the throat.

The liquid compressibility is generally negligible; thus, the liquid can be considered incompressible,

$$dv_l/dP = 0 \tag{11-56}$$

The term $(dk/dP)_t$, which is indicative of the momentum transfer rate, is difficult to evaluate and, as shown previously, has a significant effect on the system compressibility. Vogrin[39] measured axial velocity ratio profiles for the critical flow of air–water mixtures in a converging–diverging nozzle. These results indicate that the velocity ratio exhibits a minimum at the throat,

$$(dk/dP)_t = 0 \tag{11-57}$$

Like the local heat transfer rate, the rate of mass transfer at the throat can be appreciable. In Ref. 36 it was shown that if an equilibrium quality is defined as

$$x_E = (s_0 - s_{lE})/(s_{gE} - s_{lE}) \tag{11-58}$$

the exit plane mass transfer rates for steam–water flows in constant–area ducts can be correlated by

$$(dx/dP)_t = N(dx_E/dP)_t \tag{11-59}$$

where

$$N = N(x_E) \tag{11-60}$$

It is shown in Ref. 35 that in the low-quality range for which the correlation was intended, this is equivalent to

$$\left(\frac{1}{s_{g0} - s_{l0}} \frac{ds_l}{dP}\right)_t = N\left(\frac{1}{s_{gE} - s_{lE}} \frac{ds_{lE}}{dP}\right)_t \tag{11-61}$$

The derivative $(ds_g/dP)_t$ can be determined from the expression

$$T_g\, ds_g = dh_g - v_g\, dP \tag{11-62}$$

If it is assumed that the vapor behaves as a real gas following the polytropic process given in Eq. (11-54), then

$$(ds_g/dP)_t = -(c_{pg}/P_t)[(1/n) - (1/\gamma)] \tag{11-63}$$

The above approximations simplify the critical flow rate expression to

$$G_c^2 = \left\{\frac{x_0 v_g}{nP} + (v_g - v_{l0})\left[\frac{(1 - x_0)N}{s_{gE} - s_{lE}} \frac{ds_{lE}}{dP} - \frac{x_0 c_{pg}[(1/\eta) - (1/\gamma)]}{P(s_{g0} - s_{l0})}\right]\right\}_t^{-1} \tag{11-64}$$

The function N is assumed to be of the form recommended in Ref. 36, i.e., the product of a constant and the equilibrium quality. The value of the constant was determined from the steam–water data of Ref. 40 such that N can be expressed as

$$N = x_{Et}/0.14 \tag{11-65}$$

For throat qualities greater than 0.14, N is set equal to unity.

The critical flow equation is combined with the momentum equation describing the overall pressure history to obtain a solution in terms of the stagnation conditions. Under the restrictions listed above, the momentum equation can be written as

$$-[(1 - x_0)v_{l0} + x_0 v_{g0}]\, dP = d(u^2/2) \tag{11-66}$$

Equation (11-66) can be integrated between the stagnation and throat regions to give

$$\eta = \left\{\frac{[(1 - \alpha_0)/\alpha_0](1 - \eta) + [\gamma/(\gamma - 1)]}{(1/2\beta\alpha_t^2) + [\gamma/(\gamma - 1)]}\right\}^{\gamma/(\gamma-1)} \tag{11-67}$$

where

$$\eta = \frac{P_t}{P_0} \tag{11-68}$$

$$\beta = \frac{1}{n} + \left(1 - \frac{v_{l0}}{v_{gt}}\right)\left[\frac{(1 - x_0)NP_t}{x_0(s_{gE} - s_{lE})_t}\left(\frac{ds_{lE}}{dP}\right)_t - \frac{c_{pg}[(1/n) - (1/\gamma)]}{s_{g0} - s_{l0}}\right] \tag{11-69}$$

$$\alpha_0 = \frac{x_0 v_{g0}}{(1 - x_0)v_{l0} + x_0 v_{g0}} \tag{11-70}$$

$$\alpha_t = \frac{x_0 v_{gt}}{(1 - x_0)v_{l0} + x_0 v_{gt}} \tag{11-71}$$

and

$$v_{gt} = v_{g0}\eta^{-1/\gamma} \tag{11-72}$$

For specified stagnation conditions of P_0 and x_0, the transcendental expression for the critical pressure ratio (Eq. (11-67)] can be solved. Such a solution implicitly yields a prediction for the critical flow rate.

The model is compared to the two-phase carbon dioxide nozzle data of Ref. 41 in Fig. 11-12.

11.4.1.2 *Orifices and Short Tubes:* It has been experimentally demonstrated that compressible flows through sharp-edged orifices do not choke; however, they do asymptotically approach a maximum flow rate. To describe these flows, a compressible discharge coefficient can be defined as

$$C = \frac{\text{actual flow rate}}{\text{critical flow rate in ideal nozzle}} \tag{11-73}$$

For single-phase flows operating well into the compressible range, this

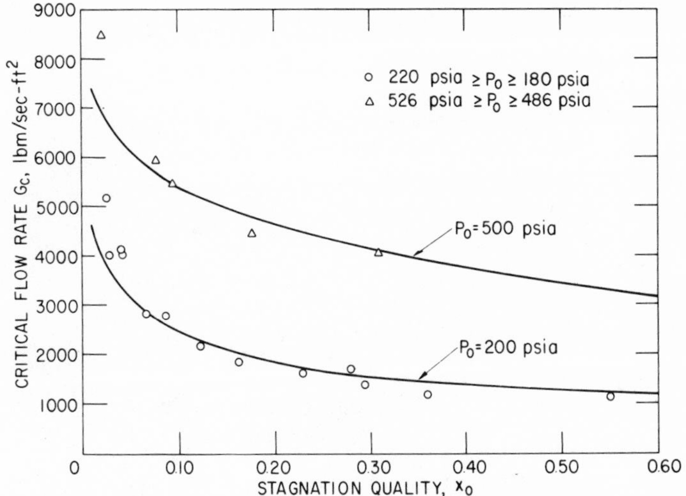

FIG. 11-12. Two-phase critical flow of carbon dioxide through a nozzle.

coefficient asymptotically approaches a value of 0.84 for orifices and short tubes.[42,43] This definition can be incorporated into Eq. (11-66) to give a formulation for two-phase compressible flows through orifices. Equation (11-67) can then be written as

$$\eta = \left\{\frac{[(1-\alpha_0)/\alpha_0](1-\eta) + [\gamma/(\gamma-1)]}{(1/2C^2\beta\alpha_t^2) + [\gamma/(\gamma-1)]}\right\}^{\gamma/(\gamma-1)} \tag{11-74}$$

As before, this transcendental expression can be solved to give a prediction for the maximum flow rate.

It is assumed that the numerical value for the single-phase discharge coefficient also characterizes two-phase flows. The orifice prediction compares very well with the CO_2 data of Ref. 41. The model is compared with the two-phase nitrogen data of Ref. 44 in Fig. 11-13.

The models described above apply to saturated vapor inlet conditions when $x_0 = 1$.

11.4.1.3 *Long Tubes:* In the literature, several models have been proposed to describe one-component, two-phase critical flow in long ducts.[36,46–48] These models give reasonably accurate predictions when based on the pressure at the plane of choking, which is generally unknown in any design calculation. If only the inlet stagnation conditions are known, it is recommended that one formulate a homogeneous equilibrium model

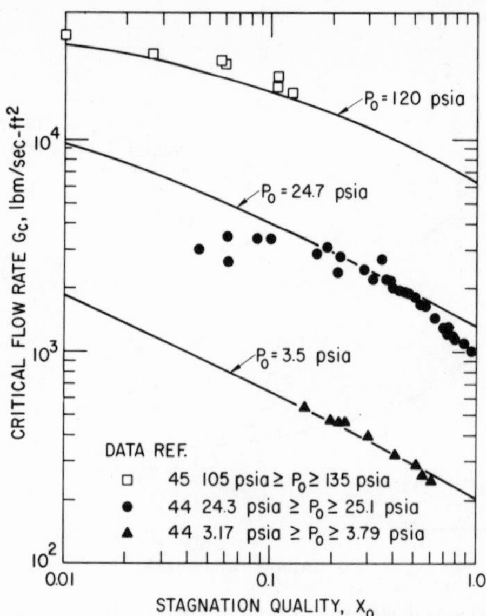

FIG. 11-13. Compressible two-phase flow of nitrogen through orifices.

using the homogeneous frictional and momentum pressure drop techniques discussed in Section 11-2 and the compressible flow relation given in Eq. (11-9).

11.4.2 Saturated and Subcooled Liquid Inlet Conditions

11.4.2.1 *Nozzles:* The model for nozzle flows can also be applied to cases where the stagnation condition is either saturated or subcooled liquid. For such cases $x_0 = 0$ and the critical flow expression is simplified to

$$G_c^2 = \left[(v_{gE} - v_{l0}) \frac{N}{S_{gE} - s_{lE}} \frac{ds_{lE}}{dP} \right]^{-1} \tag{11-75}$$

where N is given by Eq. (11-65). Since no vapor is formed until the throat is reached, it is assumed that the vapor which is formed is saturated at the local pressure. The critical pressure ratio relationship for such flows is greatly simplified

$$\eta = 1 - (v_{l0} G_c^2 / 2P_0) \tag{11-76}$$

Equations (11-75) and (11-76) can be combined to give a transcendental expression for the critical pressure ratio. As before, a solution for the critical pressure ratio also yields a prediction for the flow rate. The model is compared to the saturated and subcooled liquid as well as the saturated vapor

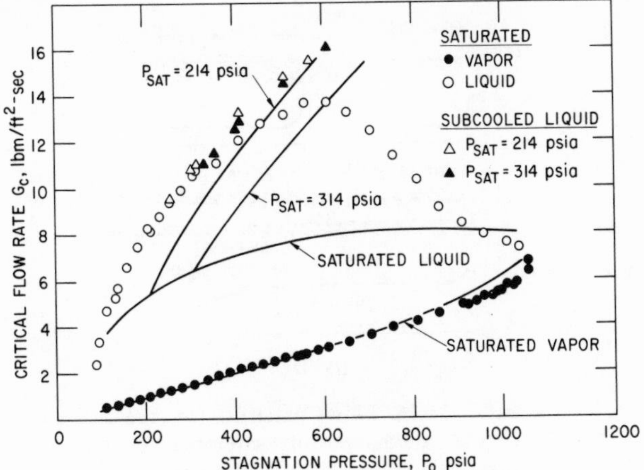

FIG. 11-14. Two-phase critical flow of saturated carbon dioxide through a nozzle.

carbon dioxide data of Ref. 41 in Fig. 11-14. Figure 11-15 compares the analytical prediction with the subcooled liquid nitrogen data of Ref. 49.

11.4.2.2 *Orifices:* Saturated or subcooled liquids discharging through a sharp-edged orifice behave in a completely metastable manner and can be predicted by the conventional incompressible relationship

$$G = 0.61(2\rho_l \Delta P)^{1/2} \tag{11-77}$$

This fact has been demonstrated numerous times for water and at least twice for cryogens.[50,51]

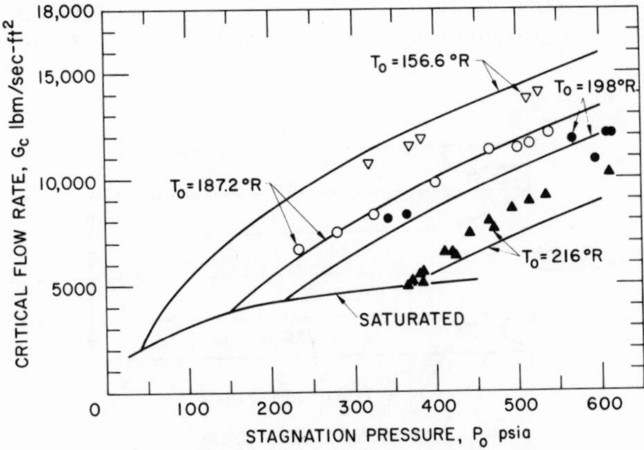

FIG. 11-15. Two-phase critical flow of subcooled liquid nitrogen through a nozzle.

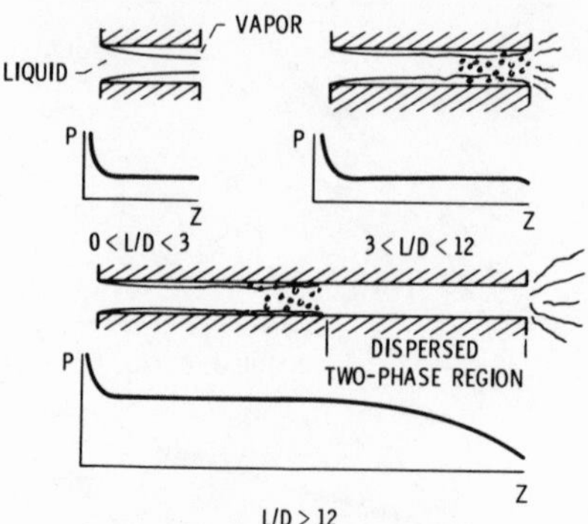

FIG. 11-16. Flow patterns for the critical discharge of saturated or subcooled liquids through short tubes.

11.4.2.3 *Short Tubes:* A compilation of experimental data with water and Freons shows that saturated and subcooled liquids flowing through sharp-edged inlet, constant-area test sections form the various flow patterns shown in Fig. 11-16.

For $0 \leqslant L/D \leqslant 3$, the flow pattern is a superheated liquid jet surrounded by an essentially saturated vapor. The momentum equation relating the stagnation and throat pressures is

$$P_0 - P_t = G_c^2 v_{10}/2(0.61)^2 \tag{11-78}$$

The mass transfer for such a flow pattern is determined by the rate of evaporation from the surface of the liquid jet where the surface temperature is the saturation value corresponding to P_t. To solve this conduction problem, a simplified jet configuration is assumed as shown in Fig. 11-17. In the analysis,

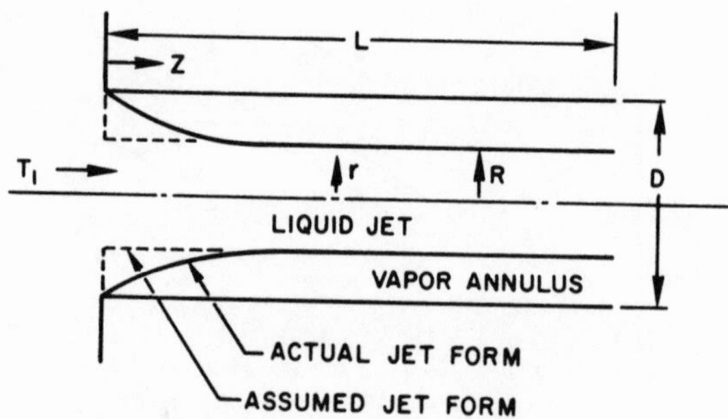

FIG. 11-17. Simplified jet configuration for short-tube model.

the flow is assumed to be one-dimensional, and axial temperature gradients are small compared to radial gradients. The resulting energy equation and accompanying boundary conditions are

$$\frac{1}{ur}\frac{\partial}{\partial r}\left(r\frac{\partial T}{\partial r}\right) = \frac{\rho_l c_l}{k}\frac{\partial T}{\partial z} \tag{11-79}$$

$$r = 0, \qquad \partial T/\partial r = 0 \tag{11-80}$$

$$r = \mathrm{R} \qquad T = T_{\text{sat}} \tag{11-81}$$

$$z = 0 \qquad T = T_1 \tag{11-82}$$

Equation (11-79) can be solved for the average liquid temperature at the throat,

$$T_t = T_{\text{sat}} + (T_0 - T_{\text{sat}}) \sum_{n=1}^{\infty} (2/\lambda_n^2) \exp[-(4\lambda_n^2 KL/c_l GD^2)] \tag{11-83}$$

The throat quality is given by

$$x_t = (h_1 - h_t)/(h_{gEt} - h_t) \tag{11-84}$$

where

$$h_1 = h_0 - \tfrac{1}{2}G^2 v_l^2 \tag{11-85}$$

$$h_t = c_l T_t \tag{11-86}$$

Differentiating Eqs. (11-83) and (11-84) with respect to the pressure shows that the mass transfer rate can be expressed as

$$\left(\frac{dx}{dP}\right)_t = -\frac{x_t}{T_0 - T_{\text{sat}}}\frac{T_{\text{sat}} v_{fgE}}{h_{fgE}} \tag{11-87}$$

As shown in Fig. 11-16, the pressure gradient within the duct is essentially zero; hence, it is assumed that the velocity ratio is unity at all times and $dk/dP = 0$.

The above assumptions reduce the critical flow expression to

$$G_c^2 = \left[x v_{gE}\left(\frac{1}{P} + \frac{T_{\text{sat}} v_{fgE}}{(T_1 - T_{\text{sat}})h_{fgE}}\right]\right)_t^{-1} \tag{11-88}$$

Equation (11-88) can be solved in conjunction with Eq. (11-78) to give a prediction for the critical flow rate and pressure ratio. The model is compared to Freon 11 data in Fig. 11-18.

As illustrated in Fig. 11-16, for the range $3 < L/D < 12$, the jet is breaking up at the surface and bubbles are being formed within. The only model available[52] for this geometry is quite complex. The short tube prediction can be used to estimate the flow in these longer geometries, in which case it will overpredict the flow rate by 15–30%.

11.4.2.4 *Long Tubes:* For length-to-diameter ratios equal to or greater

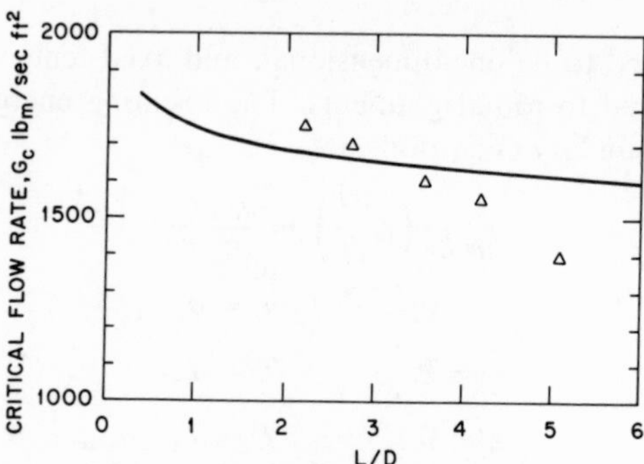

FIG. 11-18. Critical discharge of subcooled Freon-11 through short tubes.

than 12, the phases are thoroughly dispersed. In the region $0 < L/D \leqslant 12$ the pressure is essentially constant and the amount of vapor formed is small. Therefore, it is assumed that

$$P_0 - P_{(L/D=12)} = G_c^2 v_{l0}/2C^2 \tag{11-89}$$

and

$$x_{(L/D=12)} = 0 \tag{11-90}$$

It is also assumed that the flow is homogeneous ($k = 1$ and $dk/dP = 0$). For lengths longer than $L/D = 12$, flashing of the mixture generates additional pressure drop and, thus, the momentum equation relating the stagnation and throat pressures must include this additional loss. As discussed in Section 11-2 for homogeneous flashing flow in a constant-area duct, the momentum pressure drop can be expressed as

$$-dP_M = G^2\, d[(1 - x)v_l + xv_g] \tag{11-91}$$

Since the quality is assumed to be zero at $L/D = 12$, Eq. (11-91) can be combined with Eq. (11-89) resulting in

$$P_0 - P_t = G_c^2[(v_{l0}/2C^2) + x_t(v_{gt} - v_{l0})] \tag{11-92}$$

The mass transfer rate at the throat is assumed to be given by the correlation in Ref. 36, which can be stated as

$$dx/dP = N\, dx_E/dP \tag{11-93}$$

where

$$N = 20x_E, \qquad x_E < 0.05 \tag{11-94}$$

$$N = 1.0, \qquad x_E > 0.05 \tag{11-95}$$

where x_E is defined by Eq. (11-58).

Under these assumptions, the critical flow expression is then

$$G_c^2 = [(xv_g/P) - (v_g - v_{l0})N(dx_E/dP)]_t^{-1} \tag{11-96}$$

where the vapor is assumed to follow a thermodynamic equilibrium path which is closely approximated by an isothermal process,

$$dv_g/dP = -\ v_g/P \tag{11-97}$$

One needs only a knowledge of x_t to solve Eqs. (11-92) and (11-96) for the critical flow prediction. It is proposed that the two-phase system relaxes in an exponential manner from $x = 0$ at $L/D = 12$ toward the long-tube value given in Ref. 36,

$$x_t = x_{\mathrm{LT}}[1 - e^{-B[(L/D)-12]}] \tag{11-98}$$

where

$$x_{\mathrm{LT}} = Nx_E \tag{11-99}$$

and N is defined by Eqs. (11-94) and (11-95). The constant B is representative of the decay of entrance effects. In Ref. 53 critical flow rates of initially saturated and subcooled water through sharp-edged geometries were measured for length-to-diameter ratios ranging from 0 to 625. These experimental results indicate that the entrance effects are negligible for $L/D > 100$. Therefore, the constant B can be determined by

$$x_t = 0.99x_{\mathrm{LT}} \qquad \text{at} \quad L/D = 100 \tag{11-100}$$

Equations (11-92, (11-96), (11-98), and (11-100) can be combined to give a critical flow prediction for geometries in the range $12 \leqslant L/D \leqslant 100$. The model is compared to the Freon data of Refs. 32 and 54 in Fig. 11-19.

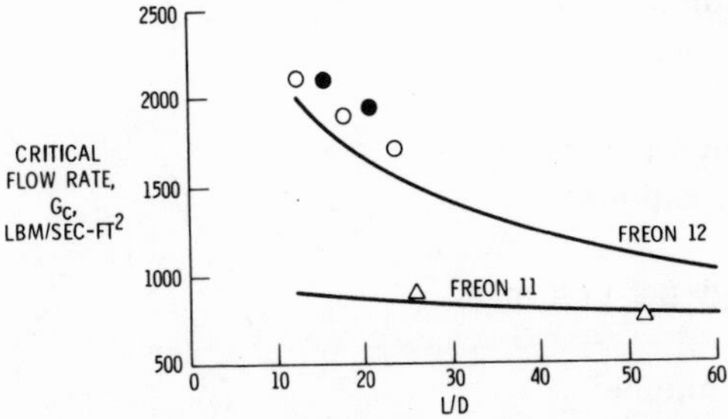

FIG. 11-19. Critical discharge of saturated and subcooled Freons through long tubes. The Freon 12 data are at $P_0 = 93$ psia, saturated; the open circles are for a diameter of 0.085 in., the solid black circles for a diameter of 0.095 in. The Freon 11 data are for $P_0 = 15$ psia, $P_s = 13.3$ psia ($T_0 = 70°\mathrm{F}$), and a diameter of 0.116 in.

References 55 and 56 give more detailed developments for the short- and long-tube, saturated liquid inlet models.

11.5 SUMMARY

The previous sections clearly demonstrate that a knowledge of the local pressure and quality is not sufficient information to analyze a two-phase system. As was demonstrated in the section on pressure wave propagation, the flow configuration can change the system compressibility by an order of magnitude. Therefore, any analysis of a flowing two-phase mixture must be predicated on a reasonable flow pattern for a given problem. The models recommended in this review cover some of the more common flow configurations and should provide a good basis for design calculations.

11.6 NOMENCLATURE

A = area
a = pressure wave propagation velocity
C = discharge coefficient
c = specific heat
D = diameter
E = kinetic energy per unit mass
G = mass flow rate per unit area
g = acceleration of gravity
h = enthalpy
K = thermal conductivity
k = velocity ratio, u_g/u_l
L = length
N = experimental parameter
n = polytropic exponent
P = pressure
q = heat energy per unit mass
R = jet radius
Re = Reynolds number
r = radial distance
s = entropy
T = temperature
u = velocity
v = specific volume
W = flow rate

x = quality, $W_g/W_g - W_l)$
z = axial distance

Greek Letters

α = void fraction, $A_g/(A_g + A_l)$
γ = isentropic exponent
η = critical pressure ratio
θ = angle from vertical
λ = eigenvalue
μ = viscosity
ρ = density
τ = wall shear stress
ϕ_l = two-phase to single-phase liquid pressure gradient ratio
χ_{tt} = Lockhart–Martinelli parameter

Subscripts

c = critical or maximum
E = equilibrium
F = friction
f = saturated liquid
f_g = difference between saturated vapor and liquid
g = vapor or gas phase
H = homogeneous
LT = long tube
l = liquid
l_0 = liquid only
M = momentum
n = polytropic exponent
p = constant pressure
sat = saturation
TP = two-phase
t = throat
0 = stagnation

11.7 REFERENCES

1. R. W. Lockhart and R. C. Martinelli, *Chem. Eng. Prog.* **45**, 39 (1949).
2. R. C. Martinelli and D. B. Nelson, *Trans. ASME* **70**, 695 (1948).
3. M. R. Hatch and R. B. Jacobs, *AIChE J.* **8**, 18 (1962).

4. R. J. Richards, W. G. Stewart, and R. B. Jacobs, in *Advances in Cryogenic Engineering*, Vol. 5, Plenum Press, New York (1960), p. 103.
5. P. S. Shen and Y. W. Jao, in *Advances in Cryogenic Engineering*, Vol. 15, Plenum Press, New York (1969), p. 378.
6. A. de La Harpe, S. Lehongre, J. Mollard, and C. Johannes, in *Advances in Cryogenic Engineering*, Vol. 14, Plenum Press, New York (1969), p. 170.
7. R. P. Sugden and K. D. Timmerhaus, in *Advances in Cryogenic Engineering*, Vol. 12, Plenum Press, New York, (1969), p. 420.
8. A. Lapin and E. Bauer, in *Advances in Cryogenic Engineering*, Vol. 12, Plenum Press, New York (1967), p. 409.
9. S. Levy, *Trans. ASME, J. Heat Transfer* **82**(5), 113 (1960).
10. S. G. Bankoff, *Trans. ASME, J. Heat Transfer* **82**(4), 265 (1960).
11. A. A. Armand, *Izv. Vses. Teplotekhn. Inst.* **1**, 16 1946; also UKAEA, *AERE Trans.* 828 (1959).
12. G. B. Wallis, *One-Dimensional Two-Phase Flow*, McGraw-Hill Book Co., New York (1969), p. 51.
13. G. B. Wallis, *One-Dimensional Two-Phase Flow*, McGraw-Hill Book Co., New York (1969), p. 325.
14. H. K. Fauske, in *Proc. Heat Transfer and Fluid Mechanics Institute, Stanford, California*, Stanford Univ. Press (1961).
15. R. W. Graham, R. C. Hendricks, Y. Y. Hsu, and R. Friedman, in *Advances in Cryogenic Engineering*, Vol. 6, Plenum Press, New York (1961), p. 517.
16. H. K. Fauske and M. A. Grolmes, "Modeling of Liquid Vapor Metal Flows with Non-Metallic Fluids," ASME Paper 70-HT-21 (1970).
17. P. A. Lottes and W. S. Flinn, *Nucl. Sci. Eng.*, **1**, 461 (1956).
18. B. L. Richardson, ANL 5949 (1958).
19. S. C. Rose, Jr. and P. Griffith, ASME Paper 65-HT-58 (1965).
20. M. R. Hatch, R. B. Jacobs, R. J. Richards, R. N. Boggs, and G. R. Phelps, in *Advances in Cryogenic Engineering*, Vol. 4, Plenum Press, New York (1958), p. 357.
21. R. C. Williamson and C. E. Chase, *Phys. Rev.* **176**(1), 285 (1968).
22. W. van Dael, A. van Iterbeek, and J. Thoen, in *Advances in Cryogenic Engineering*, Vol. 12, Plenum Press, New York (1966), p. 754.
23. R. E. Henry, M. A. Grolmes, and H. K. Fauske, in *Cocurrent Gas-Liquid Flow* (Rhodes and Scott, eds.), Plenum Press, New York (1969), p. 1.
24. H. K. Fauske, in *Proc. Symposium on Two-Phase Dynamics*, Eindhoven, Netherlands (1967).
25. R. E. Henry, *Chem. Eng. Progr. Symp. Series, Heat Transfer* **66**, 1 (1970).
26. L. J. Hamilton, Ph.D. Dissertation, Nuclear Engineering Dept., Univ. of California (1968).
27. H. B. Karplus, ARF 4132-12 (1961).
28. N. I. Semenov and S. I. Kosterin, *Teploenergetica* **11**(6), 46 (1964).
29. M. A. Grolmes and H. K. Fauske, *Nucl. Eng. Design* **11**, 137 (1969).
30. R. E. Henry, *Chem. Eng. Progr. Symp. Series, Convective and Interfacial Heat Transfer* **67**, 38 (1971).
31. W. G. England, J. C. Firey, and O. E. Trapp, *I&EC Process Design and Dev.* **5**, 198 (1966).
32. H. K. Fauske and T. C. Min, ANL-6667, Argonne National Laboratory (1963).
33. P. F. Pasqua, *Refrigerating Eng.* **61**, 1084A (1953).
34. J. C. Hesson and R. E. Peck, *AIChE J.* **4**, 207 (1958).
35. R. E. Henry and H. K. Fauske, *Trans. ASME, J. Heat Transfer* **93**(2), 179 (1971).

36. R. E. Henry, ANL-7430, Argonne National Laboratory (1968).
37. R. V. Smith, L. B. Cousins, and G. F. Hewitt, AERE-R5736, Atomic Energy Research Establishment, Harwell (1968).
38. R. F. Tangren, C. H. Dodge, and H. S. Seifert, *J. Appl. Phys.* **20**, 736 (1949).
39. J. A. Vogrin, ANL-6754, Argonne National Laboratory (1963).
40. E. S. Starkman, V. E. Schrock, V. E. Neusen, and D. J. Maneely, *Trans. ASME, J. Basic Eng.* **86-D**, 247 (1964).
41. J. C. Hesson, Ph.D. Dissertation, Illinois Institute of Technology (1957).
42. J. A. Perry, Jr., *Trans. ASME,* **71**, 757 (1949).
43. B. T. Arnberg, *J. Basic Eng.* **84**, 447 (1962).
44. H. M. Campbell and T. J. Overcamp, NASA TMX-53492 (1966).
45. F. W. Bonnet, in *Advances in Cryogenic Engineering*, Vol. 12, Plenum Press, New York (1966), p. 427.
46. H. K. Fauske, ANL-6633, Argonne National Laboratory (1962).
47. S. Levy, *Trans. ASME, J. Heat Transfer* **87-C**, 53 (1965).
48. F. J. Moody, *Trans. ASME, J. Heat Transfer* **87-C**, 134 (1965).
49. R. J. Simoneau, R. E. Henry, R. C. Hendricks, and R. Watterson, NASA TMX–67863, Lewis Research Center (1971).
50. R. J. Richards, R. B. Jacobs, and W. J. Pestalozzi, in *Advances in Cryogenic Engineering*, Vol. 4, Plenum Press, New York (1958), p. 272.
51. J. A. Brennan, in *Advances in Cryogenic Engineering*, Vol. 9, Plenum Press, New York (1963), p. 292.
52. A. K. Edwards, AHSD(S) R147, United Kingdom Atomic Energy Authority (1968).
53. H. Uchida and H. Nariai, in *Proc. of the Third International Heat Transfer Conference*, AIChE (1966,), Vol. 5, p. 1.
54. P. F. Paqua, Ph.D. Dissertation, Dept. of Mech. Eng., Northwestern Univ. (1952).
55. R. E. Henry, *Nucl. Sci. Eng.* **41**, 336 (1970).
56. R. E. Henry, to be published as an Argonne National Laboratory Report.

FORCED CONVECTION HEAT TRANSFER WITH TWO-PHASE FLOW* 12

K. D. WILLIAMSON, JR. and F. J. EDESKUTY

12.1 INTRODUCTION

Two-phase flow with heat transfer is, in the case of cryogenic applications, both frequently encountered and of considerable interest. Since the heat of vaporization is small for most cryogens (approximately 5–100 cal/g), the quality changes rapidly with heat addition. This can result in a relatively short two-phase section which may be difficult to locate if oscillations are present. In any case the quality will vary rapidly, thus complicating the computational process. For hydrogen, Bartlit [1] has developed a method to determine the proper average quality to be used in heat transfer calculations which eliminates the necessity of using an iterative procedure.

Two approaches are commonly used to correlate forced convection two-phase heat transfer data. One of these simply combines pool boiling with single-phase forced convection equations. Correlations of this type may or may not require weighting factors on each of the two terms. The second and more complicated technique utilizes the Martinelli parameter χ_{tt} to obtain a correction factor to be applied to a Dittus–Boelter-type single-phase equation.

* Work performed under the auspices of the U.S. Atomic Energy Commission.

K. D. WILLIAMSON, JR. and F. J. EDESKUTY Los Alamos Scientific Laboratory, University of California, Los Alamos, New Mexico.

Other techniques which have been used for two-phase heat transfer correlations will also be discussed briefly.

The dimensionless ratios commonly used in these correlations are the Nusselt number, the Reynolds number, the Prandtl number, and the Sterman or boiling number (the ratio of the radial mass velocity of vapor formed at the wall to the axial mass velocity in the pipe). Proper evaluation of these numbers requires use of either film or bulk fluid properties or a combination of both. While the heat transfer correlations do consider both nucleate and film boiling, the effects of specific flow regimes (see Chapter 4) have not been considered experimentally or theoretically to any extent.

12.2 PARTIAL OR SUBCOOLED NUCLEATE BOILING

Very little cryogenic data exist for this boiling regime; therefore the discussion in this section will be confined to correlations developed from noncryogenic heat transfer experiments. Equation (5-9) can be used to predict the boiling inception point. In the case of inception nucleate boiling of cryogens, $T_w - T_{\text{sat}}$ seldom exceeds 2 K, which for low-pressure systems (1.5 atm) would indicate a maximum heat flux for inception of boiling of approximately 1 W/cm^2. This appears to be a reasonable number when compared to existing cryogenic pool boiling data.

Bergles and Rohsenow[2] have developed an interpolation procedure for constructing the partial nucleate boiling curve illustrated in Fig. 12-1 and given in the following equation, where the partial nucleate boiling heat flux $(Q/A)_{\text{PNB}}$ is given as

$$(Q/A)_{\text{PNB}} = (Q/A)_{\text{FC}}\left\{1 + \left[\frac{(Q/A)_{\text{NB}}}{(Q/A)_{\text{FC}}}\left(1 - \frac{(Q/A)_{\text{B}i}}{(Q/A)_{\text{NB}}}\right)\right]^2\right\}^{1/2} = h\,\Delta T \tag{12-1}$$

Here $(Q/A)_{\text{FC}}$ is the forced convection nonboiling heat flux computed for all liquid. The heat transfer coefficient h_{FC} is calculated from

$$h_{\text{FC}} = 0.023\text{Re}_b^{0.8}\text{Pr}_b^{0.4}k_b/D = (Q/A)_{\text{FC}}/\Delta T \tag{12-2}$$

where the properties are evaluated at the bulk fluid temperature. The fully developed nucleate boiling line $(Q/A)_{\text{NB}}$ was obtained from experimental data in Bergles and Rohsenow's work. However, for cryogenic applications it is suggested that values of $(Q/A)_{\text{NB}}$ be calculated from one of the correlations given in Chapter 5. The term $(Q/A)_{\text{B}i}$ is the fully developed nucleate boiling heat flux at the $T_w - T_{\text{sat}}$ inception point. It is important to emphasize that these equations have not been experimentally verified for cryogenic

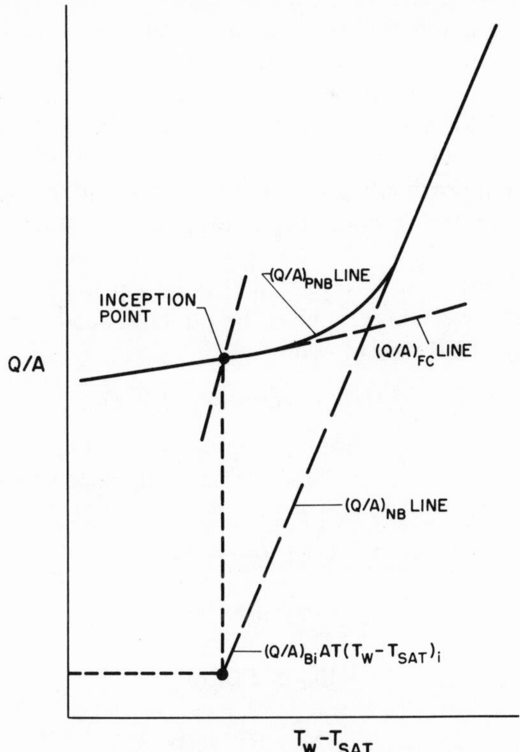

FIG. 12-1. Interpolation procedure for construction of the partial nucleate boiling curve.[2]

fluids and that in fact Eq. (12-2) without correction factors has been shown experimentally[3] to be invalid for cryogenic fluids. It is therefore recommended that equations presented in Chapter 3 and specifically developed for cryogens be substituted for Eq. (12-2). Similar interpolation procedures have been developed by McAdams *et al.*,[4] Kutateladze,[5] and Forster and Greif.[6]

12.3 FULLY DEVELOPED NUCLEATE BOILING

Several experiments in this boiling regime have been carried out utilizing cryogenic fluids. Among these are the work of Wright and Walters[7] (hydrogen); Dean and Thompson[8] (nitrogen); Richards *et al.*[9] (nitrogen and hydrogen); Sydoriak and Roberts[10] (helium); and Chi[11] (hydrogen). However, all workers but Wright and Walters utilized unusual geometries or their published reports lack complete tabulated data. For this reason the correlations in use were for the most part developed for noncryogens.

The most straightforward approach, at first glance, would appear to be the utilization of the pool boiling equations such as those of Forster and Greif,[6] Gilmour,[12] Kutateladze,[13] Labountzov,[14] and Michenko.[15] The use of such equations has been shown to be invalid except for low-flow-rate cases approximating the pool boiling situation and for high-heat-flux cases where the bubbles begin to pack the surface regardless of flow-rate.[16]

In 1953 Rohsenow[17] suggested that an equation combining a non-boiling forced convection term given by Eq. (12-2) and a pool boiling term might be useful. Thus we can write

$$(Q/A)_{\mathrm{NB}} = (Q/A)_{\mathrm{FC}} + (Q/A)_{\mathrm{PB}} \tag{12-3}$$

where $(Q/A)_{\mathrm{PB}}$ is calculated from an appropriate pool boiling equation given in Chapter 5. Giarratano and Smith[18] have developed an equation of the same type, namely

$$h_{\mathrm{NB}} = h_{\mathrm{FC}} + h_{\mathrm{PB}} \tag{12-4}$$

in which

$$h_{\mathrm{FC}} = 0.023\mathrm{Re}_l^{0.8}\mathrm{Pr}_l^{0.4}k_l/D \tag{12-5}$$

where

$$\mathrm{Re}_l = DG_{\mathrm{mix}}/\mu_l \tag{12-6}$$

and

$$\mathrm{Pr}_l = (C_p)_l\mu_l/k_l \tag{12-7}$$

and in which

$$h_{\mathrm{PB}} = 1.58\left[\frac{k_l\rho_l{}^{1.282}P^{1.75}(C_p)_l^{1.5}}{(\lambda\rho_v)^{1.5}\sigma^{0.906}\mu_l}\right]\Delta T^{1.5} \tag{12-8}$$

Equation (12-8) is the Kutateladze[13] pool boiling equation.

In place of Eq. (12-5) it might be better to use a correlation from Chapter 3 to evaluate h_{FC}. In the only reported application of Eq. (12-4) to cryogenic fluids,[18] h_{FC} was calculated from Eq. (12-5). No calculations have been made using an h_{FC} based specifically on cryogenic data.

From noncryogenic data Chen[19] has developed a weighted superposition equation to account for the interaction between the vapor bubbles and the flowing fluid which is given by

$$(Q/A)_{\mathrm{NB}} = h_{\mathrm{FC}}(T_w - T_{\mathrm{sat}})F + (Q/A)_{\mathrm{PB}}S \tag{12-9}$$

where h_{FC} is computed from an appropriate equation such as Eq. (12-2) (see Chapter 3) and $(Q/A)_{\mathrm{PB}}$ is calculated from the Forster–Zuber[20] equation:

$$\mathrm{Nu}_{\mathrm{PB}} = (Q/A)_{\mathrm{PB}}D/(\Delta T)k_l = 0.0015\mathrm{Re}^{0.62}\mathrm{Pr}_l^{0.33} \tag{12-10}$$

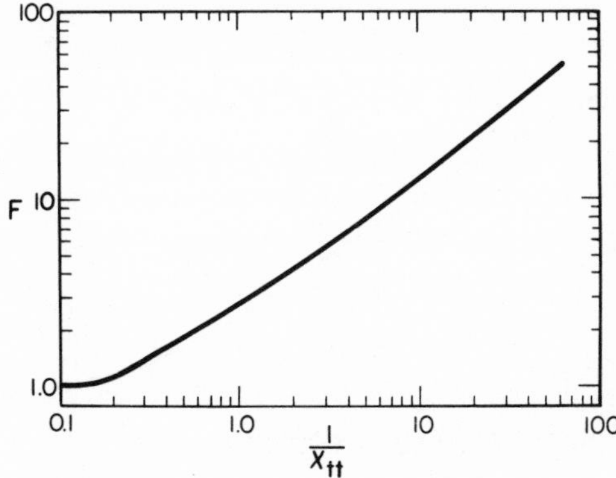

FIG. 12-2. F factor[19] used in Eq. (12-9).

which on rearrangement leads to

$$(Q/A)_{\mathrm{PB}} = 0.0015(k_l\,\Delta T/D)\mathrm{Re}^{0.62}\mathrm{Pr}_l^{0.33} \tag{12-11}$$

In this equation the characteristic dimension D is obtained from bubble dynamics and is given as

$$D = \frac{\Delta T_{\mathrm{sat}}(C_p)_l\rho_l(\pi a)^{1/2}}{\lambda\rho_v}\left(\frac{2\sigma}{\Delta P}\right)^{1/2}\left(\frac{\rho_l}{\Delta P}\right)^{1/4} \tag{12-12}$$

while the Reynolds number is given as

$$\mathrm{Re} = \left(\frac{\rho_l}{\mu_l}\right)\left(\frac{\Delta T_{\mathrm{sat}}(C_p)_l\rho_l(\pi a)^{1/2}}{\lambda\rho_v}\right)^2 \tag{12-13}$$

and the Prandtl number is

$$\mathrm{Pr}_l = (C_p)_l\mu_l/k_l \tag{12-14}$$

The correction factors F and S are shown in Figs. 12-2 and 12-3. Note that both of these factors are functions of the fluid quality X. As one approaches the all-liquid case ($X = 0$), $F = 1$ and S approaches zero for high Reynolds numbers and unity for low Reynolds numbers. This implies that pool boiling effects are negligible at high flow rates, as one would expect. Care should be taken in using these equations at high qualities since the transition to the all-gas case is not smooth. In addition, when only small volumes of liquid are present (high quality) it is doubtful that the walls are wetted; thus nucleate boiling has little meaning.

Kutateladze[21] has suggested an equation of the following form:

$$h_{\mathrm{NB}} = h_{\mathrm{FC}}[1 + (h_{\mathrm{PB}}/h_{\mathrm{FC}})^n]^{1/n} \tag{12-15}$$

Using the data of Wright and Walters,[7] Giarratano and Smith[18] investigated values of n from 0.7 to 2.0. Best results were obtained with $n = 2$.

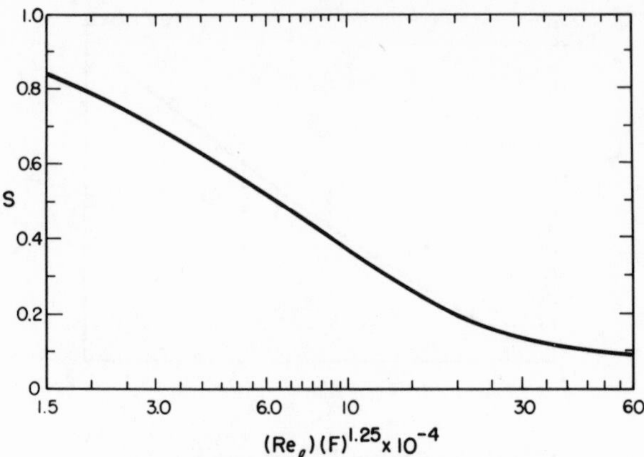

FIG. 12-3. S factor[19] used in Eq. (12-9).

In the same study a comparison of Eqs. (12-4), (12-9), and (12-15) was made. They found root mean square fractional deviations of 0.400 for Eq. (12-4), 0.415 for Eq. (12-9), and 0.338 for Eq. (12-15) when the predicted values were compared to the Wright and Walters data (see Table 12-1, p. 274).

12.4 TRANSITION BOILING

To date there has been a considerable amount of investigation of both pool boiling and forced convection at the peak nucleate boiling heat flux (see Chapter 6) as well as some investigation (noncryogenic) of the pool boiling Leidenfrost transition phenomenon.[22] No investigators have addressed themselves to the transition region for cryogenic, forced convection heat transfer. Experimentation in this region is always difficult and in the case of the lower boiling cryogens even more so since the transition covers a small ΔT region, i.e., approximately 1 K in helium, approximately 5 K in hydrogen, approximately 8 K in nitrogen, and approximately 30 K in oxygen.

12.5 STABLE FILM BOILING

As the temperature difference between the wall and bulk fluid is increased the vaporization becomes sufficiently rapid that the entire heat transfer surface becomes vapor bound. This condition (film boiling) results in a lower heat transfer coefficient. For cryogens, wall to bulk fluids ΔT's to establish film boiling are low ($< \sim 50$ K). This, coupled with the low fluid temperatures, permits experimentation with two-phase heat transfer over a wider range without equipment damage due to burnout or overheating.

Several correlations have been developed from film boiling cryogen heat transfer data. Most of these correlations utilize the Martinelli parameter χ_{tt}. Another type of correlation utilizes the superposition principle in a similar fashion to that described in the section on fully developed nucleate boiling. Another correlation presented is suitable for mist flow only (quality $> 50\%$) and depends upon considering the fluid as a single-phase material with pseudoproperties representing the mixture.

Following a method suggested by Guerriere and Talty,[23] Hendricks *et al.*[24] developed a correlation for hydrogen heat transfer data in the film boiling region. Later work[25] including additional data resulted in two updated equations. The first of these which is applicable in the subcritical region is

$$\mathrm{Nu_{FB}} = \left(\frac{1}{0.7 + 2.4\chi_{tt}} + 0.15\right)\mathrm{Nu} \tag{12-16}$$

In this equation

$$\chi_{tt} = \left(\frac{1 - X}{X}\right)^{0.9}\left(\frac{\rho_f}{\rho_l}\right)^{0.5}\left(\frac{\mu_l}{\mu_f}\right)^{0.1} \tag{12-17}$$

and

$$\mathrm{Nu} = 0.021(\rho_{f\mathrm{m}}\mu_b D/\mu_f)^{0.8}\mathrm{Pr}_f^{0.4} \tag{12-18}$$

where

$$\rho_{f\mathrm{m}} = \frac{1}{(X/\rho_f) + [(1 - X)/\rho_l]} \tag{12-19}$$

In these equations χ_{tt} and consequently $\mathrm{Nu_{FB}}$ vary with the fluid quality. In most applications, the fluid will enter at a quality of zero and approach or equal unity. To avoid an implied stepwise, calculation Bartlit[1] has developed a method for selecting an average quality which can be used to represent the entire vaporization process from an entrance quality of zero to an exit quality X_0. Though Bartlit's correlation was derived from the earliest correlation of Hendricks *et al.*,[24] it should be applicable to any equation that fits the data similarly. For an inlet quality of zero, as long as the axial wall temperature varies no more than 30 K in the two-phase section, the average quality for heat transfer equals $0.22X_0$. For wall temperature gradients greater than 30 K see Bartlit.[1] For inlet qualities greater than zero, two calculations are required. The first calculation gives the length from zero to the inlet quality and the second, from zero to the exit quality, the difference being the required length.

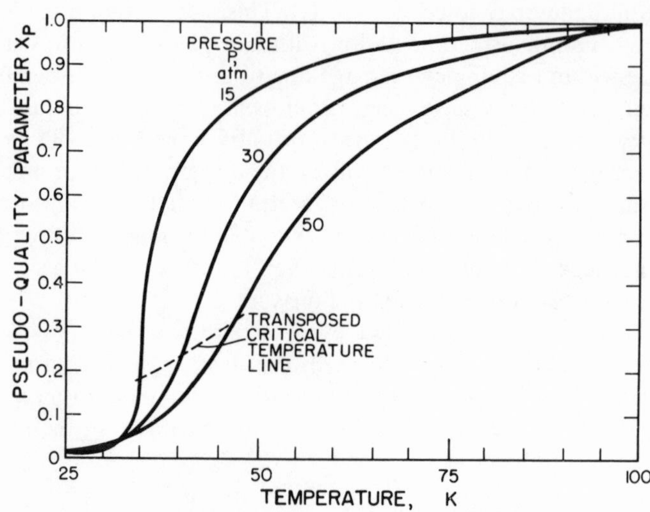

FIG. 12-4. Pseudoquality parameter X_P as a function of bulk temperature with pressure as a parameter.[25]

The second correlation of Hendricks *et al.*,[25] which is applicable for hydrogen in the vicinity of the critical point and on into the supercritical region, is

$$\mathrm{Nu_{FB}} = \left(\frac{1}{0.81 + 2.52\chi_{tt,P}} + 0.12\right)\mathrm{Nu} \tag{12-20}$$

where

$$\chi_{tt,P} = \left(\frac{1 - X_P}{X_P}\right)^{0.9}\left(\frac{\rho_f}{\rho_l}\right)^{0.5}\left(\frac{\mu_l}{\mu_f}\right)^{0.1} \tag{12-21}$$

and X_P is a pseudoquality. For hydrogen X_P as a function of temperature is given in Fig. 12-4.

Hendricks *et al.*[25] have compared these correlations with existing cryogenic data as shown in Figs. 12-5 and 12-6, where $\mathrm{Nu_{FB}/Nu}$ is shown as a function of χ_{tt}.

Giarratano and Smith[18] have developed a modified equation similar to Hendricks' original correlation[24] but with the Reynolds and Prandtl numbers evaluated at the fluid saturation temperature rather than at the film average temperature. This equation is as follows:

$$\mathrm{Nu_{FB}} = \exp[0.222 + 0.160\ln\chi_{tt} - 0.008(\ln\chi_{tt})^2]\mathrm{Nu} \tag{12-22}$$

where χ_{tt} is defined by Eq. (12-17) and Nu by

$$\mathrm{Nu} = 0.026\mathrm{Re}_v^{0.8}\mathrm{Pr}_v^{0.33}(\mu_v/\mu_w)^{0.14} \tag{12-23}$$

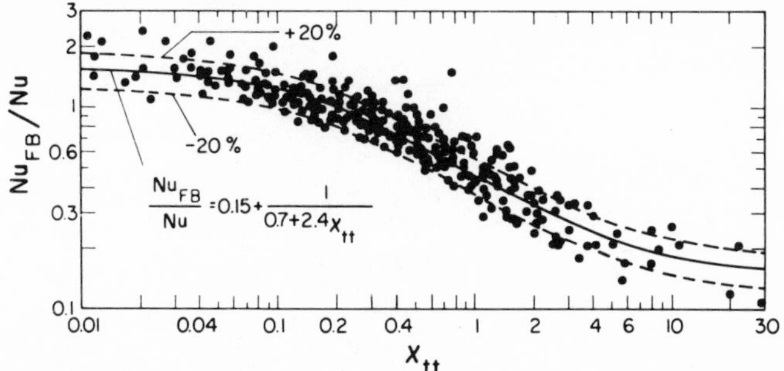

FIG. 12-5. Comparison of experimental subcritical hydrogen heat transfer data[25] with Eq. (12-16). Here χ_{tt} is the Martinelli parameter based on X.

At the loss of some generality, Eq. (12-22) has been written for hydrogen systems in simplified form in terms of quality instead of χ_{tt}. This equation is

$$\mathrm{Nu_{FB}} = \exp(-0.185 - 0.251 \ln X - 0.00767 \ln X^2)\mathrm{Nu} \qquad (12\text{-}24)$$

Further examination of Hendricks' data led von Glahn[27] to develop another correlation based on H_2, N_2, and Freon 113 data. This correlation, which presents the ratio $\mathrm{Nu_{FB}/Nu}$ as a function of a film vaporization parameter X_F and a modifying two-phase correlation factor F_{TP}, is

$$(\mathrm{Nu_{FB}/Nu})F_{\mathrm{TP}} = (1 - X_c)GD\lambda/4qL = 1/X_F \qquad (12\text{-}25)$$

where

$$F_{\mathrm{TP}} = 2.0 \times 10^{-10}\alpha(\beta)^{0.167}(\gamma)^{(1.8 - X_F^{a'})}(0.005)^{(1 - X_F^{a'})}N^{-0.667} \qquad (12\text{-}26)$$

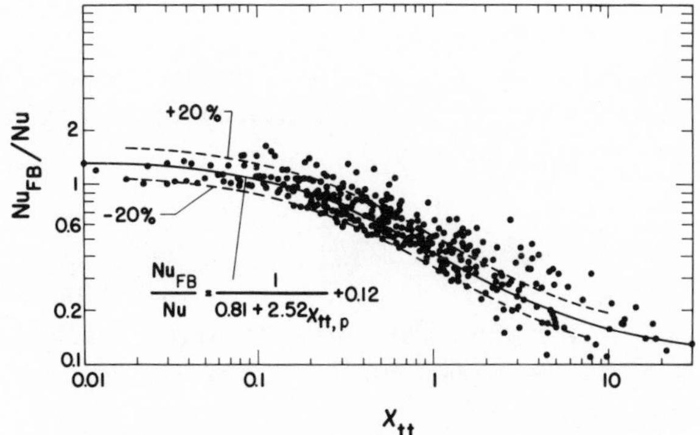

FIG. 12-6. Comparison of experimental critical and supercritical hydrogen heat transfer data[25] with Eq. (12.20). Here χ_{tt} is the Martinelli parameter based on X_P.

and

$$\alpha = \left[4.2\left(1 - \frac{\mathrm{Re}_v^{\,2}}{\mathrm{Re}_v^{\,2} + 5.85 \times 10^{11}}\right) + 0.92\right] \quad (12\text{-}27)$$

$$\beta = [g(\rho_l - \rho_v)D^2/\sigma_l] \quad (12\text{-}28)$$

$$\gamma = \mathrm{Re}_v[1 + (2500\mu_v h_{fg}/qD)] \quad (12\text{-}29)$$

$$N = \mu_v^{\,2}[g(\rho_l - \rho_v)]^{1/2}/\rho_v(\sigma_l)^{1.5} \quad (12\text{-}30)$$

$$a' = 0.5\left[1 - \frac{(L_c/D)^2}{(L_c/D)^2 + 0.05}\right] + 0.13 \quad (12\text{-}31)$$

The above were obtained from data giving values of X_F from 0.01 to 1. Outside these limits caution should be observed. If the fluid quality at the beginning of the film boiling section X_c is zero (as is frequently true), and if the heat flux is constant along the tube, then $1/X_F$ is a measure of the thermodynamic fluid quality.

Ellerbrock *et al.*[26] noted that the data of Hendricks lie on a series of curves, each curve at a constant boiling number. This implied the importance of including the boiling number in the correlation. Ellerbrock graphically presented such a correlation in which $(\mathrm{Nu_{FB}}/\mathrm{Nu})\mathrm{Bo}^{-0.4}$ was plotted against χ_{tt}. By means of a least squares fit of the data, Giarratano and Smith[18] give Ellerbrock's correlation as

$$\frac{\mathrm{Nu_{FB}}}{\mathrm{Nu}}\left(\frac{1}{\mathrm{Bo}^{0.4}}\right) = \exp[2.35 - 0.266 \ln \chi_{tt} - 0.0255(\ln \chi_{tt})^2] \quad (12\text{-}32)$$

where

$$\mathrm{Bo} = q/\lambda G_{\mathrm{mix}} \quad (12\text{-}33)$$

Here χ_{tt} is defined by Eq. (12-17) and the equation for Nu is identical to Eq. (12-19) except that the constant is 0.023.

Perroud and Rebiere[28] developed an equation similar to Eqs. (12-16) and (12-22) utilizing their H_2 data. This correlation, which also includes a boiling number correction factor, is

$$\mathrm{Nu_{FB}} = \left(\frac{1}{0.16 + 0.3\chi_{tt}} + 1.16\right)\left(\frac{q}{\lambda G_{\mathrm{mix}}}\right)^{1/3}\mathrm{Nu} \quad (12\text{-}34)$$

Perroud and Rebiere first arrived at a correlation which differed from Eq. (12-16) only in the numerical values. However, like Ellerbrock, they found, by including the boiling number correction factor, a correlation which better fit their data.

Hsu *et al.*[29] have developed a different approach on the basis of hydrogen which is applicable to the film boiling mist flow regime (high quality). The flow is treated as single phase in which the properties are synthesized by void fraction weighting of the liquid- and vapor-phase properties. In the development of the correlation several assumptions were made, including the following: (1) temperature and velocity profiles are fully developed, and (2) the droplets are moving at the same velocity as the vapor in the axial direction but can diffuse from the bulk region into the wall region where the drops impinge on the wall and evaporate. The detailed computer analysis considered eddy diffusivity, velocity profiles, and temperature profiles. For design purposes a simplified approximation to this analytical model was developed. This depends upon an empirical film coefficient C and the Dittus–Boelter equation [Eq. (12-17) with the coefficient equal to 0.023] in which the synthesized physical properties are evaluated at a calculated film temperature. The calculational procedure is as follows. First an average volume void fraction $\bar{\alpha}_v$ is computed by

$$\bar{\alpha}_v = (X/\rho_v)/(X/\rho_v + (1 - X)/\rho_l) \tag{12-35}$$

The value of $\bar{\alpha}_v$ is then used to calculate the empirical film coefficient C by

$$C = (0.964\bar{\alpha}_v - 0.9684)/(\bar{\alpha}_v - 1.02) \tag{12-36}$$

Next the reference film temperature T_f and reference void fractions $\alpha_{v,f}$ and $\alpha_{l,f}$ are calculated from

$$T_f = T_b + C(T_w - T_b) \tag{12-37}$$

$$\alpha_{v,f} = \alpha_{v,b} + C(1 - \alpha_{v,b}) \tag{12-38}$$

$$\alpha_{l,f} = 1 - \alpha_{v,f} \tag{12-39}$$

The synthesized properties can then be evaluated from

$$\varphi_f = \alpha_{l,f}\varphi_l(T_{\text{sat}}) + \alpha_{v,f}\varphi_v(T_f) \tag{12-40}$$

where φ_f is the property in question, namely ρ_f, $(C_p)_f$, k_f, or μ_f. These pseudo-properties are then used to calculate a heat transfer coefficient by

$$h_{\text{FB}} = 0.023(k_f/D)(\text{Re}_f)^{0.8}(\text{Pr}_f)^{0.4} \tag{12-41}$$

As shown in Fig. 12-7, this correlation was able to predict approximately 90% of the experimental heat transfer coefficients to within ±20% at pressure levels of less than 5 atm. At higher pressures the deviation between experiment and calculated heat transfer coefficients increases. While the reason for this deviation is not understood at present, it has been attributed to nonequilibrium, lack of information on the distribution profile of the droplets, and the influence of acceleration on turbulence.

A completely different approach to the correlation of film boiling data was made by Giarratano and Smith,[18] who applied the superposition

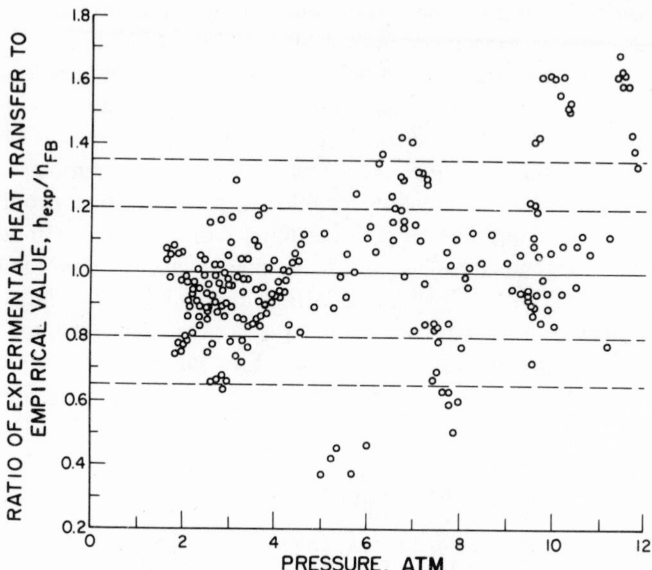

FIG. 12-7. Comparison of experimental hydrogen heat transfer data in the film boiling mist flow regime[29] with Eq. (12-41), which utilizes pseudoproperties.

theory previously used for nucleate boiling correlations. The resulting correlation is given by

$$h_{\mathrm{FB}} = h_{\mathrm{FC}} + h_{\mathrm{FPB}} \tag{12-42}$$

In this equation, h_{FC} is calculated from a modified Sieder–Tate single-phase forced convection equation:

$$h_{\mathrm{FC}} = 0.026\mathrm{Re}_v^{0.8}\mathrm{Pr}_v^{0.33}\left(\frac{\mu_v}{\mu_w}\right)^{0.14}\frac{k_v}{D} \tag{12-43}$$

$$h_{\mathrm{FPB}} = \left[4.94\,\frac{(\Delta\rho_f)^{0.375}}{\sigma^{0.125}} + 0.115\,\frac{\sigma^{0.375}}{D(\Delta\rho_f)^{0.125}}\right]\left[\frac{k_v{}^3 h'_{fg}\rho_f}{\mu_f}\right]^{0.250}(\Delta T)^{-0.25} \tag{12-44}$$

in which

$$h'_{fg} = \{[h_{fg} + 0.340(C_p)_f\,\Delta T]/h_{fg}\}^2 \tag{12-45}$$

Napadensky[30] reported the results of a Freon 12 two-phase study using an electrically heated, vertical test section. The subject of interest was annular flow consisting of a vapor core with liquid annulus. In this type of flow the principal resistance to heat transfer is across the liquid–vapor boundary. Note that the opposite type of annular flow (liquid core, vapor annulus) has been observed in cryogenic systems.[31] Napadensky's work yielded a correlation involving the boiling number, heat flux, and fluid quality:

$$\frac{h_{FB}}{h}\,\mathrm{Bo}^{0.1} = 1.67q^{0.39}\left(\frac{1+X}{1-X}\right)^{1.21q^{0.018}} \tag{12-46}$$

in which

$$h = 0.023(k_l/D_e)(D_eG_l/\mu_l)^{0.8}\mathrm{Pr}_l^{0.4} \tag{12-47}$$

$$D_e = \mathrm{D}(1 - \sqrt{\alpha}) \tag{12-48}$$

Equation (12-46) fits the Freon 12 data to $\pm 20\%$ over a range of q from 1.5 to 9.3 W/cm^2 and over a range of χ_{tt} from 0.5 to 6. Napadensky makes no claim for the generality of Eq. (12-46) for other geometries or fluids.

12.6 FORCED CONVECTION HEAT TRANSFER TO TWO-PHASE HELIUM

The only forced convection two-phase heat transfer experiments carried out with liquid helium are those of de La Harpe *et al.*[32] and Dorey.[33] The latter work involved the flow of helium over small plates rather than through channels, while de La Harpe *et al.* investigated the single-phase liquid, nucleate boiling, and film flow regions, covering qualities from approximately 0 to 0.9 in a helical coil. They found that over the quality range from 0.2 to 0.8 (film flow) their data could be correlated by

$$h_{FB} = 1.8(1/\chi_{tt})^{0.75}h \tag{12-49}$$

where

$$h = 0.023(k_lD)\mathrm{Re}_l^{0.85}\mathrm{Pr}_l^{0.4}(D/d)^{0.1} \tag{12-50}$$

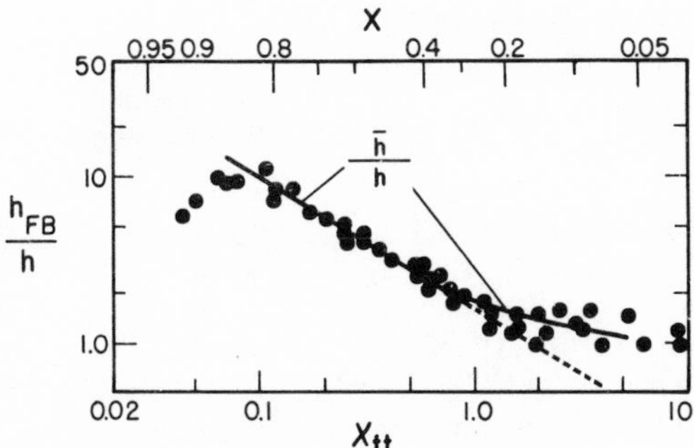

FIG. 12-8. Comparison of helium heat transfer data[32] with Eqs. (12-49) (broken curve) and (12-52) (solid curve).

In Eq. (12-50) the Reynolds number is evaluated by using liquid properties and assuming the total flow as liquid. They also found, surprisingly enough, over the quality range from 0 to 0.8, that if the Reynolds number in Eq. (12-50) was evaluated with a mixed viscosity calculated from

$$\bar{\mu} = X\mu_v + (1 - X)\mu_l \tag{12-51}$$

then the heat transfer coefficient so calculated $\bar{h}$ is equal to the experimental value (i.e., $\bar{h} = h_{\mathrm{FB}}$). This is shown in Fig. 12-8, which indicates that the correlation can be written as

$$\bar{h} = h_{\mathrm{FB}} = 0.023k_l D(DG_{\mathrm{mix}}/\bar{\mu})^{0.85}\mathrm{Pr}_l^{0.4}(D/d)^{0.1} \tag{12-52}$$

Another interesting peculiarity associated with forced flow liquid helium heat transfer is that the transition from nucleate to film boiling occurs at relatively high qualities (i.e., approximately 0.2). With other fluids this transition occurs at much lower qualities (e.g., hydrogen 0.05). This has been attributed to the fact that the density ratio of liquid to gas at T_{sat} is much lower for helium than for other fluids.[34]

12.7 CONCLUDING REMARKS

A number of correlations have been presented to represent forced convection heat transfer to two-phase cryogenic fluids. In most cases these

TABLE 12-1

Summary of the Giarratano and Smith Correlation Evaluation

Correlation[a]	Type	RMS fractional deviation
Nucleate boiling correlations		
Giarratano and Smith[18] Eq. (12-7)	Superposition	— 0.400[b]
Chen[19] Eq. (12-9)	Superposition	— 0.417[b]
Kutateladze[21] Eq. (12-15)	Superposition	— 0.338[b]
Film boiling correlations		
Hendricks *et al.*[24] Eq. (12-16)	χ_{tt}	0.605,[c] 0.259[d]
Giarratano and Smith[18] Eq. (12-22)	χ_{tt}	0.446,[c] 0.446[d]
von Glahn[27] Eq. (12-25)	X	0.681,[b] 0.175[d]
Ellerbrock *et al.*[26] Eq. (12-32)	χ_{tt}, Bo	0.209,[c] 0.158[d]
Giarratano and Smith[18] Eq. (12-42)	Superposition	0.482,[c] 0.317[d]

[a] Equations not evaluated are (12-3), (12-16), (12-20), (12-34), and (12-46).

[b] Data from Ref. 8 only; 19 points.

[c] Data from Refs. 7, 24, and 35.

[d] Data from Ref. 24 only.

were developed from limited data for a single fluid (most often hydrogen). Only one critical evaluation and comparison has been made.[18] This is summarized in Table 12-1. It should be noted that this evaluation is based only on hydrogen data and that the nucleate boiling portion is based on only 19 data points. It should also be noted that several of the correlations presented in this chapter are more recent than the review by Giarratano and Smith.[18] Specific recommendations for general use of the equations have not been made because of a lack of generality in their development and the paucity of data. To determine the applicability of any of these correlations with fluids other than those for which they were developed, much more data are required. For cryogens, this should include liquefied methane (liquefied natural gas) nitrogen, oxygen, and helium. While liquid nitrogen and liquid oxygen are already common engineering materials, the rapid growth of the technology of liquefied natural gas and liquid helium presents a real need for more engineering data.

Also, it should be pointed out that because of space program applications the effect of gravity on two-phase forced convection heat transfer should be studied. Also, the pressure effects that have been noted are not well explained and the transition region has not been studied.

12.8 NOMENCLATURE

A = area, cm^2
a = thermal diffusivity, cm^2/sec
Bo = boiling number
C = empirical film coefficient, Eq. (12-36)
C_p = specific heat, J/g-K
D = diameter, cm
d = coil diameter, cm
F = correction factor
G = mass velocity, g/sec-cm^2
g = acceleration of gravity, 980 cm/sec^2
h_{fg} = heat transfer coefficient, W/cm^2-K, latent heat of vaporization
k = thermal conductivity, W/cm-K
L = length along heated tube measured from the burnout location, cm
P = pressure, atm
ΔP = pressure difference, dynes/cm^2
$\Delta P'$ = vapor pressure difference corresponding to the superheat temperature minus the saturation temperature, dynes/cm^2
Pr = Prandtl number

Q = rate of heat flow, W
q = heat flux, W/cm^2
Re = Reynolds number
S = correlation factor
T = temperature, K
ΔT = temperature difference, $T_w - T_b$, K
ΔT_{sat} = temperature of the liquid minus the saturation temperature (degree of superheat)
X = quality, G_{vapor}/G_{total}
X_F = film vaporization parameter

Greek Letters

α = volume fraction
$\bar{\alpha}$ = average volume fraction
μ = viscosity, g/cm-sec
$\bar{\mu}$ = weighted viscosity [see Eq. (12-51)] g/cm-sec
ρ = density, gm/cm^3
σ = surface tension, dynes/cm
φ = synthesized property
χ_{tt} = Martinelli parameter defined by Eq. (12-17)

Subscripts

b = bulk conditions
c = burnout location
e = equivalent
f = film conditions
FB = film boiling
FC = forced convection single phase
fm = weighted property, see Eq. (12-19)
i = inception
l = liquid phase at saturation
mix = based upon total (vapor + liquid) flow rate
NB = fully developed nucleate boiling
P = pseudo
PB = pool boiling
PNB = partial nucleate boiling
sat = saturation
v = vapor phase at saturation
w = wall
0 = exit

12.9 REFERENCES

1. J. R. Bartlit, Rept. LA-3177-MS, Los Alamos, New Mexico.
2. A. E. Bergles and W. M. Rohsenow, *ASME Trans., J. Heat Transfer* **86,** 365 (1964).
3. K. D. Williamson, Jr., J. R. Bartlit, and R. S. Thurston, *Chem. Eng. Progr. Symp. Series* **64** (87), 103 (1968).
4. W. H. McAdams *et al., Ind. Eng. Chem.* **41**, 1945 (1949).
5. S. S. Kutateladze, *Intern. J. Heat Mass Transfer* **4**, 31 (1961).
6. K. Forster and R. Greif, *Trans. ASME, J. Heat Transfer* **81C**, 43 (1959).
7. C. C. Wright and H. H. Walters, WADC-59-423; also in *Advances in Cryogenic Engineering*, Vol. 6, Plenum Press, New York (1961), p. 509.
8. L. E. Dean and L. M. Thompson, Rept. 56-982-035, Bell Aircraft Corp. (1955).
9. R. J. Richards, R. F. Robbins, R. B. Jacobs, and D. C. Holten, in *Advances in Cryogenic Engineering*, Vol. 3, Plenum Press, New York (1957), p. 375.
10. S. G. Sydoriak and T. R. Roberts, *J. Appl. Phys.* **28**, 143 (1957).
11. J. W. H. Chi, Rept. WANL-TNR-154, Westinghouse Electric Corp. (1964).
12. C. H. Gilmour, *Chem. Eng. Progr.* **54**, 77 (1958).
13. S. S. Kutateladze, *Heat Transfer in Condensation and Boiling*, Transl. Ser. AEC-TR-3770, Tech. Info. Service, Oak Ridge, Tennessee (1952).
14. D. A. Labountzov, *Teploenerg.* **7**(5), 76 (1960).
15. N. Michenko, *Teploenerg.* **7**(6), 17 (1960).
16. J. A. Clark, in *Advances in Heat Transfer*, Vol. 5 (T. F. Irvine, Jr. and J. P. Hartnett, eds.), Academic Press, New York (1968), p. 325.
17. W. M. Rohsenow, *Heat Transfer Symposium*, University of Michigan Press, Ann Arbor, Michigan (1953).
18. P. J. Giarratano and R. V. Smith, in *Advances in Cryogenic Engineering*, Vol. 11, Plenum Press, New York (1966), p. 492.
19. J. C. Chen, ASME Paper 63-NT-34, presented at the ASME–AIChE Heat Transfer Conference, Boston, Massachusetts (1963).
20. H. K. Forster and N. Zuber, *AIChE J.* **1**, 532 (1955).
21. S. S. Kutateladze, *Fundamentals of Heat Transfer*, translated from the second revised and augmented edition by Scripts Technica, Academic Press, New York (1963).
22. K. J. Baumeister and R. C. Hendricks, NASA-TN-D-3226 (1966).
23. S. A. Guerriere and R. D. Talty, *Chem. Eng. Progr. Symp. Ser.* **18**(52), 69 (1956).
24. R. C. Hendricks, R. W. Graham, Y. Y. Hsu, and R. Friedman, NASA-TN-D-765 (1961).
25. R. C. Hendricks, R. W. Graham, Y. Y. Hsu, and R. Friedman, NASA-TN-D-3095 (1966).
26. H. H. Ellerbrock, J. N. B. Livingood, and D. M. Straight, NASA-SP-20 (1962).
27. U. H. von Glahn, NASA-TN-D-2294 (1964).
28. P. Perroud and J. Rebiere, Note ASP No. 63/10 Centre d'Etudes Nucleaires de Grenoble, Grenoble, France (1963).
29. Y. Y. Hsu, G. R. Cowgill, and R. C. Hendricks, NASA-TN-D-4149 (1967).
30. V. R. Napadensky, Doctoral Dissertation No. 4285, Swiss Federal Institute of Technology, Zurich, Switzerland (1969).
31. K. D. Williamson Jr. and J. R. Bartlit, in *Advances in Cryogenic Engineering*, Vol. 10, Plenum Press, New York (1965), p. 375.
32. Lehongre S. de La Harpe, J. Mollard, and C. Johannes, in *Advances in Cryogenic Engineering*, Vol. 14, Plenum Press, New York (1969), p. 170.

33. A. P. Dorey, *Cryogenics* **5**(3), 146 (1965).
34. R. V. Smith, in *Proc. 1968 Summer Study on Superconducting Devices and Accelerators*, Part 1, 249 (Rept. BNL-50155), Brookhaven National Laboratory, Upton, Long Island, New York (1968).
35. T. C. Core, J. F. Harkee, B. Misra, and K. Sato, WADD-60-239 (1961).

13 TRANSIENT CONDITIONS IN BOILING AND TWO-PHASE DISCHARGE

G. H. NIX and R. I. VACHON

13.1 INTRODUCTION

Transient effects are of importance in the analysis of two-phase flow and boiling heat transfer. Such information has application to boiler and reactor safety, venting of cryogenic systems, and space vehicle designs.

The first part of this review focuses attention on analyses and experimental data related to transient effects on boiling heat transfer. The latter portion is devoted to an analysis of transient flow boiling and pressurized discharge.

13.2 TRANSIENT BOILING

Closely connected with the goal of analyzing the behavior of a pool boiling system is the task of predicting heat transfer fluxes with varying pressure.

G. H. NIX Optimal Systems, Inc., Atlanta, Georgia.
R. I. VACHON Auburn University, Auburn, Alabama.

Early work concerned with the effect of pressure on boiling systems was performed by Cichelli and Bonilla,[1] who showed the displacement of the boiling curve with pressure. Corty and Foust[2] presented the equation

$$P_{vp} - P_{ext} = (2\sigma/r)\cos(\tfrac{1}{2}\phi - \theta) \tag{13-1}$$

for the radius of curvature of an idealized nucleating cavity. Since surface tension decreases with increasing temperature and the contact angle probably increases, Corty and Foust assumed the right-hand term did not change greatly with increasing temperature. However, the slope of the vapor pressure–temperature curve becomes greater as the temperature is increased. Thus, they explained that less superheat would be required for the necessary excess vapor pressure $P_{vp} - P_{ext}$ and the boiling curve would be shifted toward a lower wall superheat by increasing system pressure. Similarly, Griffith and Wallis[3] developed a relation between the cavity radius and the wall superheat. An inverse relationship was predicted between system pressure and wall superheat necessary to sustain nucleation of a given size cavity. This conclusion was also obtained by Corty and Foust.

Bankoff[4] presented an equation which predicts the minimum wall superheat for pressures less than the critical pressure. Bankoff[5] also proposed that the increase in heat transfer with pressure for a given wall superheat is caused by the increase in the number of active sites.

Other investigators[6–8] postulated that the wall superheat for a given heat flux is primarily a function of system pressure. Cryder and Finalborgo[6] found that for a given liquid–surface combination, increased wall superheat is required when the system pressure is reduced at constant heat flux. Kreith and Summerfield,[7] in an experimental study, found that the wall superheat at constant heat flux varies as the reciprocal of the absolute pressure to the three-fourths power. Raben *et al.*[8] conducted an investigation of nucleate boiling at pressures below atmospheric pressure and concluded that the mechanisms of heat transfer become less effective with reduced pressure because of (1) decreased number of active sites, (2) reduced contribution of convective transfer, and (3) reduced vapor density. The reduced effectiveness of heat transfer dictates an increase in wall superheat with reduced pressure to maintain a constant heat flux.

Critical heat flux data for benzene, diphenyl, and benzene–diphenyl mixtures boiling under pressures from 13.5 to 488.5 psia were obtained by Huber and Hoehne.[9] Lienhard and Schrock[10] correlated the peak and minimum heat fluxes with pressure for a variety of fluids. They showed that variations in pressure and geometry have a significant effect on the peak and minimum flux. A later investigation by Lienhard and Watanabe[11] concluded that the effects of geometry and pressure on the peak and minimum nucleate boiling heat fluxes are separable.

More recently, Lienhard and Schrock[12] proposed a generalized correlation for the displacement of the nucleate boiling heat flux curve with pressure. The basis of their correlation is the hypothesis that the superheat for any configuration and heat flux is directly proportional to the van der Waals maximum superheat. Justification for the hypothesis is by comparison with experimental data.

13.3 IMPOSED FLUCTUATIONS

Imposed pressure oscillations have been utilized to alter the heat transfer characteristics of a boiling system. A sizable influence on the heat transfer has been obtained in some instances at frequencies ranging from a few cycles per second through the kilocycle range. Such results are of great interest since, in addition to being imposed artificially, pressure fluctuations may be produced by instabilities in hydraulic systems, from system vibrations, or by other means.

The employment of boiling for the cooling of nuclear reactors and rocket motors entails heat release rates per unit volume which are very large, possibly of the order of 10^9 Btu/hr-ft^3. In many such cases, the heat generation rate is essentially constant and not adjustable; thus, if the cooling is inadequate, the surface will fail by melting or by rapid corrosion at high temperatures. Because of the importance in this and other applications, much attention has been given to the alteration of heat transfer rates by imposing oscillating pressures.

One of the first researchers in this field was Isakoff,[13] who experimentally demonstrated that an ultrasonic field can cause the temperature difference at burnout to increase by about 10°F and can also induce the critical heat flux to increase by about 60%. It was also observed that the ultrasonic field could cause film boiling to revert to nucleate boiling. Other investigators have reported that the influence of sonic and ultrasonic vibrations did not result in an appreciable increase in burnout flux, but did promote an increase in the critical temperature difference. This is true of the study by Markels, Durfee, and Richardson.[14] A sonic field was imposed over the entire boiling range by Gibbons,[15] increasing the heat transfer in each region, with the most notable increase in the free convection regime. He also reported that the application of the proper vibrations to the system could cause film boiling to revert to nucleate boiling, or prevent its occurrence completely, in accordance with Isakoff.[13]

In a more recent investigation, Wong and Chon[16] obtained increases in heat transfer of up to 800% in the free convection regime with an imposed ultrasonic field for subcooled methanol and subcooled water. The effect in

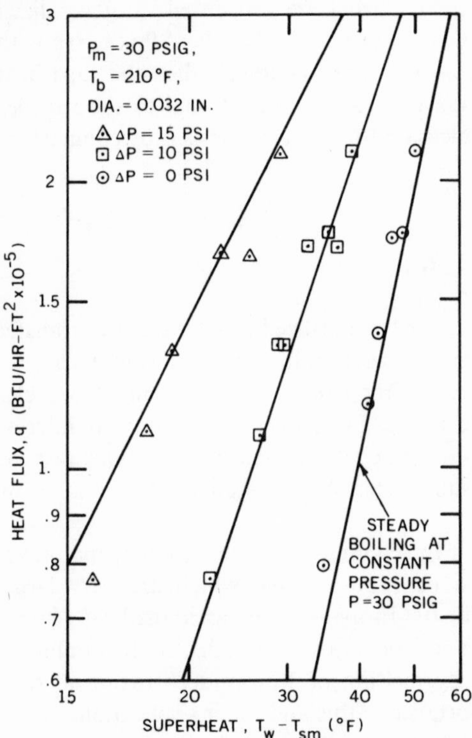

FIG. 13-1. Effect of pressure amplitude on nucleate boiling for a frequency of 15 cps.[20]

the nucleate region was reported as negligible. Large pressure pulsations applied to the surrounding fluid increased heat transfer by over 100% in film boiling in a study by DiCicco and Schoenhals.[17]

Several analytical studies of imposed pressure oscillations have shown the value of the technique. Carson[18] studied the case in which pressure was increased in a ramp fashion in the range of 1–2 atm, and calculated that the wall heat flux to water could be increased from 10 to 50 times the steady-state value. More recently, Burmeister and Schoenhals[19] predicted threshold values of amplitude and frequency which separate the region of negligible influence and the region of substantial increase due to dynamic effects.

An eightfold increase in nucleate pool boiling heat flux was observed for a sinusoidally varying pressure by McCoy, Schoenhals, and Winter.[20] A corresponding quasisteady analysis predicted only a $2\frac{1}{2}$-fold increase for the experimental conditions. The discrepancy was attributed to dynamic effects, and interpreted in the light of the naturally occurring bubble release frequencies. The investigation is one of the first to examine the effect of high-amplitude, low-frequency pressure oscillations on nucleate boiling.

The study[20] yielded a quasisteady model to estimate the time-averaged heat flux in the absence of dynamic influences, with the restriction of validity for very low frequencies. The authors stated the basis for their model as follows:

> "The effect of an oscillating pressure on nucleate boiling, if dynamic effects are absent, can be predicted by averaging the instantaneous heat flux over a complete cycle of oscillation. When the pressure is changing relatively slowly, a bubble can form, grow, and break away while the pressure is essentially constant, so the instantaneous heat flux is not significantly affected by the fact that the system pressure is varying. However, if the oscillation frequency is of the order of magnitude of the bubble release frequency, the pressure changes significantly while each bubble is forming and growing. In this situation, it can be expected that the instantaneous heat flux might be significantly affected by the dynamic effects occurring under these conditions. The equations of McFadden and Grassman[21] and Semeria[22] predict bubble release frequencies for water of 44 per second at 1 atm and 66 per second at 5 atm."

Typical results obtained by McCoy *et al.*[20] are shown in Figs. 13-1 and 13-2, representing data for various pressure amplitudes at frequencies of 15 and 30 cps. All the data exhibit an increasing heat flux with increasing amplitude

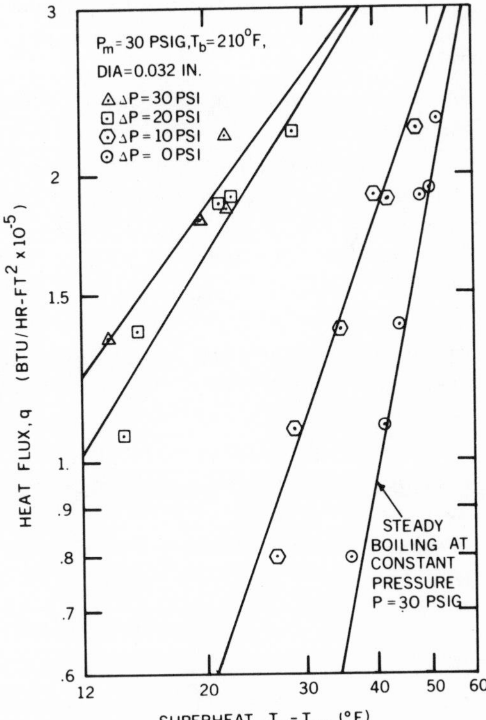

FIG. 13-2. Effect of pressure amplitude on nucleate boiling for a frequency of 30 cps.[20]

for any given value of the superheat. All the measured data lie well above the predictions of the quasisteady analysis. This could be expected since the frequency range (15–40 cps) was of the same order as the predicted bubble release frequencies. Thus, the existence of significant dynamic effects appears to be reasonable. Two major conclusions were drawn from the study:

1. At very low frequencies, where quasisteady conditions apply, the time-averaged heat flux increases with pressure oscillation amplitude because of the nonlinear relationship between heat flux and superheat.
2. When the pressure oscillation frequency is of the order of magnitude of the bubble release frequencies, much larger increases can be obtained as a result of dynamic effects.

13.4 QUENCHING EXPERIMENTS

Heat transfer under quenching conditions is highly illustrative of transient boiling, and is of interest in both metallurgy and heat transfer work. The transient temperature distribution in the piece undergoing quenching can be predicted if the surface heat transfer coefficient is known. Similarly, a knowledge of heat transfer rates from particles at high flux levels, which can readily occur when normal temperature particles are immersed in a cryogen, has become increasingly significant to the design of energy transfer systems now being developed in the space industries.

Heat transfer analyses of such systems have, in the past, been dependent on data obtained under steady-state conditions. Recently, however, more attention has been given to the experimental determination of the required heat transfer data under transient conditions. Bergles and Thompson[23] examined available quench data for water and found that the generally assumed equivalence of steady-state and quench data was in error. The object of their study was to investigate the conditions under which steady-state pool boiling data may be reliably used to generate temperature histories for quenching operations.

Results of the study are shown in Fig. 13-3. The nucleate and film boiling data are in good agreement with saturated pool boiling data for a $\frac{5}{8}$-in.-OD horizontal copper tube obtained by Flynn *et al.*[25] Shown for comparison is the correlation of Frederking and Clark,[26] which is frequently used with turbulent film boiling from various geometries in saturated pools:

$$\mathrm{Nu} = 0.15(\mathrm{Ra}^*)^{1/3} \tag{13-2}$$

where

$$\mathrm{Ra}^* = \frac{D^3 \rho_{\mathrm{vf}}(\rho_l - \rho_{\mathrm{vf}}) g (h_{fg} + 0.5 C_{P\mathrm{vf}}\, \Delta T)}{\mu_{\mathrm{vf}} k_{\mathrm{vf}}\, \Delta T}$$

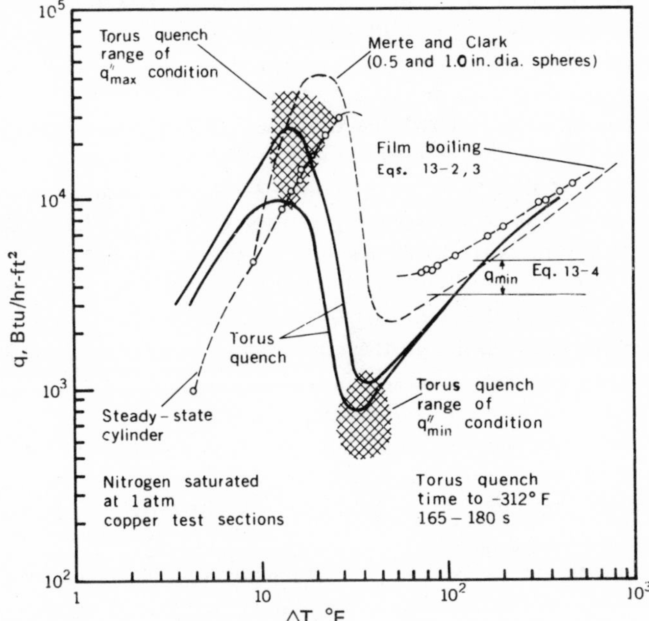

FIG. 13-3. Comparison of quench and steady-state data for copper test sections in nitrogen.[23]

is valid for Ra* > 5 × 10⁷. The total heat transfer coefficient is estimated using Bromley's[27] suggestion for the radiation correction

$$h \simeq h_{co} + 0.75h_r \tag{13-3}$$

with

$$h_r \simeq \sigma(T_w{}^4 - T_{sat}^4)/\Delta T$$

The relationship for the minimum heat flux in film boiling q_{min} developed by Zuber on the basis of hydrodynamic stability theory is generally considered valid for large heaters:

$$q_{min} = Ch_{fg}\rho_v[\sigma g(\rho_l - \rho_v)/(\rho_l + \rho_v)^2]^{0.25} \tag{13-4}$$

where C is given as 0.177,[28] 0.13,[29] or 0.09.[30] The steady-state film boiling data in Fig. 13-3 are somewhat higher than predicted by Eqs. (13-2) and 13-3; however, q_{min} is within the range suggested by Eq. (13-4).

Shown in Fig. 13-3 are data derived from typical quench runs, where the shaded areas indicate the range of q_{max} and q_{min} for 27 tests. Indicated for comparison is the boiling curve of Merte and Clark[24] for quenching of 0.5- and 1.0-in.-diameter copper spheres. The use of smaller spheres by

Merte and Clark explains the fact that they obtained less deviation between runs due to localized inception of nucleation. The data of Fig. 13-3 point out significant differences between the steady-state and quench data. During quenching the film boiling is much more stable; accordingly, the onset of transition boiling occurs at reduced heat flux and wall superheat. The data of Merte and Clark indicate a similar shift in the $q_{\min}$ condition.

The results obtained by Bergles and Thompson indicate that actual quenching times will be considerably shorter than would be predicted by conventional boiling correlations used to calculate the temperature history. Specimens to be quenched usually have surface contamination which acts to destabilize film boiling, reducing the characteristic slow cooling portion of the quench. The results further imply that the transient calorimeter technique[24] is not generally suited for obtaining reliable boiling curve information. The transition and nucleate boiling regions may be greatly distorted unless surface conditions are strictly controlled. The differences between quench and steady-state data are sufficient to indicate that, even with a cryogen, the quenching experiment may not provide accurate transition and nucleate boiling data.

There are numerous instances when steady-state data may not suffice to explain transient phenomena. Witte *et al.*[31] conducted an experimental study of heat transfer from extremely high-temperature particles to liquid sodium. In their study, data were obtained with a transient quenching technique in which a tantalum sphere attached to a swinging-arm apparatus was passed through a molten pool of sodium. A corresponding analysis of film boiling[32] revealed that for large amounts of subcooling in liquid sodium, the contribution of sodium vaporization to the net heat transfer was negligible. This implied that the vapor film was very thin (of the order of 10^{-6} in.) and that practically all of the energy transferred from the sphere was transferred to the liquid sodium. Film thicknesses computed from the experimental data were even smaller, leading to the conclusion that the vapor film thickness must "adjust" to accommodate the energy being transferred through it.

The results of their study are shown in Fig. 13-4 and are correlated by the expression

$$q = K[U_\infty(\rho k C_p)_L/D]^{1/2}(T_L - T_B) \tag{13-5}$$

where $T_L = T_w$, and $K = 0.675$. Also shown is the forced-convection heat transfer expression developed by Sideman:[33]

$$q = 1.13[U_\infty(\rho k C_p)_L/D]^{1/2}\,\Delta T \tag{13-6}$$

The value of 0.675 in Eq. (13-5) is obtained by replacing the constant of Sideman[33] by a factor K, and selecting K to obtain the best agreement with

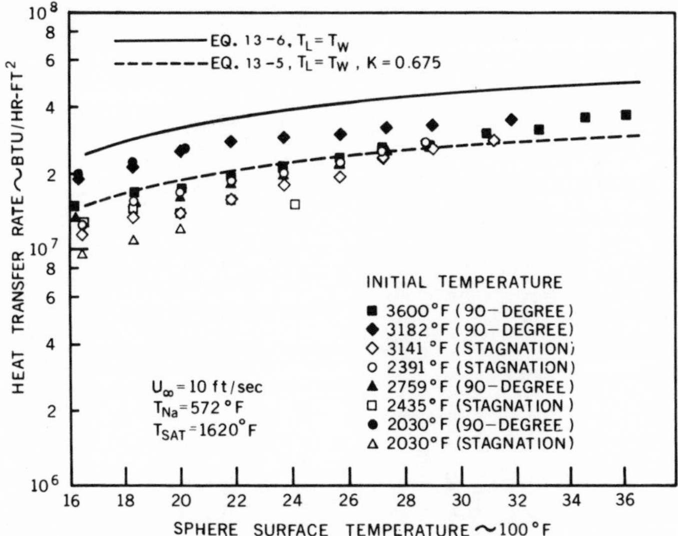

FIG. 13-4. Heat transfer rate vs. surface temperature for ½-in.-diameter tantalum spheres in liquid sodium.[31]

the experimental results. In both cases, the temperature difference ΔT, normally given as $T_l - T_b$, where l and b refer to liquid and bulk, is taken for the case $T_l = T_w$, where T_w is the temperature of the sphere wall. This implies no contact resistance at the interface and indicates that the sodium can be superheated to extremely high temperatures. The phenomenon of superheating of liquid metals during heat transfer has been observed by others.[34] Actually, the value of $K = 0.675$ represents a modified superheat theory which might be the result of a contact resistance at the solid–liquid interface. It appears that complete verification of the transient data obtained might require additional heat transfer and hydrodynamic measurements.

13.5 TRANSIENT CONDITIONS IN TWO-PHASE FLOW AND DISCHARGE

Considerable material has been published on the subject of two-phase flow, especially in the area of flow boiling in a pipe. Tong[35] and Kepple and Tung[36] have prepared comprehensive reviews of this topic. Boure *et al.*[37] present an excellent review of two-phase flow instability and identify the causes and mechanisms of instabilities in boiling flow. Table 13-1 taken from Ref. 37, presents a synopsis of flow instabilities. Basically, instabilities

TABLE 13-1
Classification of Flow Instability[37]

Class	Type	Mechanism	Characteristics
1. Static instabilities			
Fundamental (or pure) static instabilities	Flow excursion or Ledinegg instability	$\partial(\Delta p)/\partial G\|_{\text{int}} \leqslant \partial(\Delta p)/\partial G\|_{\text{ext}}$	Flow undergoes sudden, large-amplitude excursion to a new, stable operating condition
	Boiling crisis	Ineffective removal of heat from heated surface	Wall temperature excursion and flow oscillation
Fundamental relaxation instability	Flow pattern transition instability	Bubbly flow has less void but higher ΔP than that of annular flow	Cyclic flow pattern transitions and flow rate variations
Compound instability	Bumping, geysering, and chugging	Periodic adjustment of metastable condition, usually due to lack of nucleation sites	Period process of superheat and violent evaporation with possible expulsion and refilling
2. Dynamic instabilities			
Fundamental (or pure) dynamic instabilities	Acoustic oscillations	Resonance of pressure waves	High frequencies (10–100 Hz) related to time required for pressure wave propagation in system
	Density wave oscillations	Delay and feedback effects in relationship between flow rate, density and pressure drop	Low frequencies (1 Hz) related to transit time of a continuity wave
Compound dynamic instabilities	Thermal oscillations	Interaction of variable heat transfer coefficient with flow dynamics	Occurs in film boiling
	BWR instability	Interaction of void reactivity coupling with flow dynamics and heat transfer	Strong only for a small fuel time constant and under low pressures
	Parallel channel instability	Interaction among small number of parallel channels	Various modes of flow redistribution
Compound dynamic instability as secondary phenomena	Pressure drop oscillations	Flow excursion initiates dynamic interaction between channel and compressible volume	Very low-frequency periodic process (0.1 Hz)

are related to the geometry of the system, system operating conditions, and boundary conditions. Instabilities are classified in Table 13-1 as either static or dynamic. A static instability exists when a small change in flow condition from the original steady-state condition causes a deviation to an unsteady condition or another steady-state condition other than the original condition. Steady-state laws are used to predict static instabilities. Dynamic instabilities are generated by inertia and other feedback phenomena. Compound instabilities are combinations of elementary mechanisms that cannot be studied separately.

The review here is concerned mainly with discharging analyses as opposed to the general case of flow in a pipe as is the case for Ref. 37.

Numerous investigators have studied the discharging of a single-phase fluid from tanks and vessels. These efforts will be discussed briefly before the more general and difficult problem of two-phase discharge is considered.

13.5.1 Single-Phase Discharge

The quasisteady analysis of rapid discharge from a vessel is a classical problem in thermodynamics. However, many discrepancies arise which can not be treated by the strict classical thermodynamic approach. Giffen[38] analyzed the rate of pressure decrease of a vessel discharging to the atmosphere through a rapidly opened port. Using a numerical technique, Giffen found that the internal pressure change did not decrease in a continuous fashion, as predicted by the classical methods, but varied in a stepwise fashion. He concluded that the speed of sound within the gas is the controlling parameter. When the orifice or port diameter is small with respect to the vessel, the rate of pressure variation within the vessel is negligible compared to the rate at which disturbances are propagated throughout the vessel. Thus, the pressure can be assumed to be constant through the vessel, and the velocity of the approach to the discharge port can be neglected.

Numerous other authors[39–42] considered the problem of nonsteady discharge of a gas, basing their calculations on the method of characteristics. Progelhof[42] found that the mass versus time relationship for sonic discharge through a nozzle can be approximated by application of the velocity-of-approach correction factor to the quasi-steady result. However, for sonic discharge through a nozzle, accuracy depends on the estimated average discharge coefficient for the process. For a subsonic discharge, Progelhof found that the pressure in the vessel may fall below the pressure of the surroundings. This is a further discrepancy not predicted by the classical quasisteady approach.

Giffen[38] gives some general conclusions concerning the nature of the discharge process:

1. The extent of pressure variation in a vessel is determined by (a) the rate of change of pressure near the port, and (b) the time required for changes at the port to be propagated throughout the vessel.
2. Unless the port opens with a speed greater than the sonic velocity, area variations of the port will be accompanied by instantaneous variations in the rate of discharge.
3. Resistance to the pressure waves by the surrounding medium is most important when a pipe is connected to the discharge port.
4. A pressure depression will usually occur when a vessel is discharged rapidly to the atmosphere.

The previous analyses consider only a single-phase flow, which is somewhat simpler than the case of one-component, two-phase flow. Two-phase (vapor–liquid) discharge is complicated by the flashing of liquid to vapor as a result of a pressure decrease. However, certain aspects of the single-phase problem may be carried over into the investigation of two-phase discharge. For instance, it is likely that the sonic velocity explanation of discharge rate variations offered by Giffen[38] may be used to help explain the pressure spikes[43–46] noticed for a two-phase pressure decay.

13.5.2 Two-Phase Discharge

Recently, a number of investigators have been concerned with the problem of two-phase flow through various apertures[47–51] and with the mechanisms and interface relations[52–55] involved with boiling and two-phase flow. Also, emphasis was placed on the prediction of the critical flow rate in two-phase flow. The models of Fauske,[56] Levy,[57] and Moody[58] are perhaps the best known of these analyses.

Timmerhaus[51] discusses flow and heat transfer phenomena in liquid transfer lines during cooldown of the lines from ambient to cryogenic temperatures. The pressure history in a transfer line has been observed to exhibit an initial "spike" due to the vaporization of the initial charge of the cryogen as it enters the line. The pressure then drops with some instabilities as the flow goes from the two-phase regime to the single-phase regime as the transfer line reaches the temperature of the cryogen. An equation for the cooldown time for transfer lines is presented by Timmerhaus, who also discusses other problems of transfer line cooldown.

The problem of two-phase discharge has been the object of a number of other studies.[43,45,46,59–65] Pollard[45,65] studied the history of liquid superheat in a pool boiling system subjected to a sudden pressure release.

A 304 stainless steel heater surface in contact with distilled, degassed water was monitored during system pressure decay through a 1-in. orifice to the atmosphere. The data indicated that the liquid superheat reached a maximum during the initial transient pressure phase and became negative during the final phase with a release time of approximately 12 sec. Figure 13-5a shows some typical results.

Pollard also found a "spike" in the pressure versus time trace during the initial phase. The spike (Fig. 13-5b) was similar to those observed by Howell and Bell,[44] Ordin, Weiss, and Christenson,[46] and Moody[43] and somewhat similar to that reported in Ref. 51. Ordin *et al.*[46] presented results of an experimental investigation concerning temperature stratification and pressure rise of liquid hydrogen contained in an aircraft-type tank exposed to atmospheric turbulence conditions during flight. Pressures and temperatures were monitored at various positions in the tank during pressurization and venting conditions. They found—as did Pollard—that immediately after venting, the liquids are superheated and very little boiling takes place during the initial drop in pressure. This initial period is marked by a sharp drop in pressure with gas outflow from the tank. The authors found that the decay in tank pressure is a function of liquid temperature, vapor temperature, vaporization rate, ullage, and line and valve size.

Nuclear reactor containment and emergency cooling systems are designed to cope with the transient and maximum pressure buildups arising from loss of coolant accidents. The expulsion of the coolant from pipe and vessel breaks in high-pressure water systems involves metastable and two-phase flow phenomena (including the critical two-phase flow rate). Adequate predictive models have not been developed for the discharge of saturated water through short length apertures. Numerous experimental studies have attempted to characterize the important parameters. One such study is the large-scale LOFT (Loss of Fluid Test) program conducted by Philips Petroleum Company at the National Reactor Testing Station in Idaho Falls, Idaho.

Objectives of the LOFT program include the generation of data for verification of analytical models and scaling considerations used to predict and extrapolate LOFT behavior. Detailed information as a result of this study includes:

1. The overall thermal behavior of the coolant in the subcooled region (where acoustic effects predominate), through the two-phase saturated region, and in the post-blowdown convection region.
2. The transient and semi-steady-state hydraulic loading applied to various system components and structures.
3. The importance of system configuration and initial conditions; i.e.,

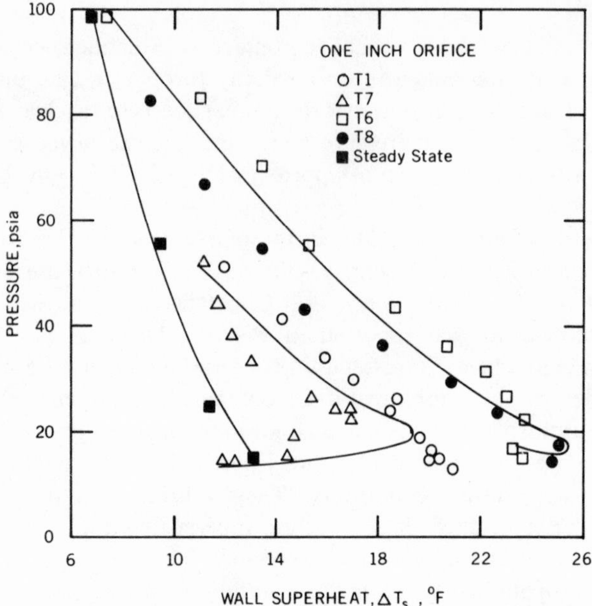

FIG. 13-5a. Pressure vs. wall superheat for discharge from different initial pressures with $q = 29{,}000$ Btu/hr-ft^2.[45]

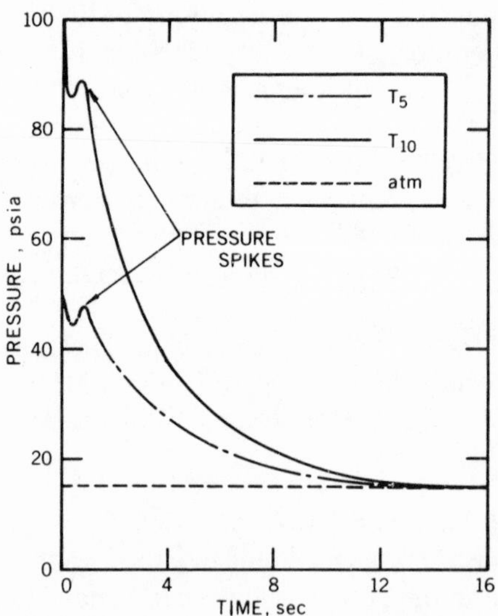

FIG. 13-5b. Pressure vs. time for discharge through a 1-in. orifice from initial pressures of 50 and 100 psia.[45]

break size and location, hot and cold leg temperature differences, initial flow, and break duration.

4. The downstream shock pressure generation.
5. The amount and location of water remaining in the system after the break.

Plans included simulation of core heat input to determine its influence on blowdown behavior. The blowdown program consisted of three phases. Phase I included tests using an unscaled empty vessel; phase II included tests using a quarter-scale model of the LOFT vessel with a simulated reactor core and internal components; and phase III included tests using a quarter-scale model of the integral LOFT reactor and coolant system.

The LOFT phase I blowdown facility is shown in Fig. 13-6. The vessel is attached to an I-beam frame through thrust load cells and weight-loss cells and can be utilized for top or bottom blowdowns. Initial tests[62] were performed at subcooled conditions with test pressures ranging from 600 to

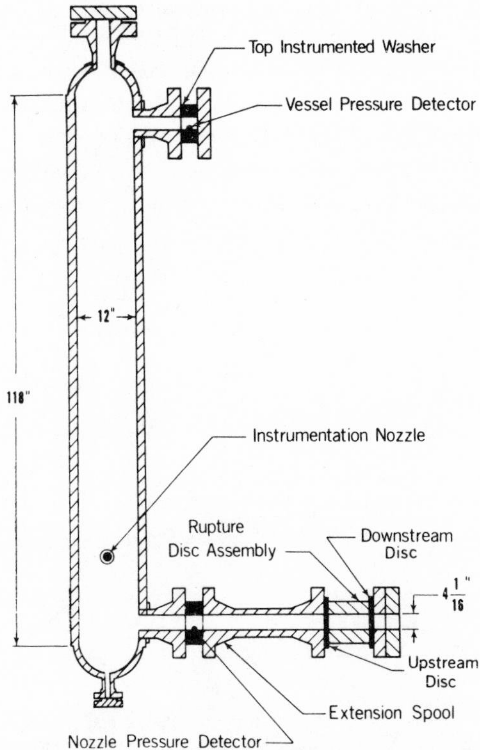

FIG. 13-6. LOFT Phase I blowdown facility.[62]

1600 psi and at room temperature to investigate the behavior of the subcooled fluid in the semiscale vessel when subjected to rapid decompression. Several tests were performed using a liquid-filled vessel to obtain data which should be similar to the subcooled portion of decompression during LOFT blowdowns.

The second series of tests[63] in the LOFT program was concerned with bottom blowdown tests at elevated pressure and temperature conditions. Tests were conducted at fluid temperatures of 400, 445, 495, and 540°F with corresponding pressures of 600, 1270, 1700, and 2300 psi. The enthalpy stored in the vessels was varied to investigate enthalpy effects on blowdown phenomena. It was found that the durations of the subcooled portions of the decompression were essentially identical to the duration of a comparable subcooled test at ambient temperature. This implied that over the temperature range considered the acoustic relaxation of the system was primarily dependent upon the disturbance propagation time in the exit nozzle. The blowdown time was observed to decrease with an increase in the stored enthalpy of the fluid.

The facility shown in Fig. 13-6 was also utilized for top blowdown tests.[64,65] Partial pipe breaks were simulated by using a sharp-edged orifice in the top blowdown nozzle to determine the effect of break size on such phenomena as subcooled decompression, blowdown time, and fluid remaining in the vessel. Separation of the fluid phase was enhanced by the top blow-

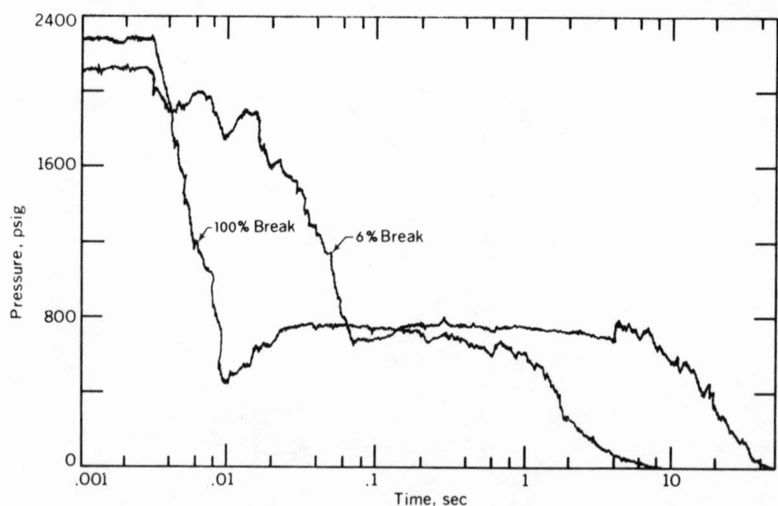

FIG. 13-7. LOFT vessel pressure for top blowdown test—effect of break size on subcooled decompression behavior.[65] Initial pressure 2330 psig; initial temperature 520°F.

down, and liquid entrainment was reduced. Separation of the fluid phases was promoted by a reduction in break size, and the increased blowdown time allowed more time for the fluid to reach thermodynamic equilibrium. The break size markedly affected the subcooled decompression, as shown in Fig. 13-7. Smaller break sizes increased the subcooled decompression and could result in oscillatory pressure forces coincident to the natural frequency of a reactor core.

Top blowdown tests at elevated temperature and pressure conditions were conducted with break sizes corresponding to 2, 6, 10, 30, 60, and 100% of the full pipe area.[66] It was found that the residual water in the vessel depended not only on break size but also on the location of the blowdown nozzle. Also, it was found that the steam exiting from the top nozzle was of a higher quality than that of comparable bottom blowdown tests. The conclusion was reached that the elevation difference between the nozzles accounted for the difference in quality. In the top blowdown configuration, the enthalpy stored in the fluid was dissipated by expanding or flashing the fluid to high-quality steam which escaped through the blowdown nozzle. In the bottom blowdown, the expanding fluid in the vessel maintained a fluid pressure which forced a fluid mixture out the bottom nozzle. Results of the top blowdown with small simulated breaks indicated that an appreciable amount of residual water might remain in the vessel.

Moody[43,67] developed a theoretical blowdown model to predict two-phase blowdown for various steam–water reference systems. The saturated system blowdown model used in developing the analysis is shown in Fig. 13-8. The model consisted of an adiabatic, constant-volume system which contained an equilibrium mixture of liquid and vapor. Mass and energy escape through a single pipe at rates W and $h_E W$, respectively, where h_E is the stagnation enthalpy of the fluid in the immediate pipe neighborhood. Moody obtained the following expressions for system mass, energy, and pressure rates:

$$dM^*/dt^* = -G \tag{13-7}$$

$$dU^*/dt^* = -(h_e/u_i)G \tag{13-8}$$

and

$$\frac{dP_0}{dt} = -\frac{[h_E + (u_{lv}/v_{lv})v_l - u_l]G}{\{(v_i/M^*)(u_{lv}/v_{lv})' - [(u_{lv}/v_{lv})v_l]' + u_l'\}M^*} \tag{13-9}$$

The value of G is determined as a function of stagnation pressure, enthalpy, and the quantity $\bar{f}L/D$. Moody presents a series of graphs giving G_M in the form

$$G_M(P_0, h_0, \bar{f}L/D) = 0 \tag{13-10}$$

The theory requires an estimation of $\bar{f}L/D$ to predict the blowdown rates.

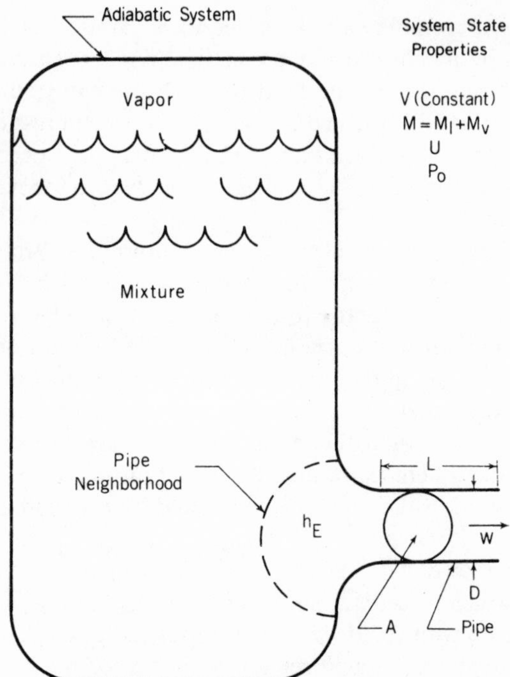

FIG. 13-8. Saturated system blowdown model of Moody.[67]

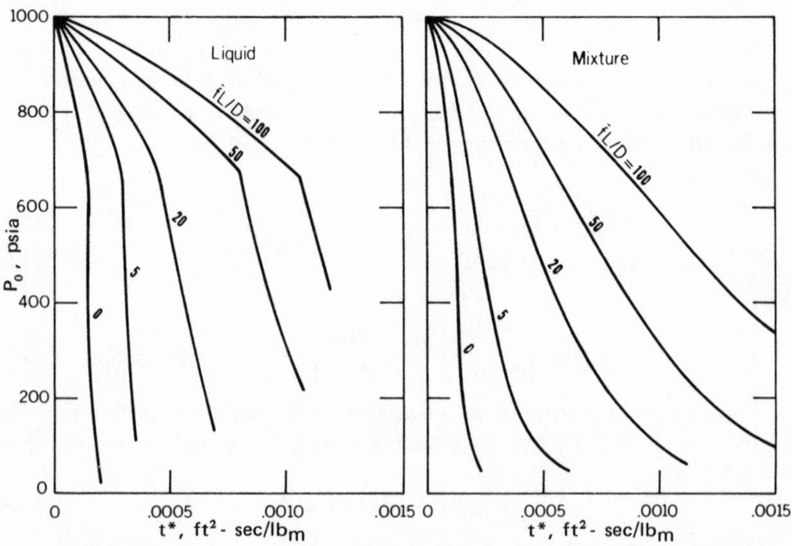

FIG. 13-9. Blowdown from 1000 psia steam/water reference system.[67]

The value of h_E depends on P_0 and on the liquid–vapor action in the system. Moody considers three characteristic types of blowdown:

1. Saturated liquid blowdown, characterized by

$$h_E = h_l(P_0) \tag{13-11}$$

2. Homogenized mixture blowdown, characterized by

$$h_E = h_l(P_0) + \frac{h_{lv}(P_0)}{v_{lv}(P_0)}\left[\frac{v_i}{M^*} - v_l(P_0)\right] \tag{13-12}$$

3. Saturated vapor blowdown, characterized by

$$h_E = h_v(P_0) \tag{13-13}$$

Integration of Eqs. (13-7)–(13-9) yields P_0, M^*, and U^* in terms of t^*. Sample results are shown in Figs. 13-9 and 13-10 for an initially saturated system at 1000 psia. The three characteristic blowdowns are shown for $\bar{f}L/D$ values from 0 to 100. The results in Figs. 13-9 and 13-10 can be used to estimate system pressure, mass, and energy at various times during blowdown.

Liquid blowdown corresponds to mass loss from a low point on the system if vapor entrainment is minor. Mixture blowdown applies to rapid mass loss from the system when vapor is formed faster than it can separate from the liquid. Vapor blowdown occurs from a high point on the physical

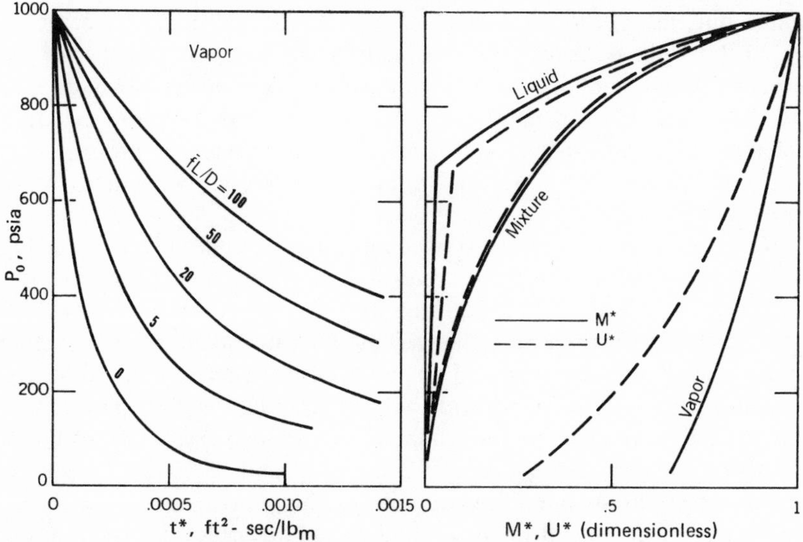

FIG. 13-10. Blowdown from 1000 psia steam/water reference system.[67]

TABLE 13-2
System Data for Blowdown Tests Used by Moody[43]

Description	Bodega 40	Bodega 22	Bodega 16	Humboldt 22
Vessel initial pressure, psia	1250.0	1250.0	1250.0	1250.0
Saturation temperature °F	572.4	572.4	572.4	572.4
Saturation enthalpy, Btu/lb_m	578.6	578.6	578.6	578.6
Initial water temperature °F	537.4	572.4	572.4	572.4
Initial water enthalpy, Btu/lb_m	532.5	578.6	578.6	578.6
Initial water subcooling, Btu/lb_m	46.1	0	0	0
Flashing pressure, psia	940.0	1250.0	1250.0	1250.0
Nozzle throat area, in.^{2}	8.25	8.25	—	—
Orifice throat area, in.^{2}	—	—	20.6	0.95
Back pressure, atm	1.0	1.0	1.0	1.0
Upstream pipe area, in.^{2}	115.0	115.0	115.0	26.0

system, which is slow enough for vapor separation without liquid entrainment.

The model results were compared with blowdown test data from a full-scale, $\frac{1}{112}$ segment of the Bodega Bay atomic power plant[68] and a full-scale, $\frac{1}{48}$ segment of the Humboldt Bay plant.[69] A list of conditions for these tests is given in Table 13-2. The vessel initially contained saturated steam–water at 1250 psia. The test results are shown in Figs. 13-11 and 13-12. As pointed out earlier, the value of $\bar{f}L/D$ must be estimated for a meaningful comparison with the theory. This estimation can be made by either of two methods: summation of standard single-phase geometric loss coefficients plus the actual $\bar{f}L/D$ components associated with the system tested, or calculation of an equivalent $\bar{f}L/D$ from measured irreversible pressure drop at a known cold-water flow rate. Moody chose the latter method for his comparison.

The theory with $\bar{f}L/D = 1.0$ is compared with Bodega Test 21 in Fig. 13-11. A sharp initial dip in test pressure which essentially recovers in 1 sec appears in all saturated blowdown tests. This phenomenon, not explained by the model, is similar to that noticed by other investigators.[44,46] Moody explains the characteristic as a combination of two effects: initial discharge of slightly subcooled liquid which is not restricted by the two-phase mechanism and the delay time for vapor bubbles to form and expand in the liquid.

Also evident in the test data of Figs. 13-11 and 13-12 is a "knee" or sudden increase in the rate of pressure drop. This is predicted by the theory, as shown in Figs. 13-9 and 13-10, when saturated water blowdown is followed by saturated vapor blowdown. The much larger blowdown area of Test

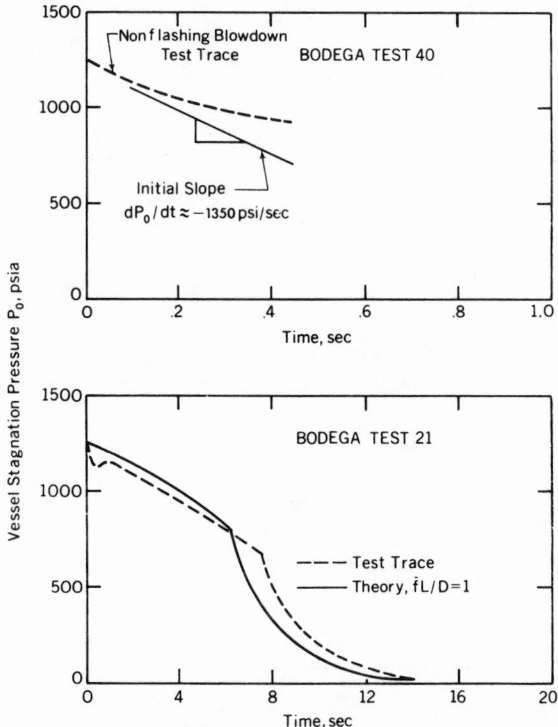

FIG. 13-11. Comparison of Moody's theory with experimental blowdown results.[67] For Bodega test 40, the initial subcooled water was used to estimate the nonmaximum water blowdown rate to obtain the equivalent $\bar{f}L/D$; initial water temperature 537°F; flashing pressure 940 psia; nozzle area 8.25 in^2. For Bodega test 21, the saturated blowdown nozzle area was 8.25 in^2.

16 (Fig. 13-14) nearly caused the disappearance of the knee. A large nozzle might cause such rapid vapor formation that a nearly homogeneous mixture would fill the vessel and the knee would disappear completely.

Further, it was concluded that it is unlikely that a high concentration of vapor would occupy the vessel lower region during blowdown while liquid was still present. Therefore, steam/water action in the vessel for all Bodega and Humboldt tests should lie somewhere between a homogeneous mixture filling the vessel and completely separated phases with water occupying the vessel lower region until fully expelled.

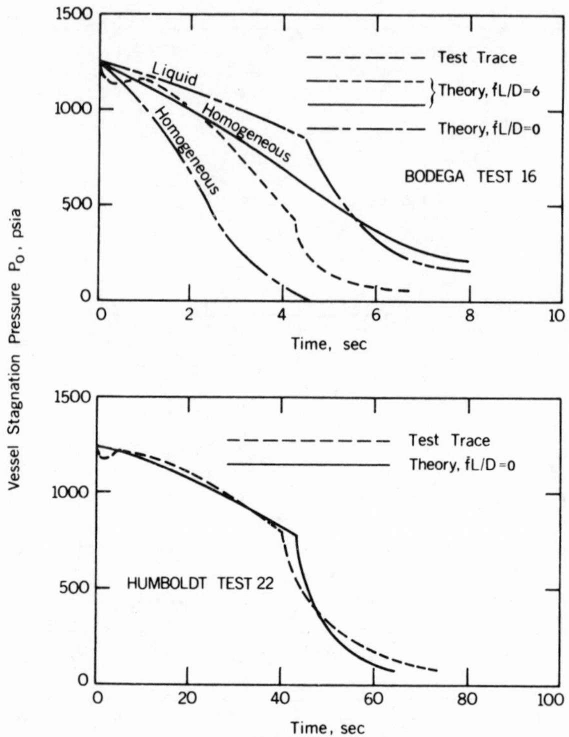

FIG. 13-12. Comparison of Moody's theory with blowdown results illustrating the effect of blowdown area.[67] For Bodega test 16, the saturated blowdown orifice area was 20.6 in.^{2} For Humboldt test 22, the saturated blowdown orifice area was 0.95 in.^{2}

13.6 SUMMARY

A number of recent studies[70–78] have attempted to simulate the liquid/vapor behavior of a full-scale system during discharge. The advantages of modeling are obvious. Grolmes and Fauske[70] used Freon 11 to simulate the expulsion of sodium from complex geometries. Nix and Vachon[71] performed blowdown tests from 100 psia with water in a small-scale system, and obtained many similar features of full-scale systems. Results of a mathematical model were compared favorably to those of Moody.[43] Other mathematical models[72,78] have contributed to the overall understanding of the blowdown process.

Other large-scale tests have been recently completed on the Containment

Systems Experiment (CSE) recently completed at Battelle Northwest for the U.S. Atomic Energy Commission.[79,82] These studies were conducted on a design basis reactor piping accident to check the safety analysis methods applied to large reactors for licensing purposes. Theoretical analyses of the CSE reactor system revealed that empirical coefficients were necessary to adjust the predictions to agree with the observed results.

From the large number of recent studies in this field, and from the varying results that have been obtained, the reader may gain an understanding of the complexities of two-phase flow and pressurized discharge. Clearly, a great deal remains to be done to coordinate the enormous amount of experimental data and the theories that have been forwarded. This survey has been able to discuss only a few of these investigations in detail, and the authors are indebted to those investigators from whom we have borrowed freely. For those who desire to enter the field, the authors recommend detailed study of the list of references.

In view of the paucity of cryogenic data on the aforementioned transient two-phase flow phenomena, the preceding discussion has through necessity considered mainly high-boiling-temperature fluids. The conclusions drawn can therefore give only qualitative information regarding cryogenic fluids. Until more data are experimentally obtained the designer is cautioned to apply the knowledge cited in this chapter with a certain degree of engineering intuition.

13.7 NOMENCLATURE

C_P = specific heat
$C_{P\mathrm{vf}}$ = specific heat of vapor evaluated at film temperature
D = diameter, hydraulic diameter
$\bar{f}$ = average Darcy friction factor
G = mass velocity per unit area
h = heat transfer coefficient
h_{co} = convective coefficient in film boiling
h_E = blowdown escape value of enthalpy
h_e = specific enthalpy at exit condition
h_{fg} = latent heat of vaporization
h_r = radiative coefficient in film boiling
K = thermal conductivity
k_{vf} = thermal conductivity of vapor evaluated at film temperature
L = length
M = mass
M_l = mass of liquid

M_v = mass of vapor
N_w = Weber number
ΔP = pressure oscillation amplitude, psi
P_{ext} = external pressure
P_m = mean pressure
P_{vp} = vapor pressure
P_0 = pressure at incipience of boiling
r = radius of cavity mouth
ΔT = difference between surface temperature and saturation temperature
T_b = bulk temperature
T_l = temperature of liquid
T_{sat} = saturation temperature
T_{sm} = mean saturation temperature, °F
T_w = surface temperature, °F
t = time
U = internal energy
U_∞ = sphere velocity
u_i = initial value of specific internal energy
u_l = specific internal energy of liquid
u_{lv} = specific internal energy of liquid vapor
v_l = specific volume of liquid
v_{lv} = specific volume of liquid vapor
$(\,)^*$ = ratio of quantity to initial value

Greek Letters

μ_{vf} = dynamic viscosity of vapor evaluated at film temperature
θ = contact angle
ρ_l = density of liquid
ρ_{vf} = density of vapor evaluated at film temperature
σ = interfacial surface tension
ϕ = conical angle of cavity

13.8 REFERENCES

1. M. T. Cichelli and C. F. Bonilla, *AIChE Trans.* **41**, 755 (1945).
2. C. Corty and A. S. Foust, *Chem. Eng. Progr. Symp. Series* **51**(17), 1 (1955).
3. P. Griffith and J. D. Wallis, *Chem. Eng. Progr. Symp. Series* **56**(30), 49 (1960).
4. S. G. Bankoff, *Chem. Eng. Progr. Symp. Series* **55**(29), 87 (1959).
5. S. G. Bankoff, *AIChE J.* **8**(1), 63 (1962).
6. D. S. Cryder and A. C. Finalborgo, *AIChE Trans.* **33**, 346 (1937).
7. F. Kreith and M. J. Summerfield, *ASME Trans.* **71**, 805 (1949).

8. I. A. Raben, R. T. Beaubouef, and G. E. Commerford, *Chem. Eng. Progr. Symp. Series* **61**(57), 249 (1965).
9. D. A. Huber and J. C. Hoehne, *ASME Trans. J. Heat Transfer* **85C**, 215 (1963).
10. J. H. Lienhard and V. E. Schrock, *J. Heat Transfer, ASME Trans.* **85C**(3), 261 (1963).
11. J. H. Lienhard and K. Watanabe, *J. Heat Transfer, ASME Trans.* **88C**(1), 94 (1966).
12. J. H. Lienhard and V. E. Schrock, *Intern. J. Heat Mass Transfer* **9**(4), 355 (1966).
13. S. E. Isakoff, in *Ninth Heat Transfer and Fluid Mechanics Institute*, Reprints of Papers, Stanford Univ. Press (1956), p. 15.
14. M. Markels, R. Durfee, and S. Richardson, NYO-9500, Atlantic Research Corporation, Alexandria, Virginia (1960).
15. J. H. Gibbons, Ph.D. Dissertation, Univ. of Pittsburgh (1961).
16. S. W. Wong and W. Y. Chon, *AIChE J.* **15**(2), 281 (1969).
17. D. A. DiCicco and R. J. Schoenhals, *ASME Trans.* **86C**(3), 457 (1964).
18. J. L. Carson, Ph.D. Dissertation, Purdue Univ. (1963).
19. L. C. Burmeister and R. J. Schoenhals, *Intern. J. Heat Mass Transfer* **2**, 371 (1969).
20. K. E. McCoy, R. J. Schoenhals, and E. R. F. Winter, in *Proceedings of the 1970 Heat Transfer and Fluid Mechanics Institute*, Stanford Press (1970), p. 4.
21. P. W. McFadden and P. Grassman, *Intern. J. Heat Mass Transfer* **5**, 169 (1963).
22. R. L. Semeria, *Proc. Inst. Mech. Eng.* **1962**, 57.
23. A. E. Bergles and W. G. Thompson Jr., *Intern. J. Heat Mass Transfer* **13**, 55 (1970).
24. H. Merte, Jr. and J. A. Clark, *ASME Trans., J. Heat Transfer* **86C**(3), 351 (1964).
25. T. M. Flynn, J. W. Drager, and J. Roos, in *Advances in Cryogenic Engineering*, Vol. 7, Plenum Press, New York (1962), p. 539.
26. T. H. Frederking and J. A. Clark, in *Advances in Cryogenic Engineering*, Vol. 8, Plenum Press, New York (1963), p. 501.
27. L. A. Bromley, *Chem. Eng. Progr.* **46**, 221 (1970).
28. N. Zuber, AECU-4439, Physics and Mathematics (1959).
29. N. Zuber, *ASME Trans., J. Heat Transfer* **80C**, 711 (1958).
30. P. J. Berenson, *ASME Trans., J. Heat Transfer* **83C**, 351 (1961).
31. L. C. Witte, L. Baker, Jr., and D. R. Haworth, *ASME Trans., J. Heat Transfer* **90C**, 394 (1968).
32. L. C. Witte, *ASME Trans., J. Heat Transfer* **90C**, 9 (1968).
33. S. Sideman, *Ind. Eng. Chem.* **58**(2), 54 (1966).
34. A. I. Krakoviak, ORNL TM-618 (1963).
35. L. S. Tong, *Boiling Heat Transfer and Two-Phase Flow*, John Wiley and Sons, New York (1965).
36. R. R. Kepple and T. V. Tung, "Two-Phase Gas-Liquid System: Heat Transfer and Hydraulics—An Annotated Bibliography" ANL-6734, USAEC Report (1963).
37. J. A. Boure, A. E. Bergles, and L. S. Tong, ASME paper 71-HT-42.
38. E. Giffen, *Engineering* **150**, 134–136, 154–155, 181–183 (1940).
39. J. Kestin and J. S. Glass, *Aircraft Engineering* **23**, 300 (1951).
40. R. Progelhof and J. A. Owczarek, *AIAA J.* **1**, 2182 (1963).
41. J. Kestin and J. S. Glass, *Proc. Inst. Mech. Eng.* **161**, 250 (1949).
42. R. C. Progelhof, *AIAA J.* **2**, 137 (1964).
43. F. J. Moody, *ASME Trans. J. Heat Transfer* **88C**(3), 285 (1966).
44. J. R. Howell and K. J. Bell, *Chem. Eng. Progr. Symp. Series* **59**, 88 (1963).
45. G. E. Tanger, R. I. Vachon, and R. B. Pollard, in *Proc. Third Intern. Heat Transfer Conference*, Vol. IV (1966), p. 38.

46. P. M. Ordin, S. Weiss, and H. Christenson, in *Advances in Cryogenic Engineering*, Vol. 5, Plenum Press, New York (1960), p. 481.
47. H. S. Isbin and G. R. Gavalas, in *Proc. 1970 Heat Transfer and Fluid Mechanics Institute* (1962), p. 126.
48. J. K. Ferrell and J. W. McGee, "Two-Phase Flow Through Abrupt Expansions and Contractions," Final Report, Volume III on USAEC Contract At-(40-1)-2950, North Carolina State Univ. Raleigh, North Carolina (June 1966).
49. H. K. Fauske, ANL-6633, Argonne National Laboratory, Argonne, Illinois (October (1962).
50. H. K. Fauske, ANL-6779, Argonne National Laboratory, Argonne, Illinois (October 1963).
51. K. D. Timmerhaus, *Applied Cryogenic Engineering*, John Wiley and Sons (1962).
52. W. Bornhorst and A. Shavit, "Study of the Boundary Conditions at a Liquid-Vapor Interface Through Irreversible Thermodynamics," Quarterly Progress Report, Massachusetts Institute of Technology, Cambridge, Massachusetts (September 1965).
53. P. Griffith and G. Snyder, "The Mechanism of Void Formation in Initially Subcooled Systems," Report 9041-26 on Contract Nonr-1841 (39), Massachusetts Institute of Technology, Cambridge, Massachusetts (September 1963).
54. W. A. Olsen, NASA TN-D-3219 (March 1966).
55. K. D. Coughren, *Pressurizing Vessel Performance Equations*, BNWL-116, Battelle-Northwest, Richland, Washington (August 1965).
56. H. K. Fauske, *Chem. Eng. Progr. Symp. Series* **59**(61), 210 (1965).
57. S. Levy, *ASME Trans., J. Heat Transfer* **87**(1), 53 (1965).
58. F. J. Moody, *ASME Trans., J. Heat Transfer* **87C**(1), 134 (1965).
59. L. S. Tong, in *Proc. Fourth Annual Symposium on Heat Transfer*, Idaho Falls, Idaho (May 1967).
60. F. W. Albaugh, J. J. Fuquay, H. Harty, E. E. Voiland, and D. C. Worlton, "Nuclear Safety Quarterly Report, " BNWL-433, Battelle-Northwest, Richland, Washington (July 1967).
61. R. T. Allemann, (Ed.), "Reactor Safeguards Experimental Unit," Battelle-Northwest, Richmond, Washington, Private communication (1967).
62. K. A. Dietz (Ed.), "Quarterly Technical Report STEP Project: (October 1965–December 1965)," IDO-17167 (September 1966).
63. K. A. Dietz (Ed.), "Quarterly Technical Report STEP Project: (January 1966–March 1966)," IDO-17186 (November 1966).
64. K. A. Dietz (Ed.), "Quarterly Technical Report STEP Project: (April 1966–June 1966)," IDO-17187 (January 1967).
65. R. B. Pollard, MS Thesis, Auburn Univ., Auburn, Alabama, August 1965 (unpublished).
66. K. A. Dietz (Ed.), "Quarterly Technical Report STEP Project: (July 1966–September 1966)," IDO-17213 (April 1967).
67. F. J. Moody, "Maximum Two-Phase Vessel Blowdown from Pipes," APED-4827, 65APE4, General Electric Company (April 1965).
68. "Preliminary Hazards Summary Report, Bodega Bay Atomic Park Unit No. 1," Pacific Gas and Electric Company (December 1962).
69. C. H. Robbins, "Tests of a Full Scale 1/48 Segment of the Humboldt Bay Pressure Suppression Containment," GEAP-3596, (November 1960).
70. M. A. Grolmes and H. K. Fauske, ASME paper 70-HT-24 (1970).
71. G. H. Nix and R. I. Vachon, Report X, Contract NAS8-11234, (March 1969).

72. H. K. Fauske and M. A. Grolmes, ASME paper 70-HT-21 (1970).
73. H. K. Fauske, *Reactor and Fuel-Processing Technology* **11**(2), p. 84 (1968).
74. F. J. Moody, ASME paper 69-HT-31 (1969).
75. R. E. Henry, *Nucl. Sci. Eng.* **41**, 336 (1970).
76. F. J. Moody, *Trans. ASME, J. Eng. for Power* **91**(1), 53 (1969).
77. A. N. Nahavandi, *Nucl. Sci. Eng.* **36**, 159 (1969).
78. F. J. Moody, *Trans. ASME, J. Heat Transfer* **91**(3), 371 (1969).
79. "Dissolved Gas Influence on a PWR Design Basis Decompression," AEC Research and Development Report, BNWL-SA-2746 (January 1970).
80. "Nuclear Safety Quarterly Report, November, December 1969, January 1970, for USAEC Division of Reactor Development and Technology," BNWL-1315-1, UC-41 (March 1970).
81. N. P. Wilburn, AEC Research and Development Report, BNWL-1295, UC-80 (April 1970).
82. "Experimental High Enthalpy Water Blowdown From a Simple Vessel Through a Bottom Outlet," AEC Research and Development Report, BNWL-1411, UC-80 (June 1970).

PART III

RADIATION AND HELIUM II HEAT TRANSPORT

RADIATIVE PROPERTIES 14

K. E. TEMPELMEYER and J. A. ROUX

14.1 INTRODUCTION

The radiative properties of many different types of materials and surfaces have been catalogued over the years for use in making heat transfer calculations. References 1–3 give summaries of much of this information together with detailed discussions of radiative heat transfer procedures and methods. In general, the solution of engineering problems concerning the radiative interchange between surfaces is characterized by a number of idealizing assumptions:

1. Surface reflection is either perfectly specular or perfectly diffuse.
2. Surface radiant emission is only diffuse.
3. The spectral or monochromatic emissivity or reflectivity is independent of wavelength.
4. No polarization effects occur in the reflected radiation.

Although these assumptions are very broad, they result in an acceptable accuracy for most radiative heat transfer calculations.

However, if the temperature of a surface is about 4°C below that specified by the vapor pressure curve of a surrounding gas, the solid phase of a gas is formed on the surface (a cryodeposit) and the radiative properties of the substrate and the thin solid-phase layer begin to change. As a result, the

---

K. E. TEMPELMEYER The University of Tennessee Space Institute, Tullahoma, Tennessee.
J. A. ROUX Arnold Engineering Development Center, Arnold AFS, Tennessee.

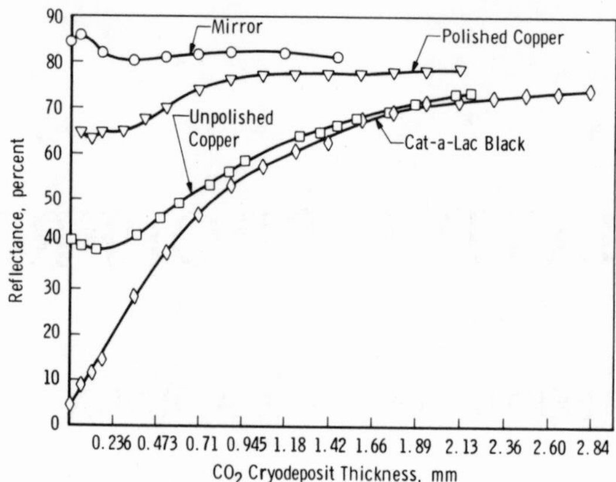

FIG. 14-1. Reflectance of various cryopanel materials with and without CO_2 frost. Angle of incidence 30°.[5]

surface radiative characteristics may begin to deviate from the above assumptions, and radiative heat transfer analysis and calculations become much more complex.

Over the past few years experimental and theoretical work has been carried out to determine the effects of cryodeposits on the reflectance of surfaces.[4–10] Most of the work to date has concerned the solid phase of CO_2 and H_2O on a variety of substrates cooled to 77 K by liquid nitrogen. Figure 14-1 illustrates some of the changes in total reflectance as a CO_2 cryodeposit forms. Additional measurements of the radiative properties of water frost have also been carried out, notably by Wood *et al.*[8,11] These results cover the UV and near-IR wavelength range and are quite similar to those illustrated here for CO_2. Much less attention has been given to the radiative properties of CO_2 and H_2O at temperatures below 77 K. However, Tempelmeyer has shown[12] that the solid phase of CO_2 goes through a structural change at about 30–35 K and Wood *et al.*[11] subsequently detected a similar effect for H_2O as its solid phase was cooled to about 145 K. Consequently, the radiative properties of solidified gases are also dependent on their temperature. This complication is discussed further below. Little attention has been given to a detailed measurement of the radiative properties of the solid phase of gases other than CO_2 and H_2O, but they have been observed to possess many of the properties which are described here for the CO_2 solid state.

14.2 EFFECTS OF THICKNESS OF THE SOLIDIFIED LAYER

Initially, when the solid layer is very thin, it is relatively transparent and the *total reflectance* is dominated by the substrate (Fig. 14-1). As the CO_2 cryodeposit reaches a thickness of about 2 mm, the effect of the substrate material becomes negligible and the surface radiative properties are governed by those of the solidified gas. Consequently, effective wall reflectance in a cryogenically pumped chamber becomes a function of deposit thickness, which probably is not uniform throughout the chamber. In addition, the energy reflected from a cryodeposit also exhibits an angular variation. Figure 14-2 shows the angular dependence for the angular-hemispherical reflectance.

The angular-hemispherical reflectance ρ_{ah} is defined as the ratio of the flux reflected from an infinitesimal element of area dA collected over the entire hemispherical space to the incident flux, which is a beam oriented at a specific angle relative to the surface normal. At all cryodeposit thicknesses the reflectance is obviously not diffuse; therefore, the equations for radiant interchange between diffuse reflecting surfaces are not applicable. The increase in reflectance with cryodeposit thickness is attributed to internal scattering.

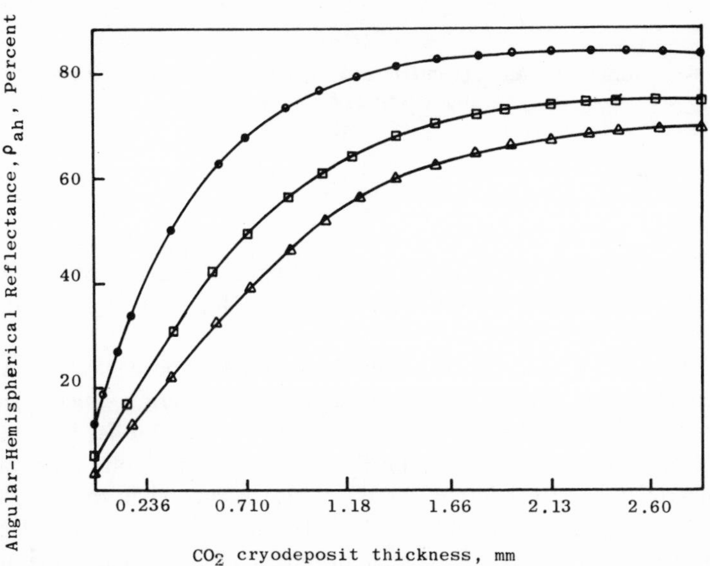

FIG. 14-2. Variation of the angular-hemispherical reflectance of CO_2 cryodeposits with difference angles of incidence of a light source. Substrate: Cat-a-lac black. Angles of incidence 0–60°: □—40°; △—10°.

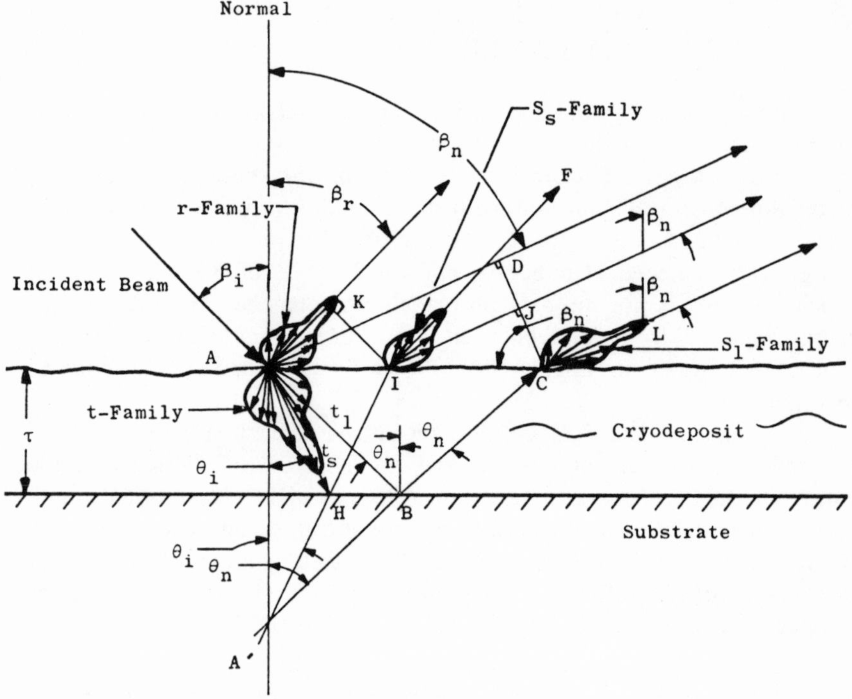

FIG. 14-3a. Sketch of scattering interference model. (See Ref. 13 for analytical model and meaning of notation).

The greater reflectivities at the larger viewing angles are due to increased scattering attendent to longer path lengths of light passing through the solidified gas.

The scattering of radiation within the solid cryodeposit has been studied by Tempelmeyer *et al.*[13] One form of internal scattering which occurs is sketched in Fig. 14-3. This internal scattering phenomenon has been utilized to develop a means of determining the thickness of the solid layer and its rate of growth. Another interesting effect also occurs when the solid layers are very thin, less than about 0.2 mm. The initial decrease in reflectance shown in Fig. 14-4 is due to a change in refractive index. Before the deposit forms, the radiant intensity travels from a medium which has a refractive index of unity (vacuum) directly to the substrate. After a very thin deposit forms, the intensity must cross a thin film which has a refractive index

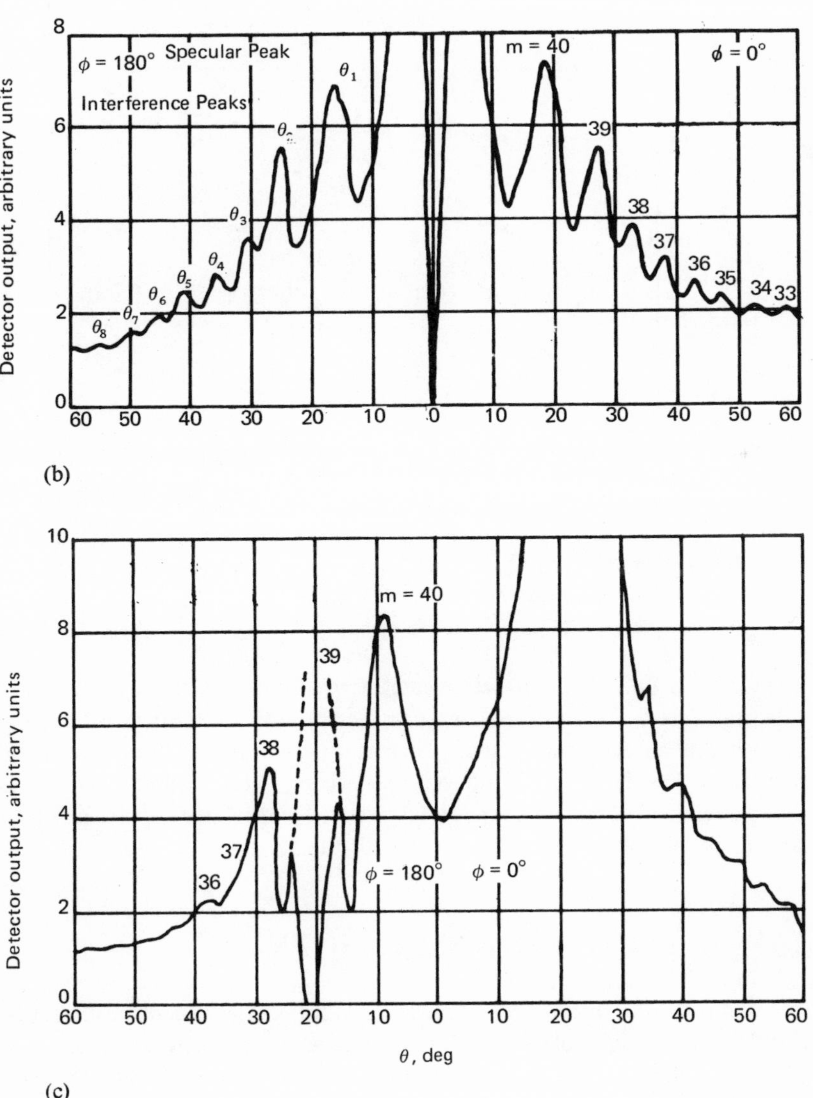

FIG. 14-3b, c. Scattering interference pattern. $p = 4 \times 10^{-4}$ torr; $\zeta = 180°$; $\dot{\tau} = 2.25$ μ/min; $\lambda = 0.9\ \mu$; $n = 1.38$. (b) $\psi = 0°$, $\tau = 13.34\mu$; (c) $\psi = 21.5°$, $\tau = 13.2\ \mu$.

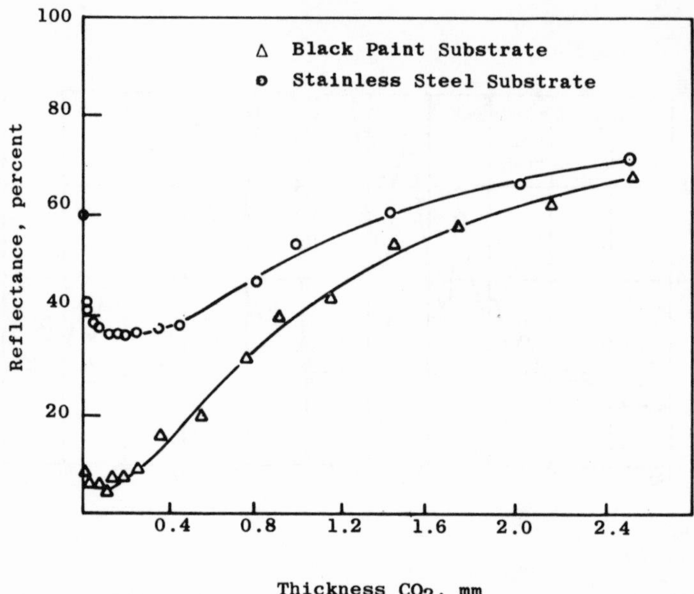

FIG. 14-4. Variation of the total reflectance of CO_2 cryodeposit as a function of its thickness. $\theta = 20°$, $\lambda = 0.75\ \mu$.

greater than unity.[6] This relative refractive index change at the deposit–substrate interface causes a sharp drop in the wall reflectance. As the cryodeposit thickness increases, internal scattering within the cryodeposit begins to become prominent. This internal scattering produces an effective increase in wall reflectance because energy is scattered out of the deposit before it is able to penetrate into the absorbing substrate.

14.3 EFFECTS OF WAVELENGTH

Figure 14-5 indicates the spectral variation of the effective reflectance of a CO_2 deposit on a black substrate for various deposit thicknesses. Again the initial decrease due to the relative refractive index change is evident. At larger thicknesses the reflectance increases due to scattering. In some wavelength regions the deposit is seen to be highly absorbing. However, the net effect over all wavelengths for solar irradiation is a reflectance increase as shown in Ref. 8 for a black substrate. Similar data for solid CO_2 deposited on a stainless steel substrate are given in Fig. 14-6. As shown in Fig. 14-1, the total reflectance in the visible wavelength range becomes independent

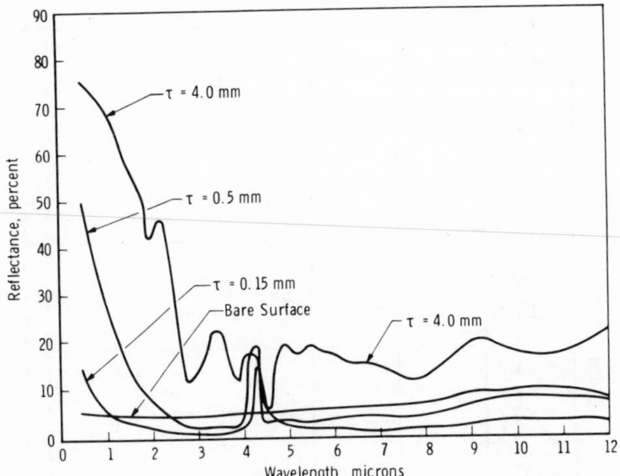

FIG. 14-5. Infrared reflectance of CO_2 cryodeposits on black epoxy paint. $\theta_v = 10°$.[8]

of the substrate upon which the solidified gas is formed as the deposit thickens. Comparing Figs. 14-5 and 14-6 illustrates, however, that the spectral reflectance is wavelength dependent, particularly in the near infrared. At wavelengths greater than 5 μm the spectral reflectance of a 4-mm-thick, solid CO_2 layer on a stainless steel surface is about $2\frac{1}{2}$ times greater than that on a black surface. Thus, in the near infrared the radiative properties of the

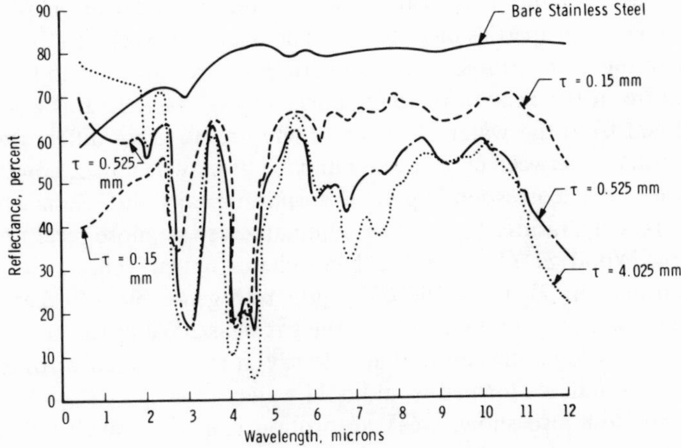

FIG. 14-6. Infrared reflectance of CO_2 cryodeposits on stainless steel. $\theta_v = 10°$.[8]

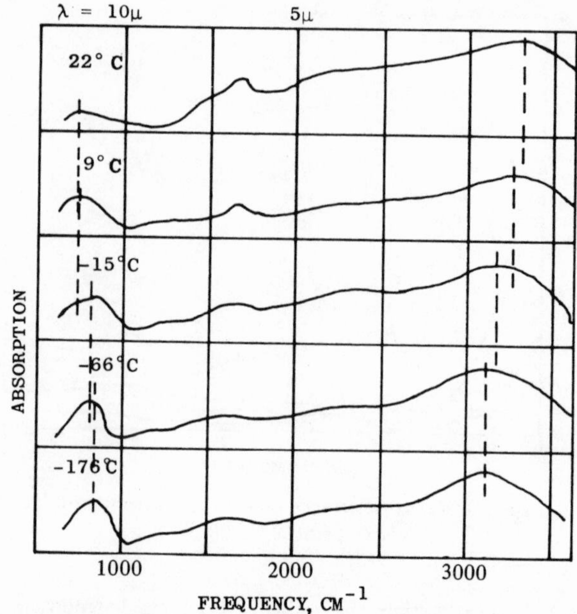

FIG. 14-7. Effect of temperature on the absorption spectra of thin films of H_2O ice condensed on a cold plate.[14]

solid gas phase are not only wavelength dependent, but they also are dependent to some degree on the substrate material.

Figures 14-7 and 14-8 also summarize the dependence on wavelength of the radiative properties of water ice. The relative absorption of water and then ice in the near infrared at atmospheric pressure and close to the freezing temperature in the near infrared is shown in Fig. 14-7. In this case the ice was formed from the water phase and there are no gross differences in the IR absorptivity between chilled water and ice. It may be noted, however, that the wavelengths corresponding to absorption peaks shift somewhat with decreasing temperature. Figure 14-8 summarizes some more recent measurements by Wood *et al.*[11] of the hemisphere-angular reflectance of H_2O cryodeposited on 77 K (−195°C), liquid nitrogen-cooled surface. In this case the ice was formed directly from the gas phase. Water ice formed in this way has a very low reflectance, that is to say, it is more highly absorptive of IR radiation than ice formed from liquid water.

Figure 14-8 also shows weak absorption bands at wavelengths of 1.05 and 1.25 μm and strong bands near 1.5, 2.0, and 3.0 μm. A broad region of strong absorption is also observed between 3 and 12 μm. Figure 14-6 shows

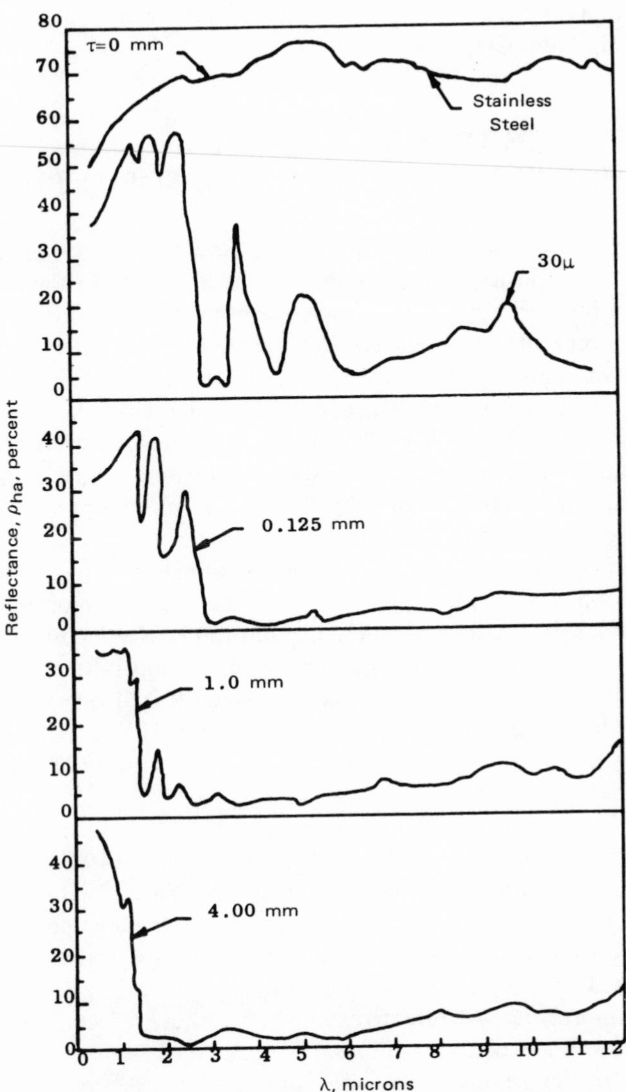

FIG. 14-8. Spectral reflectance of amorphous H_2O cryodeposits formed on a polished stainless steel substrate. Deposition pressure 2.8×10^{-2} torr; $\dot{\tau} = 0.825$ μ/sec; $\theta = 10°$.[11]

that CO_2 ice exhibits a weak absorption band at about 2 μm and strong bands near 2.5 and 4.0 μm. In general, these ranges of strong absorption of electromagnetic radiation are very similar to the ranges of absorption of the gas phases of H_2O and CO_2.

14.4 EFFECTS OF STRUCTURAL CHARACTERISTICS OF THE SOLID PHASE

In addition to the angular, spectral, substrate, and wavelength dependences of cryodeposits, further complications arise due to their structure. Besides the high-pressure forms of ice, there are three additional forms of ice which occur at low pressures. Ice consists of two crystalline forms, hexagonal and cubic, and a third form which is referred to as amorphous (i.e., no crystalline structure). The conditions under which each of the three types can be formed have been the subject of much discussion and a wide range of data have been gathered by many investigators. These investigations are summarized in tabular form by Sugisaki *et al.*[14] For ice formed at atmospheric pressure from liquid, only hexagonal ice has been observed. The general consensus is that water deposits formed in a vacuum from the vapor state on surfaces that are 115 K or colder will be amorphous. If the test surface is between approximately 115 and 150 K, the deposit formed will be of cubic crystalline form.[8] For test surface temperatures higher than 150 K the frosts will form in the hexagonal state. It is also generally agreed that these temperatures of formation are not rigid. Hence, mixtures of the different crystalline ice forms can be observed in the vicinity of the temperature-dependent transition.

The differences in the radiative properties of water ice formed in different ways may be inferred by comparison of Figs. 14-7 and 14-8 and are due to the different structural characteristics of the ices. Those data given in Fig. 14-7 were obtained from water ice formed from the liquid state resulting in a crystalline structure, probably hexagonal, while those results in Fig. 14-8 were obtained from an amorphous formation. In general, it is reasonable to expect the amorphous structures to exhibit more diffuse reflectance properties, but they are still not perfectly diffuse reflectors.

It is also important to note that structure changes may occur as an amorphous frost is allowed to warm. In this case an amorphous frost or ice may assume a variety of crystalline structures. Tempelmeyer[12] was perhaps the first to note this phenomenon, while investigating the sorption capacities of amorphous forms of the solid phase of CO_2 .The fact that H_2O deposits have various structures means that the hemispherical-angular reflectance of

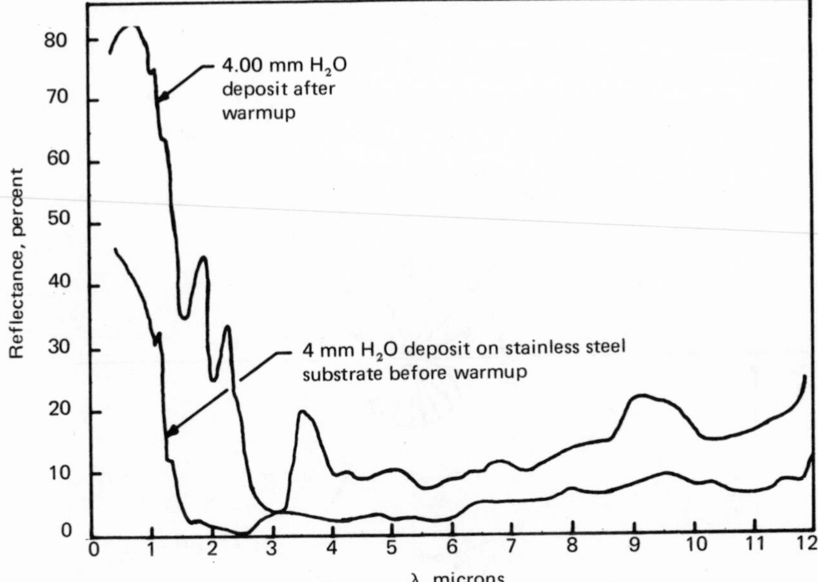

FIG. 14-9. Comparison of reflectances of a 4-mm H_2O deposit on stainless steel before warmup (amorphous structure) and after warmup (hexagonal structure). *Steps in the experimental procedure*: (1) After prewarmup data were obtained, the system was pressurized to 740 torr with bone-dry nitrogen for about 1 hour; (2) the chamber was evacuated to 10^{-4} torr; (3) liquid nitrogen was turned off and the test surface allowed to warm up to above 200 K; (4) the test surface was cooled back down to 100 K; (5) spectral reflectance measurements were made.

these deposits can vary quite strongly. Wood *et al.*[11] have documented the change of reflectance of a H_2O cryodeposit as its structure changes due to warming. Their results are given in Fig. 14-9. In this particular instance the hemispherical reflectance of the H_2O frost layer almost doubled as the amorphous frost took on a more compact crystalline form. A change in structure can cause a change in the number and size of the scatterers. The radiant energy is believed to be scattered by voids within the cryodeposit, because Muller[9] has reported amorphous ice to have a specific gravity of 0.81.

14.5 THEORETICAL DEVELOPMENTS

All of the experimental results indicate that many factors affect the radiative properties of the solid phase of gases. Thus, theoretical models predicting their radiative properties are difficult to formulate. Nevertheless

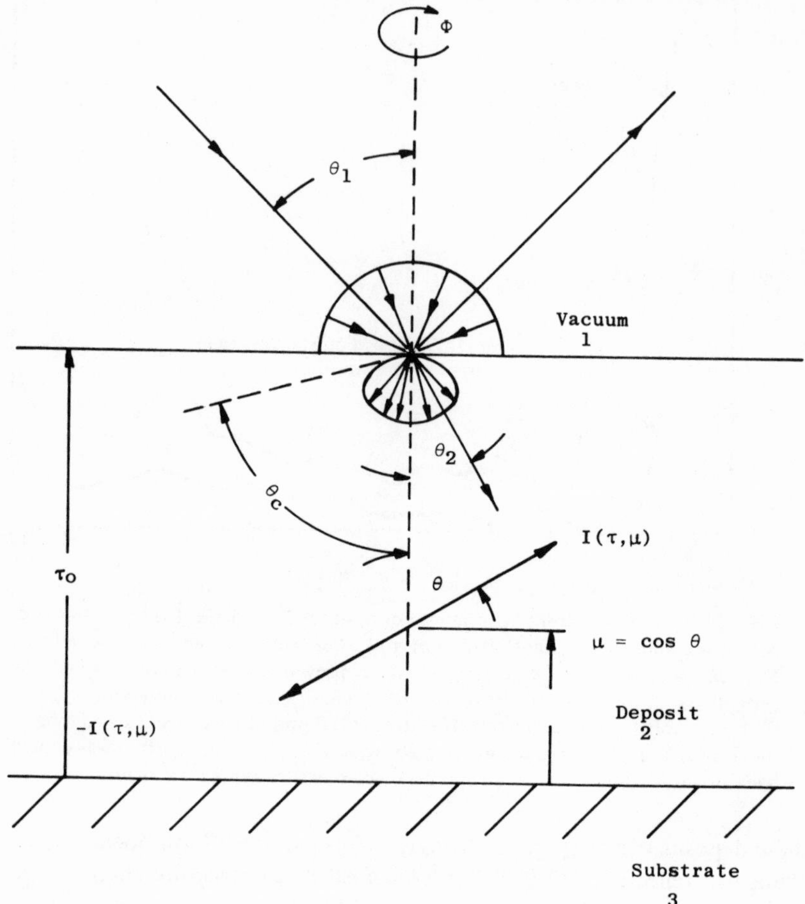

FIG. 14-10. Definition of notation with Merriam's model.

some progress has been made. Merriam[10] analytically investigated the radiative characteristics of condensed gas deposits. He considered an absorbing and scattering medium bounded by a specular surface. The geometry shown in Fig. 14-10 was used. The top interface was assumed to reflect and transmit radiant energy in accordance with Fresnel's equations. The radiative transport equation for isotropic scattering was solved,

$$\mu = dI/d\tau = -I + \tfrac{1}{2}\omega \int_{-1}^{1} I(\tau, \mu')\, d\mu' \tag{14-1}$$

where $\mu' = \cos\theta$ and $\omega = \tau/(\tau + k)$, τ is the monochromatic scattering coefficient, and k is the monochromatic absorption coefficient. The boundary conditions used by Merriam were

$$\begin{aligned} I(\tau_0, -\mu) &= [1 - \rho_{12}(\mu_1)]m_2^2 + \rho_{21}(\mu)I(\tau_0, \mu) \\ I(0, \mu) &= \rho_{23}(\mu)I(0, -\mu) \end{aligned} \tag{14-2}$$

where ρ_{12} represents the Fresnel reflectance in going from vacuum to the solid phase deposit, ρ_{21} the Fresnel reflectance in going from deposit to vacuum, and ρ_{23} the specular reflectance in going from deposit to substrate, and μ_1 and μ are related by Snell's law. The hemispherical-angular reflectance is given by

$$\rho_{ha}(\mu) = \rho_{12}(\mu_1) + \frac{1 - \rho_{12}(\mu_1)}{n_2^2} \frac{I(\tau_0, \mu)}{I_0} \tag{14-3}$$

where I_0 is the diffuse incident intensity and n_2 is the refractive index of the deposit. The solution is given in terms of the eigenvalues and eigenvectors of the resulting coefficient matrix. Results of this solution are shown in Fig. 14-11, which indicates the same trends as indicated by the experimental results shown in Fig. 14-4. The $\omega = 0.4$ curves would correspond to a highly

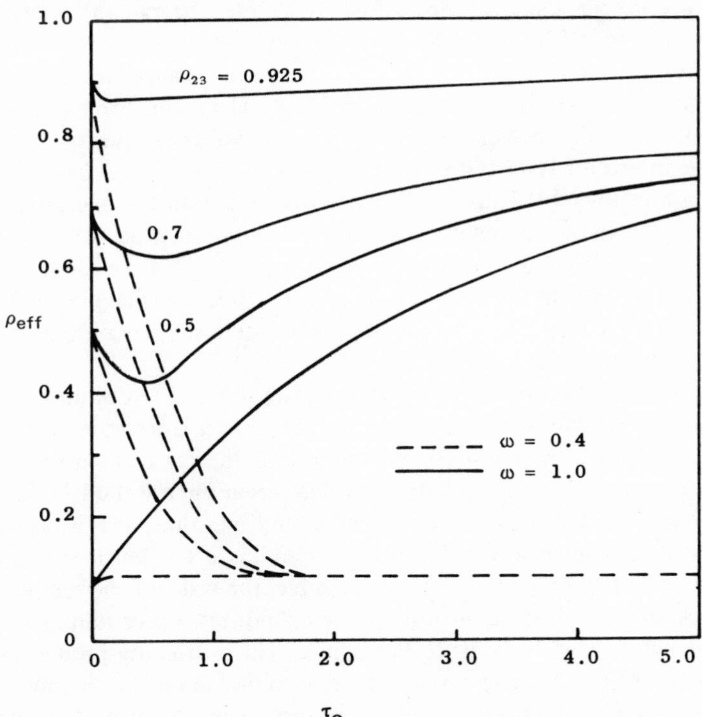

FIG. 14-11. Calculated reflectances using Merriam's model.

absorbing deposit such as H_2O in the infrared region. The $\omega = 1.0$ curves would correspond to a H_2O or CO_2 deposit in the visible region ($\lambda = 0.75\mu m$).

The decrease in reflectance for the high and moderately reflecting substrates as the optical thickness increases ($0 \leqslant \tau_0 \leqslant 1$) for $\omega = 1.0$ is due to the trapping of internally scattered energy incident upon the cryodeposit–vacuum interface at angles greater than the critical angle. This critical angle effect at the top interface is associated with Snell's law and Fresnel's reflectance law for smooth dielectrics. The $\omega = 0.4$ curves approach a value independent of the substrate. The reflectance decreases sharply due to the highly absorbing deposit.

14.6 APPLICATIONS

The angular, spectral, structural thickness, and substrate effects of a cryodeposit formed on a surface can be quite significant. All of these effects must be considered in computations requiring a high degree of accuracy. Also, since cryodeposits are not diffuse reflectors, the radiant interchange among surfaces with film deposits is difficult to compute for complex geometries. Even for simple geometries, the radiative properties of the cryodeposit are important for calculation purposes. The monochromatic absorption and scattering coefficients are not known for cryodeposits. Also, the refractive indices of H_2O and CO_2 cryodeposits are only available in limited wavelength regions.[6,9] Thus, reliable radiative heat transfer calculations can only be made in ranges where experimental data (such as some of those shown here) are available. As a result Mills and Smith[4] have carried out an interesting experiment to measure the overall effect of changing reflective properties of cryodeposits on the temperature of a simple configuration in a thermal-vacuum test.

In the space environment, the thermal radiation emitted by and reflected from a space vehicle does not return to the vehicle. Correct simulation of these conditions during environmental testing in space chambers would require that the amount of radiation incident upon the test model caused by the presence of the chamber walls be minimized; thus, the walls of the chamber should have low reflectance. To simulate the black background of space as closely and practicably as possible, the walls of thermal vacuum chambers are cooled to liquid nitrogen temperatures, hence minimizing the radiation emitted by the walls to the vehicle. The outgassing products (such as H_2O and CO_2) of the test model readily condense on the chamber walls and thus change the effective wall reflectance. The presence of condensed gases, as we have seen, can significantly affect the thermal balance of the

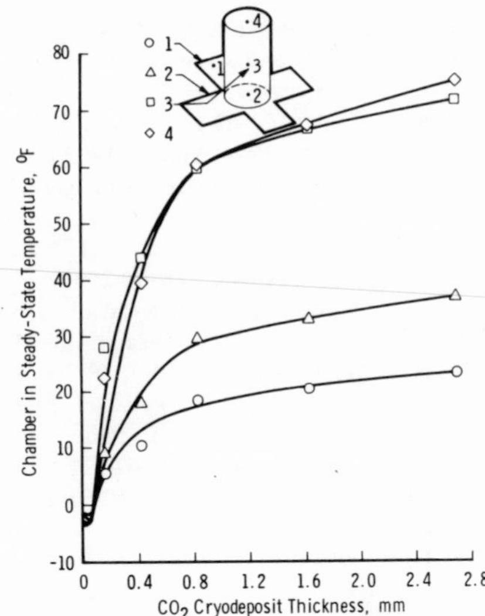

FIG. 14-12. Effect of CO_2 cryodeposit thickness on the thermal balance of a cylinder–paddlewheel spacecraft model.[4]

test model. Since the chamber walls absorb more energy when the reflectance is reduced by the presence of a thin film, the temperature of their test model, as shown in Fig. 14-12, initially undergoes a temperature decrease. The increased wall absorption allows less radiation to reflect back to the test model and hence the heat load on the model is reduced. This initial decrease in the equilibrium temperature of the test model is a direct result of the decrease in wall reflectance illustrated by some of the data presented in this chapter for very thin cryodeposits.

The radiative properties of the solid phase of gases not only may have a significant effect on the thermal control of space vehicles, but also on the long-time storage of cryogenic fluids. The presence of cryodeposits can affect experimental apparatus requiring cryogenically cooled mirrors, windows, or lenses; also, the thin deposits can affect low-temperature black bodies which are used as calibration standards. In performing thermal balance calculations under vacuum conditions, an engineer should not overlook the possible effects of cryodeposits due to outgassing. A good summary of other surface effects, such as roughness and polarization, is contained in Ref. 15.

14.8 REFERENCES

1. E. M. Sparrow and R. D. Cess, *Radiation Heat Transfer*, Brooks/Cole Publishing Company, Belmont, California (1966).
2. H. C. Hottel and A. F. Sarofim, *Radiative Transfer*, McGraw-Hill Book Co., New York (1967).

3. D. C. Hamilton and W. R. Morgan, NASA TN 2836 (1952).
4. D. W. Mills and A. M. Smith, *Spacecraft and Rockets* **7**, 374 (1970).
5. B. A. McCullough, B. E. Wood, and J. P. Dawson, Arnold Engineering Development Center TR-65-94, Arnold Air Force Station, Tennessee (August 1965).
6. K. E. Tempelmeyer and D. W. Mills, *J. Appl. Phys.* **39**(6), 2968 (1968).
7. A. M. Smith, K. E. Tempelmeyer, P. R. Muller, and B. E. Wood, *AIAA J.* **7**(12), 2274 (1969).
8. B. E. Wood, A. M. Smith, B. A. Seiber, and J. A. Roux, Arnold Engineering Development Center TR-70-205, Arnold Air Station, Tennessee (1970).
9. P. R. Muller, Ph.D. Dissertation, Univ. of Tennessee (June 1969).
10. R. L. Merriam, Ph.D. Dissertation, Purdue Univ. (June 1968).
11. B. E. Wood, A. M. Smith, J. P. Roux, and B. A. Seiber, *AIAA J.* **9**(9), 1836 (1971).
12. K. E. Tempelmeyer, *J. Vac. Sci. Tech.* **8**, 575 (1971).
13. K. E. Tempelmeyer, B. E. Wood, and D. W. Mills, Arnold Engineering Development Center, AEDC, TR-67-226 (December 1967).
14. M. Sugisaki, H. Suga, and S. Seki, *Physics of Ice*, Plenum Press, New York (1969), p. 329.
15. D. K. Edwards, *J. Heat Transfer* **91**, 1 (1969).

HEAT TRANSPORT IN LIQUID HELIUM II 15

R. K. IREY

15.1 INTRODUCTION

Few physical phenomena illustrate quantum effects on a macroscopic scale as dramatically as superfluidity.[1] Two examples are known. One is the superconducting transition which occurs in many metals at relatively low temperatures (less than 20 K). The second is the transition of the normal isotope of helium ^{4}He into a superfluid state known as liquid helium II. In each case, the unique properties of the superfluid hold promise of significant technological exploitation. In the case of superconductivity this exploitation has already begun[2] and many commercial applications are under development. The superfluidity of liquid helium II is equally unique physically. At present, the development of commercial uses which depend upon the unusual thermomechanical properties of liquid helium II is not as advanced. This is partially due to its inconveniently low temperature. Thus the greatest promise for commercially exploiting the properties of liquid helium II is in conjunction with systems involving superconductivity. In fact, it is so used in its single application to date.[3]

15.2 PROPERTIES OF LIQUID HELIUM II

The two isotopes of helium, ^{3}He and ^{4}He, due to their electronic structure and low mass, have extremely weak van der Waals binding forces.[4]

R. K. IREY Dept. of Mechanical Engineering, University of Florida, Gainesville, Florida.

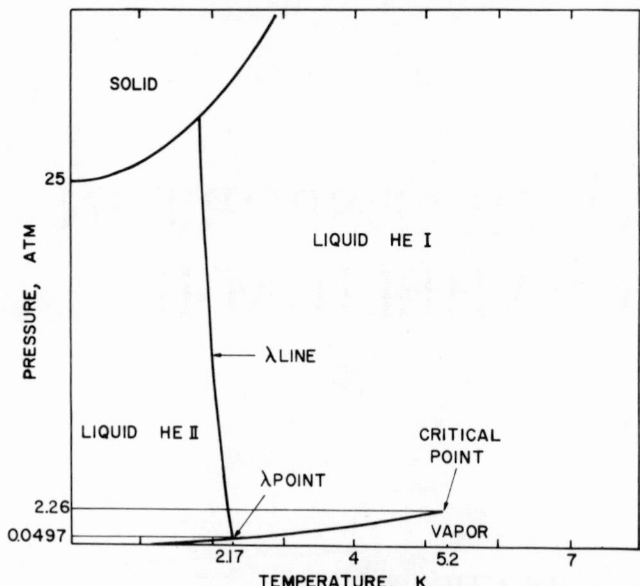

FIG. 15-1. Phase diagram for ^{4}He (schematic).

These weak intermolecular forces, in conjunction with large zero-point energies, combine to give helium extremely low critical and normal boiling points. In addition, they do not condense into a solid phase except at very high pressures (Fig. 15-1).

Helium-4 is a boson, and exhibits two liquid phases, liquid helium I and II, the latter being the superfluid phase. While the development of a satisfactory statistical mechanical theory has proven to be difficult, there seems little doubt that the origin of superfluidity in liquid helium II is essentially due to a Bose degeneracy. In fact, this assumption provides a basis for Tisza's two-fluid hypothesis, which is well supported by experiment.[5] The Tisza model divides liquid helium II into two distinct, interpenetrating components. A Bose degeneracy provides for such a division and allows that particles in the zero-point energy state in helium II comprise one component, called the superfluid. The particles in excited states are the other component, called the normal fluid. The total density of the fluid is thereby the sum of the densities of the superfluid and normal fluid components,

$$\rho = \rho_s + \rho_n \tag{15-1}$$

where the concentration of each component is a unique function of temperature (Fig. 15-2).

More than the description of helium II as a degenerate Bose system,

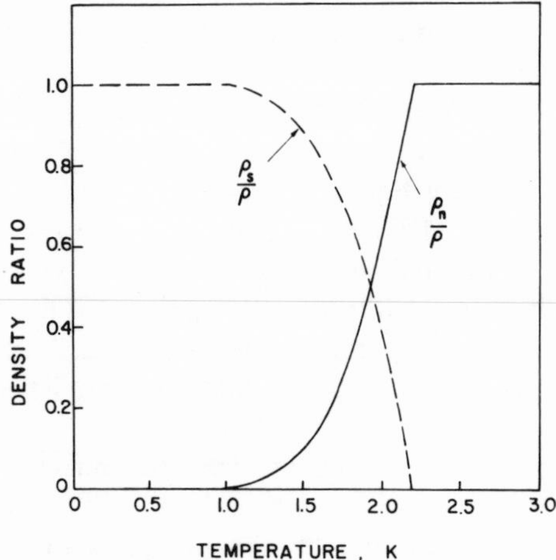

FIG. 15-2. Density ratio for liquid helium II.

however, is required to explain the superfluid effects of most significance to heat transfer processes. Scarcity of low-lying excited states in the energy spectrum allows the superfluid transport properties to be exhibited; for, then, energetic interaction between the superfluid and its environment becomes improbable. The result is that the superfluid component displays an inviscid character. Measurements of viscosity by Poiseuille techniques, which exclude normal fluid flow, give vanishing values of the viscosity of the superfluid component. In contrast, in other procedures, such as the damping of an oscillating disk which measures the bulk viscosity, the latter does not vanish because of the presence of the viscous normal fluid.[5,6]

Similarly, the peculiar energy spectrum of helium II at low temperatures strongly influences the rate of heat transfer. Under the influence of a temperature gradient, phonon transport does proceed through the normal component. In addition, heat transfer is accomplished by a peculiar internal convection discussed below. The temperature gradient sets up a thermomechanical driving potential which is related to the temperature gradient by the London equation

$$\nabla P = \rho s \, \nabla T \tag{15-2}$$

The potential induces a flow of the viscous normal fluid in the direction of the temperature gradient. The normal fluid carries with it energy in amount*

$$\bar{q} = \rho s T \bar{v}_n \tag{15-3}$$

where $\bar{v}_n$ is the local bulk velocity of the normal fluid.

* Equation (15-3) has been written in terms of the total entropy per unit volume ρs instead of for the normal component $\rho_n s_n$. This is valid since the entropy of the superfluid is zero,

$$\rho s = \rho_n s_n + \rho_s s_s{}^0$$

The system is constrained to conserve mass, so that

$$\dot{m} = \oiint (\rho_s \bar{v}_s + \rho_n \bar{v}_n) \cdot d\bar{s} \tag{15-4}$$

The condition imposed by Eq. (15-4) requires a counter flow of superfluid toward the heat source. The superfluid component is at zero entropy, thus its flow transports no thermal energy. At sufficiently low local bulk velocities of the superfluid, it does not interact with the normal fluid or the surroundings. Thus a peculiar internal convection of normal fluid from, and superfluid toward, a heat source ensues and a most efficient heat transport mechanism is established.

One further point is essential in order to understand a fundamental limitation on superfluid heat transport in liquid helium II. The efficiency of the transport process is limited by the noninteraction of the superfluid component with the normal fluid component or the surroundings. In either flow or heat transfer processes, the superfluid component acquires momentum and mechanical energy which can be sufficient to create excitation in the superfluid and allow interaction with the normal fluid or the boundaries. Thus there is a mechanical energy threshold which fixes a maximum velocity for dissipationless flow and heat transport. Excitations which develop in the superfluid component above this threshold are often viewed as associated with the onset of turbulence in the superfluid component.[7] A second critical situation is also frequently observed, and appears as a result of ordinary turbulence in the normal fluid.

The above paragraphs provide a brief introduction to some of the phenomena in liquid helium II which relate to heat transport. Much more detailed expositions are available, including a review of heat transport[7] and several summary texts.[1,4–6,8–10]

15.2.1 Kapitza Conductance

In spite of the large effective thermal conductivity of liquid helium II, in early experiments Kapitza[11] found that heat transfer from a solid into the liquid was inhibited by appreciable thermal resistance. Most early researchers ascribed this effect to processes within the superfluid.[12–14] More recently, the effect has been found in liquid ^{3}He,[15,16] between two solids,[17] and between solids and helium gas.[16,18] Indeed, Snyder[19] in a recent review argues that Kapitza resistance is present at all interfaces, but that its magnitude is much more significant at low temperatures. Evidence that such is the case may be provided by considering the phonon radiation limit of Jones and Pennebaker.[20]

Energy transport by phonons in any direction within a solid in equilibrium can be shown[21] to be

$$\bar{q}_{ph} = \rho u_{ph} C_s/4 \tag{15-5}$$

where C_s is the speed of sound and u_{ph} is the internal energy of the phonons.

At low temperature the expression for phonon energy on the basis of a Debye model[22] is

$$u_{ph} = \tfrac{1}{2}R\theta_D[1 + (6\pi^2/5)(T/\theta_D)^4] \tag{15-6}$$

where θ_D is the Debye temperature. Substituting Eq. (15-6) into Eq. (15-5), one obtains not only the rate of phonon transport in any given direction in a low-temperature crystal in equilibrium, but also the maximum rate of phonon emission from a low-temperature solid surface into another medium:

$$\bar{q}_{max} = \tfrac{1}{8}\rho R\theta_D C_s[1 + (6\pi^2/5)(T/\theta_D)^4] \tag{15-7}$$

This quantity is analogous to blackbody photon emission.* Although heat may be transported by other carriers, notably electrons, in the solid, heat transfer from the solid into the liquid is limited essentially by Eq. (15-7).† Further, since phonon excitation is a simple process in liquid helium II, one can use Eq. (15-7) as a limit on phonon absorption as well. This gives a maximum for the net heat transport in the limit of low-temperature differences as

$$q_{max} = (2\pi^4 k^2/4\theta_D{}^2\hbar)(3n/4\pi)^{2/3}T^3\,\Delta T \tag{15-8}$$

where n is the number density of the crystal. The coefficient of ΔT is the upper limit for the Kapitza conductance.‡

If one compares the largest Kapitza conductance data available for any given material§ with the phonon radiation limit, one finds for nonmetals that several samples are higher than 70% of the limit|| with others as low as 25%. A similar comparison made with metals finds a few samples as high as 25%, with 10–15% being more typical.

* Unlike photon emission, maximum phonon emission is dependent upon the medium doing the emitting.

† Photon emission is negligible and Little has considered electron tunneling[23] and emission by surface waves "epiphonons"[24] and found their influence to be small.

‡ In order to give a quantitative feel for the limitation on heat transfer imposed by Kapitza resistance, Snyder[19] compares the limit imposed by Eq. (15-8) for a low-resistance metal, lead, and liquid helium II at its maximum effective conductivity temperature, 1.9 K. In this case, a linear length of liquid helium II of 175 km is required to get an equivalent resistance. Although this comparison is somewhat misleading, since the thermal conductivity of liquid helium II depends on the dimensions and heat current density of the channel through which the heat is flowing, it does illustrate how significant the Kapitza resistance can be.

§ Since Kapitza resistance is strongly dependent on surface treatment, it is assumed that the samples with the largest conductance exhibit this because their treatment has approached phonon emission ideality most closely.

|| One sample[25] exceeds the limit by about 7%.

Some criticism may be leveled at Eq. (15-8) on the grounds that a Debye continuum model is inadequate to handle either phonon energies and/or phonon velocities in this case.[17] Further, some samples, for instance, mercury,[26] do not satisfy the low-temperature approximation imposed by Eq. (15-6) at liquid helium II temperatures. Nevertheless, Eq. (15-8) appears to give a very realistic upper bound.

There are many reasons for a given sample to have a lower Kapitza conductance than the limit. Among these would be reflections of potentially emitted phonons at the interface. Causes of these reflections can be internal to the solid, relating to phonon–phonon scattering, phonon–electron scattering, or, most significantly, phonon–defect scattering with the defects associated with the surface itself and with their reflective characteristics being strongly affected by surface treatment, mechanical, or chemical. Reflections can be caused externally in oxide-absorbed surface layers or in the liquid itself. Phonon transport from the liquid to the solid is limited by similar reflective processes (some like electron scattering being omitted and others such as roton* scattering being included). All such processes tend to reduce the rate of heat transport below the limit imposed by Eq. (15-8).

Further, if the magnitude of the phonon mean free path is of the order of or smaller than the thermal boundary layer thickness in liquid helium II, then an appreciable percentage of phonons emitted by the solid will be absorbed at temperatures higher than that of the bulk liquid. Similarly, an appreciable number of phonons absorbed by the solid will have been emitted in the liquid at temperatures higher than that of the bulk liquid temperature. The above effect represents the transition of the transport mechanism from the phonon–free molecular region on which Eq. (15-8) is based, toward a phonon continuum. The net effect is to reduce the rate of phonon transport below that predicted by Eq. (15-8).

The temperature dependence of the Kapitza conductance predicted by Eq. (15-8), T^3, is in reasonable agreement with experiment. When fit to a function of the form

$$h_K = AT^n \tag{15-9}$$

most experimental data yield n between 2.5 and 3.5.† It should be recognized that failure of the data to follow a T^3 dependence does not of itself require an inadequacy of Eq. (15-8). The deviation may be due to a temperature dependence of the scattering mechanisms which reduces the heat flow below the limit imposed by Eq. (15-8).

* The name roton was used by Landau to describe very short-wavelength excitations which correspond to the minimum in the excitation spectrum.[5,6,9]

† Some of the exceptions, such as the data on mercury,[24] $h_K = A - BT$, deviate from the prediction of Eq. (15-8) for well-recognized reasons.

The predicted dependence of Kapitza conductance upon the Debye temperature and the number density,

$$h_k \propto \theta_D^{-2} n^{2/3} \tag{15-10}$$

does not correlate the data nearly as well. In fact, θ_D^{-2} seems to fit the data somewhat better.[19]

The development of a general theory for Kapitza resistance is complicated by the strong dependence on surface condition. Clean, strain-free surfaces give conductances on the order of 1.5–10 times larger than the same material when strained and dirty.[19] For those who tend to begin any cleaning procedure with a machine operation or a mechanical polishing process, it is particularly noteworthy that this procedure tends to lower the Kapitza conductance substantially, unless adequate time for recrystallization of the surface or chemical etching to remove some of the strained material is allowed subsequently.[19]

Electronic contributions to Kapitza conductance in metals have been investigated in part by making measurements on superconductors and then repeating the measurement after driving the material normal by the application of a high magnetic field. In analogy with the discussion of liquid helium II in the introduction, the super-electron pairs will not readily interact with the lattice or phonons and hence do not participate in the heat transfer process. Indeed many experimenters[25–31] have observed an increase in Kapitza resistance in soft superconductors, up to factors of 10 or 15 in lead[31] and 1.3 in mercury.[26] The effect in hard superconductors is much less pronounced; Gittleman and Bozowski[25] reported 1.1 and 1.06 in Sn and In, respectively.* In addition to providing insight on the importance of electrons, existence of a superconducting effect and its magnitude are of considerable practical significance. This is true since, as noted in the introduction, practical applications of liquid helium II are very likely to be to cool superconductors. Unfortunately, whether the effect as evidenced by the data is really due to a direct electron effect or is rather a secondary influence of strain[33] is still in some doubt.[19]

Khalatnikov[9,34] has presented a model which corrects Eq. (15-8) to account for acoustic mismatch. Acoustic mismatch refers to the difficulty of satisfying the momentum and energy equation at arbitrary angles for phonon–phonon collisions when the acoustic speeds of the two media are significantly different. The correction predicted, however, amounts to two or more orders of magnitude for ideal surfaces and as a result overcorrects Eq. (15-8)

* Some speculation[24] has been made that the effect may actually be due to a superconducting transition in the soft superconductor in the solder, particularly since Wey-Yen[32] reported no effect for tin.

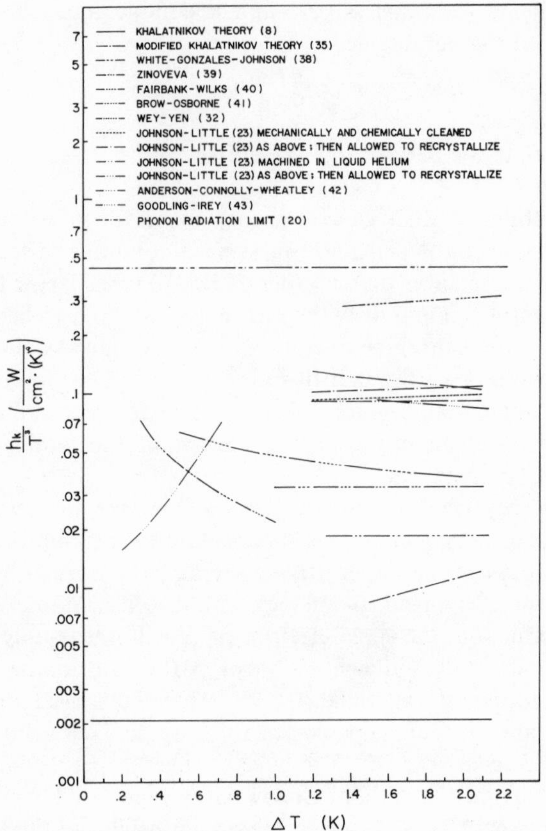

FIG. 15-3. Kapitza conductances for copper surfaces.

sufficiently to be in worse agreement than is the original form.* Several attempts to correct the Khalatnikov model to account for real surface effects,[35–37] the increased density of the liquid near the interface,[36] and absorbed layers[23] have had little success in improving the correspondence between prediction and experiment.

To illustrate the range of variation with experiment and correspondence with theory, Fig. 15-3 shows a representation of the results of several investigators in the form of $h_K T^{-3}$ for copper. Also shown are the predictions of Khalatnikov, modified Khalatnikov results, and the phonon radiation limit. Much more complete catalogs of the Kapitza conductance data and

* The Khalatnikov model is, nevertheless, very well known; probably because it incorporated the characteristic T^3 behavior, whose origin is the low-temperature T^4 dependence of the Debye phonon energy, Eq. (15-6).

many additional references can be found in three recent reviews.[17,19,37]

15.2.2 Kapitza Conductance when $\Delta T \sim T$

As the temperature difference increases, one might expect that other thermal resistance mechanisms, in addition to the phonon radiation limit imposed by Eq. (15-8), would become significant. One method of recognizing the occurrence of such additional mechanisms is through the deviation of the temperature dependence of the conductance from that expected from the phonon radiation mechanism. Equation (15-8) has been linearized to first order in ΔT. The complete equation can be obtained by a multiplicative correction factor

$$f = 1 + \tfrac{3}{2}(\Delta T/T_b) + (\Delta T/T_b)^2 + \tfrac{1}{4}(\Delta T/T_b)^3 \qquad (15\text{-}11)$$

Figure 15-4 presents representative data of Clement and Frederking[44] for silver, Madsen and McFadden[45] and Holdredge and McFadden[46] for a nickel–iron alloy, and Lyon[47],* and Goodling and Irey[43] for platinum, with the latter's results including copper. Superimposed upon the data is a family of curves representing the temperature dependence of Eq. (15-11),

$$h_k = C(T_b)f(T_b, \Delta T)\,\Delta T \qquad (15\text{-}12)$$

where T_b is the temperature of the liquid helium II bath, and the constant C has been selected for best fit to the low-ΔT limit in each case.

The data of Clement and Frederking, Lyon, and Goodling and Irey provide reasonably close agreement with Eq. (15-12), but show an increasing negative departure (up to 50%) at increased temperature differences. The data reported by McFadden and co-workers[45,46] show the same trend with somewhat larger and more irregular deviation. The deviation from Eq. (15-12) could be caused by temperature dependence of the phonon emission characteristics in the solid, but this does not appear likely.

The nature of the mechanism or mechanisms which are responsible for the deviation has not been determined. Among the mechanisms proposed are those of Kronig and co-workers,[12,13,37] who had earlier considered resistive mechanisms in the bulk fluid associated with dissipative processes in the normal fluid, and Gorter *et al.*[14] who include the effect of rate of conversion of normal fluid to superfluid. The breakdown or partial breakdown of superfluidity associated either with a critical heat current or with the thermodynamic equilibrium state of the liquid in the vicinity of the heated

* Curves are used to represent the data of Lyon, which have been calculated from the heat flux vs. ΔT figures contained in Ref. 47.

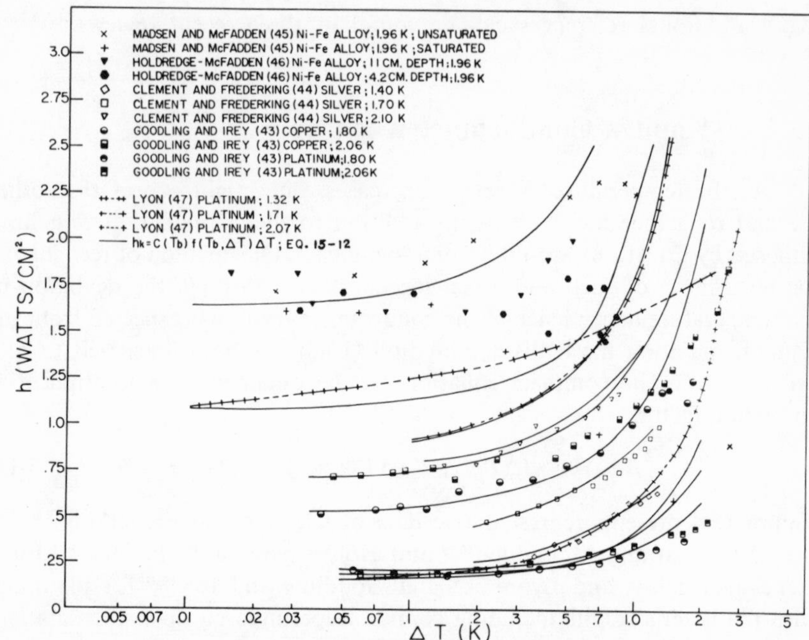

FIG. 15-4. Heat transfer coefficients in the nonboiling region.

surface may be important. As the temperature is increased, assuming that the liquid near the heated surface is a continuum, it crosses an extension of the λ line from the metastable liquid helium II into the metastable liquid helium I region* (Fig. 15-1). Frederking[17] has pointed out that the beginning of the negative deviation from Eq. (15-12) is relatively coincident with crossing the λ line in the metastable region, adding some credence to that mechanism. This places special significance on unsaturated bath data.

In early experiments, Peshkov[48,49] studied heat transfer to unsaturated baths. His experimental procedure included the use of a Mach–Zhender interferometer which allowed direct observation of the density gradients in the liquid. He presents photographs which display a liquid helium I–II interface. Peshkov's photographs show what appear to be vortex eddies of liquid helium I propagating into the liquid helium II at the interface. Unfortunately, one cannot interpret the deviation from the form of Eq. (15-12) in terms of the results given.

* As is the case in ordinary liquids in pool boiling of a saturated bath, the nucleation process is not initiated until the surface temperature is well into the gas-phase region. In the case of liquid helium II, except for exceptionally large standing liquid depths or saturated baths quite close to the λ point, the liquid near the boiling surface crosses the liquid vapor equilibrium curve before crossing the λ line in the metastable region.

More recently, Madsen and McFadden[45] reported results for both saturated and unsaturated baths (Fig. 15-4). In both cases their results are in general agreement with Frederking's comment, although the quantitative correspondence is not as close. Madsen and McFadden's data show that at relatively small temperature difference, $\Delta T < 0.1$ K, the data for saturated and unsaturated baths are the same. However, as the temperature difference is increased, they find a pronounced enhancement of heat transfer due to the elevated pressure. This enhancement reaches a magnitude of about two as the peak heat flux is approached. Adronikashvili and Mirskaia[50] had earlier reported a similar effect of magnitude up to ten or so as the pressure was increased to 8 atm.

It will be noted in subsequent sections that both peak flux and film boiling characteristics in helium II are strongly influenced by the test section submersion depth. Whether this influences the rate of heat transfer before peak flux is of interest. Holdredge and McFadden's[46] results (Fig. 15-4) indicate that depth does influence the heat transfer rate significantly prior to peak flux by as much as a factor of two, at temperature differences above about 0.4 K (larger heat transfer rates are associated with deeper immersion depth). In contrast, in very similar experiments, Goodling and Irey[43] observed no depth effect below peak flux within an experimental precision estimated as less than 2.5%.[51] The major difference between the two experiments was the test section construction. The Holdredge and McFadden experiment was conducted on a test surface which utilized a thin film of Ni–Fe alloy vacuum-deposited on the surface of glass rod, with the thermometer in the interior. Accuracy of surface temperature measurement is in this case dependent upon uniformity of the electrical dissipation in the thin film, the conductivity of the interior being too low to restore uniformity. Goodling and Irey utilized high-conductivity metal as their test specimen, thus assuring a high degree of temperature uniformity. An earlier version of the Holdredge and McFadden test surface had been employed by Irey, McFadden, and Madsen.[52] In that case, the test surface had apparently failed to maintain the degree of temperature uniformity required, particularly near points of transition, due to locally high heating rates. Although the test section was improved before subsequent studies, it is also true that local transition to film boiling would alter the data to show a depth effect in the fashion evidenced. Neither Clement and Frederking[44] nor Lyon[47] reported a depth effect on their horizontal test surfaces up to peak flux.

These two investigations[43,46] also obtained different results for a size effect. Each studied the change in heat transfer rate resulting from altering test surface diameter by a factor of about two. Again Holdredge and McFadden[46] reported a significant change. (Reducing the diameter reduced the heat transfer rate by the same order.) Goodling and Irey[43] reported no effect. It was noted above that in the Kapitza resistance range, $\Delta T \to 0$, the nature of the interface, not its geometry, determines the Kapitza conductance. Therefore, the fact that the diameter influence observed by Holdredge and McFadden is sustained undiminished even to vanishing temperature differences suggests strongly that the effect is actually a difference in surface preparation.

Finally, Goodling and Irey[43] reported that a reorientation of the test surface from a horizontal to a vertical cylinder had no measurable effect up to peak flux.

15.3 PEAK FLUX

Although no nucleate boiling occurs in liquid helium II, as the heat flux is increased, a gaseous film forms, insulating the liquid from the test surface (that is, film boiling is initiated). The nonboiling regime is separated from the film boiling regime by a range of temperature differences corresponding to unstable film boiling. The departure point is called peak flux q^*. For practical design of systems to be cooled by liquid helium II, peak flux is of particular significance.

Like the nonboiling region, peak flux is affected by bath temperature and the material and condition of the test surface. In addition, peak flux is affected by the geometry of the test section and the liquid depth. Experimentally, peak flux is difficult to investigate. Since the dissipation flux is large, up to 10 W/cm^2,[9] it is convenient experimentally to use small test surfaces. The smallest test surfaces are fine wires, for which it is impractical to use a surface thermometer other than the wire itself. Except for some unusual materials, like phosphor-bronze,[50] metallic resistance thermometers lose their thermometric sensitivity as the temperature is reduced. Even with high-purity platinum, the thermometric sensitivity is so low at helium temperatures as to make temperature measurements imprecise. This is not too serious, since the heat flux can still be measured accurately. In addition, however, the small wire test surfaces are potentially subject to quite significant surface temperature nonuniformity. The steady-state longitudinal temperature nonuniformity due to end effects is estimated to be relatively insignificant. Significant error can result, however, when there is nonuniform heating in the wire. Steed,[53] with respect to film boiling measurements, and more recently Leonard and Lady,[54] with respect to peak flux, have shown that a local high-resistance region (associated with impurities, cold working, or mechanical necking down) causes relatively extreme temperature nonuniformity due to a tendency toward thermal runaway.† In the case of peak

† As the temperature increases locally, the resistance does also, further increasing the temperature and so on.

flux the film blanket will form first on the localized hot spot and then rapidly spread to cover the wire. Leonard and Lady* reduced the diameter of one test specimen locally by 31% and found a 60% reduction in measured peak flux. Therefore, extreme care in wire test section construction and handling is required to get accurate peak flux data. In addition, test section material should be selected with consideration to alloying for strength and insensitivity of electrical resistance to cold working or local impurity.

Test surfaces made of high-conductivity metals to establish uniform surface temperature, on the other hand, must be relatively large in order to contain a separate heater and thermometer (generally a semiconductor).[43,44,47,56] The associated large net heat transfers have limited the range of data obtainable with steady-state conditions,[44] and have led to compromises on length-to-diameter ratio,[43] large corrections for end effects,[47,56] and limitation on the lowest bath temperatures which could be investigated.[43]

In spite of the various experimental difficulties noted above, considerable data have been collected on peak flux. The majority of the data has been collected on horizontal cylinders of small diameter (wires). However, in these experiments, the temperature difference is either unreported or of low resolution.[57] In addition, probably largely as a result of the hot spot difficulties noted above, one finds rather large scatter in the peak flux data (about half an order of magnitude). Since the principal sources of error all tend to reduce peak flux, this reviewer has a strong bias toward the higher points under given conditions.

Essentially all of the investigators[53–55,57–61] who have performed peak flux measurements on wires have reported on some aspect of the general characteristics relating to bath temperature, wire diameter, and depth. Since a rather successful correlation has just been reported, no attempt will be made to reproduce the wire peak flux data with any degree of completeness. Rather, some of the results of Frederking and Haben[55] will be presented to dramatize the nature of the variation of three principal variables, depth (Fig. 15-5), bath temperature, and wire diameter (Fig. 15-6).

Much fewer data are available on cylinders of larger diameter. McFadden and Holdredge[62] and Madsen and McFadden[45] reported peak flux data on composite glass cylinders of 0.145 and 0.245 cm[62] and of 0.179 cm,[45] respectively. Goodling and Irey[43] reported peak flux data on platinum and copper cylinders of 0.56 cm diameter as well as on a copper cylinder of 0.371 cm diameter. In this case, the test surfaces have too low a ratio of length to diameter to be considered approximately infinite (length was

* This investigation[54] also looked into the possibility of dewar size affecting the data as had earlier been suggested.[55] It was found that a significant change in local dewar geometry did not affect the wire data.

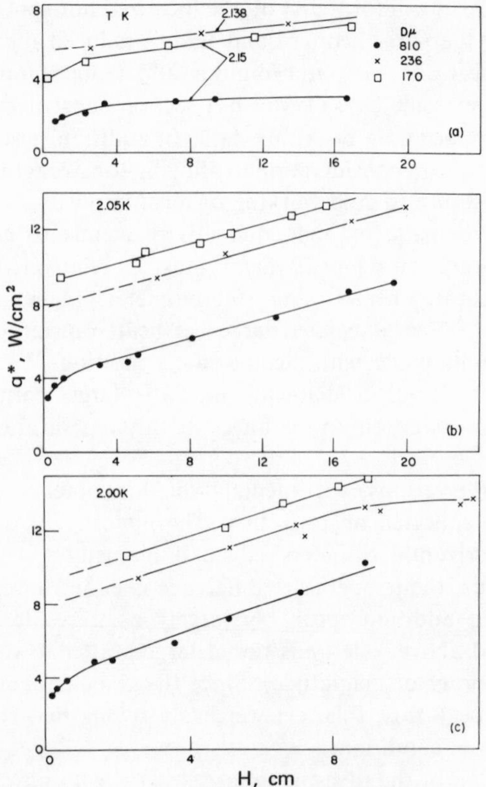

FIG. 15-5. Influence of depth on peak flux.[55]

about 0.53 cm). Although the number of bath temperatures reported is limited, all of these investigators' data show a variation with bath temperature consistent with Fig. 15-6. Both Goodling and Irey[43] and McFadden and Holdredge[62] confirmed the effect of depth indicated by Fig. 15-5.* Madsen and McFadden reported that peak flux is increased by a factor of approximately two in an unsaturated bath at 1 atm.

Goodling and Irey showed a generally small† but not insignificant reduction in peak flux for platinum in comparison to their copper surfaces.

* Madsen and McFadden's experiment did not report on a parametric variation of depth.

† The puzzling exception is at a 1.8 K bath temperature, where the difference is somewhat more than a factor of two.

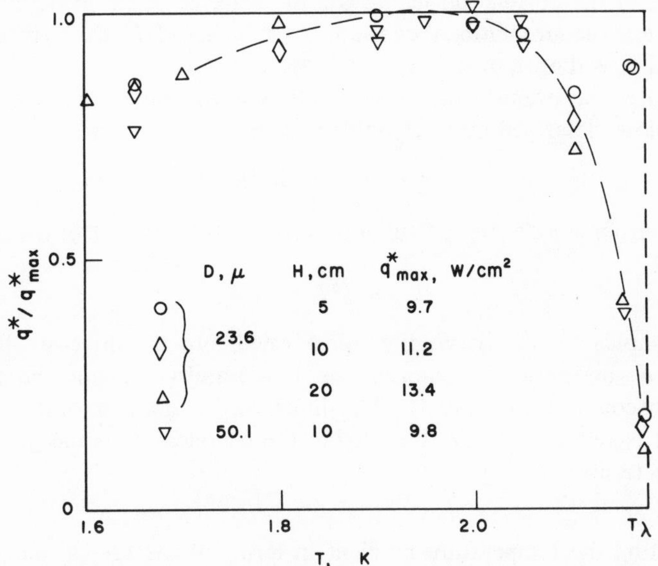

FIG. 15-6. Peak flux dependence upon bath temperature and wire diameter.[55]

Such a variation due to material has also been observed in a wire study.[47] In that case it could be a result of premature transition caused by hot spots, as discussed above. Hot spots are not probable for large, high-conductivity test surfaces such as that employed in Ref. 43. Yet, if peak flux is influenced by material, then a correlation such as proposed by Frederking *et al.*[63] cannot be truly adequate.

The studies of Goodling and Irey and of Holdredge and McFadden also showed a diameter effect consistent with that indicated by Fig. 15-6. As noted above in the case of Holdredge and McFadden, this diameter effect continues undiminished at lower temperature differences. On the copper samples of Goodling and Irey, the effect occurs only at peak flux.

Frederking *et al.*[63] have developed a correlation relating peak flux, expressed as a dimensionless heat flux

$$N_y = \rho q^* D / \rho_s s T_b \mu_n \tag{15-13}$$

where D is the cylinder diameter and μ_n is the viscosity of the normal fluid, to a dimensionless driving potential

$$N_x = \left(\frac{\rho D^3}{\mu_n{}^2}\right)\left(\frac{\rho_s}{\rho}\right)\left(\frac{\rho s T_b}{\rho_v h_{fg}}\right)\left(\frac{\rho g H}{L\zeta}\right) \tag{15-14}$$

where $L\zeta$ is the correlation length for the state of order in liquid helium II,[64–66] g is the gravitational constant, H is the liquid depth, h_{fg} is the latent heat, and ρ_v is the vapor density.

Briefly, the premise of the correlation is to relate the dimensionless viscous flow, described by a Reynolds number

$$\text{Re} = \rho(|\bar{v}_s| - |\bar{v}_n|)D/\mu_n \tag{15-15}$$

to a dimensionless driving potential in terms of the gradient of the chemical potential

$$N_{\nabla\mu} = \rho^2 D^3 \nabla\mu/\mu_n{}^2 \tag{15-16}$$

The Reynolds number is readily transformed into the dimensionless heat transfer N_y by the introduction of Eqs. (15-1) and (15-3) and the zero net mass flow condition, Eq. (15-4).† The dimensionless gradient of the chemical potential is reducible to N_x by relating the chemical potential gradient to the temperature gradient

$$\nabla\mu|_P = -s\,\nabla T \tag{15-17}$$

then writing the temperature gradient in terms of the critical temperature difference over the correlation length

$$\nabla T = \Delta T^*/L\zeta \tag{15-18}$$

Assuming that after transition the liquid–vapor interface is the local saturation temperature, then ΔT^* can be written in terms of the hydrostatic head and the Clausius–Clapeyron equation as

$$\Delta T^* = \rho g H T_b/\rho_v h_{fg} \tag{15-19}$$

Substituting Eqs. (15-18) and (15-19) into Eq. (15-16) gives N_x.

All of the parameters appearing in N_x and N_y are relatively well known‡ except the correlation length, $L\zeta$, and of course the peak flux q^*. Frederking *et al.* suggest that $L\zeta$ be treated as a constant.

A general form of the correlation is

$$N_y = f(N_x) \tag{15-20}$$

On the basis of the data, the authors adopted the specific form

$$N_y = C_n N_x{}^n \tag{15-21}$$

and two such curves are drawn through the data (Fig. 15-7), $C_n = 1.48 \times 10^2$ and $n = 0.257$, and $C_n = 1.21 \times 10^2$ and $n = 0.257$, where the first represents a mean of the points corresponding to higher peak flux and the

† Equation (15-4) expressed the condition on a global basis. In this case it is to be applied at a point or $J = \rho_s\bar{v}_s + \rho_n\bar{v}_n = 0$.

‡ The reported data for the viscosity of the normal fluid μ_n show considerable scatter.

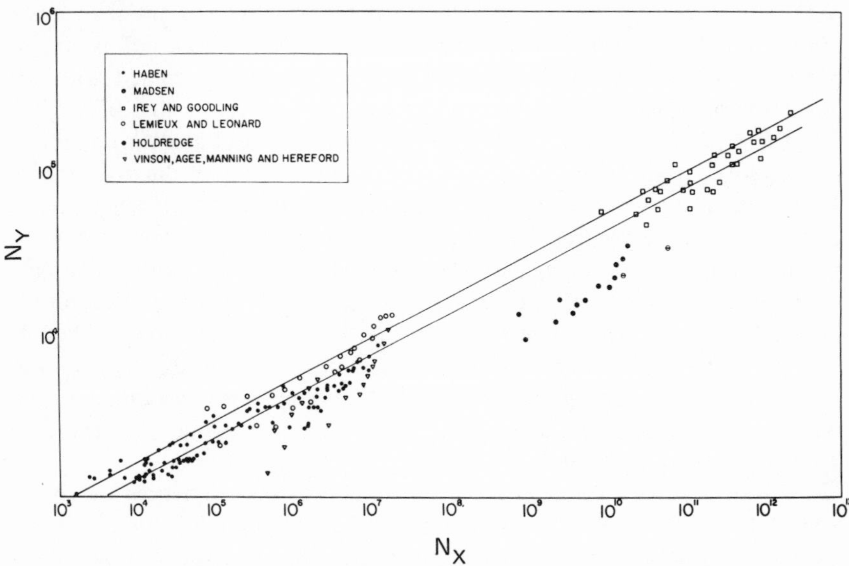

FIG. 15-7. A dimensionless correlation for peak flux.

second of all the points. It is evident from Fig. 15-7 that, although significant experimental scatter is present, the correlation groups have at least come quite close to describing the behavior of peak flux on horizontal cylinders. In particular, the parametric variation with bath temperature and wire diameter described by the correlation is quite close to that observed experimentally. The parametric dependence on depth which the correlation suggests is also quite satisfactory, with the exception that q^* goes to zero in the limit as depth goes to zero. This difficulty can be corrected within the general expression of Eq. (15-20) by choosing a different form for Eq. (15-21).

Peak flux data for other than horizontal cylinders have not yet been correlated. Goodling and Irey[43] reported peak flux data on one of their copper cylinders reoriented in the vertical position. These data display the same characteristic behavior as the data for horizontal orientation but the magnitudes are reduced on the order of 20%.

Lyon[47,56] reported peak flux values on flat platinum surfaces oriented in the horizontal, face-up position. In contrast to the results discussed above, Lyon found no effect on peak flux due to depth from 3 to 30 cm. The reported dependence upon bath temperature is similar to that reported by other investigators. However, the magnitude of peak flux is about 25% greater than the largest of those reported by other investigators.[43] Lyon also reported a few points in the horizontal, face-downward position; in this case peak flux is slightly decreased. Lyon also reported a substantial decrease in peak flux for a surface coated with ice. This decrease could well be due to hot spots appearing where the surface is not as thickly coated with the insulating material and thus causing premature transition to film boiling.

Clement and Frederking[44] reported peak flux from a silver horizontal surface located at the end of a tube a little more than two diameters deep. Their data display the characteristic dependence on the bath temperature. The authors, however, reported that depth dependence was not evident. In this investigation, pump capacity was not adequate to maintain steady state at the peak flux condition, except just below the λ point. This caused additional uncertainty in the data. A second factor which might well have masked or influenced the dependence of peak flux on depth in this study was the location of the test surface at the bottom of a tube. However, an inspection of the data of Clement and Frederking shows that the majority of the data display a depth effect consistent with that previously discussed.

Other peak flux studies have been completed on geometries which are difficult to categorize[67,68] or where it is difficult to establish the maximum current-carrying capacity of superconducting wires; see, for example, Refs. 69–71. These will not be commented upon in detail.

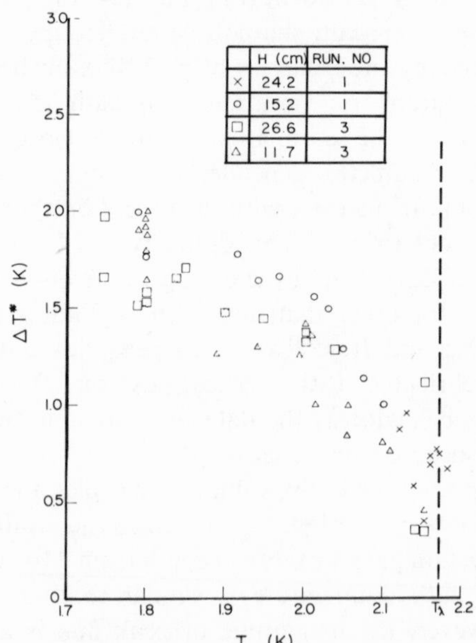

FIG. 15-8. Peak excess temperatures as a function of bath temperature.[44]

Data on the temperature difference at which peak flux occurs, ΔT^*, are much more sparse. It was noted above that the many studies on wires forego this aspect. The available data are therefore limited to the larger test surfaces.[43,44,46,47] (See Fig. 15-8.) If (1) peak flux is independent of material, as is indicated by the Frederking *et al.* correlation,[63] and (2) non-film boiling is describable in terms of the material surface condition and bath temperature by a curve of the form described by Eq. (15-12) (accounting for the negative departure indicated above), then ΔT^* is uniquely determined by the intersection of these two curves. This, one might note, is the procedure adopted in nucleate boiling of ordinary fluids. Although as suggested by Frederking,[17] there may be an upper limit in terms of the limit of metastability of the liquid adjacent to the surface, present results appear to support the above procedure at least for design purposes.

15.4 FILM BOILING

Superficially, film boiling in liquid helium II is similar to film boiling in liquid helium I or other conventional liquids. Indeed there is no discontinuity in the heat flux as the bath temperature is reduced below the λ point. There is, however, a discontinuity in $dq/d(\Delta T)$.[63] In addition, at least two distinct phenomena set liquid helium II film boiling apart from that which occurs in other liquids. These are the depth effect, discussed in relation to peak flux above, and a "noise" regime which occurs under certain experimental conditions.†

As in the case of peak flux, it is convenient experimentally to reduce the total energy dissipation by using small test surfaces. Therefore, the majority of the film boiling measurements are reported for wire test sections.[53,57–60,72–74] Unlike peak flux, however, it is essential in film boiling studies that surface temperature be reported. Whereas the problem of non-uniform wire temperature, hot spotting, affected the peak flux data as a premature transition, the same problem in film boiling affects the indicated wire surface temperature directly. Calibration of wire resistance as a function of temperature, insofar as possible, is carried out near thermodynamic equilibrium with the test wire at each calibration point as uniform in temperature as possible. Both end effects and particularly, locally high electrical resistance regions will result in nonuniform wire temperature during data acquisition. The reported wire temperature is necessarily a mean value based on total wire resistance.

† In addition, the absence of bubbles leaving the film region makes it appear different from film boiling in ordinary liquids.

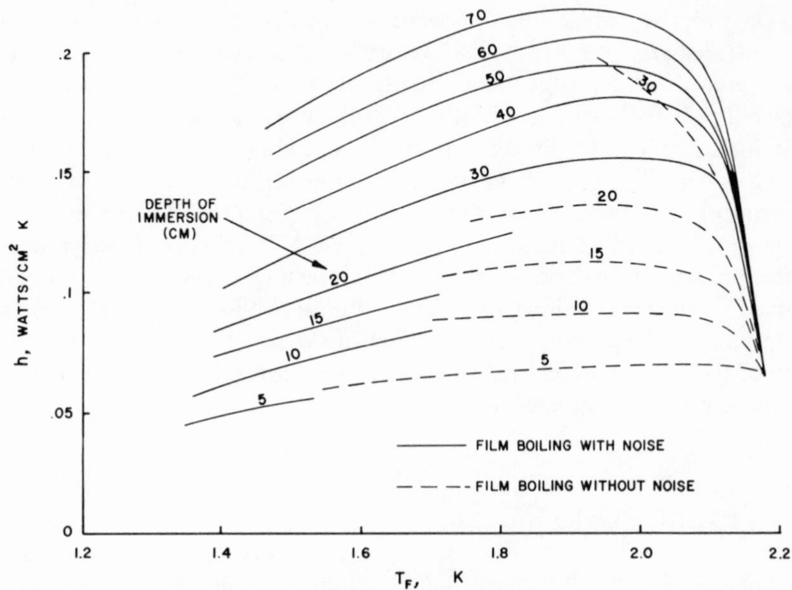

FIG. 15-9. Heat transfer coefficient as a function of bath temperature and depth of immersion (T_S = 80 K).[72]

With respect to thermometric sensitivity at low temperatures, it is desirable that the test material be a pure, low-resistivity element. Unfortunately, these materials are most susceptible to high-resistance regions caused by local impurities and to large resistance change due to local cold working. Both of these effects become more pronounced as the temperature is reduced. For the case of pure platinum* Steed[53] has estimated that cold working could easily give a local temperature region as high as 40 K with an average temperature of 25 K. The effect of cold working, exclusive of necking down, becomes less and less pronounced as the temperature is increased; in platinum, its influence is essentially gone above 80 K.[75] The use of alloyed wires reduces these difficulties, but also reduces thermometric sensitivity and decreases the accuracy of interpolation formulas to aid in calibration.

As an indication of the parametric dependence of film boiling on depth and bath temperature, smoothed curves representing† the data of Leonard and Lemieux[72] obtained on 0.003-in., 90% platinum–10% rhodium wires are represented in Fig. 15-9 for 80 K surface temperature and in Fig. 15-10 for a 400 K surface temperature. This investigation has been found to be

* Pure platinum maintains sufficient thermometric sensitivity to enable data to be obtained down to the minimum point of stable film boiling.

† Leonard and Lemieux do not present the data points in their publication.

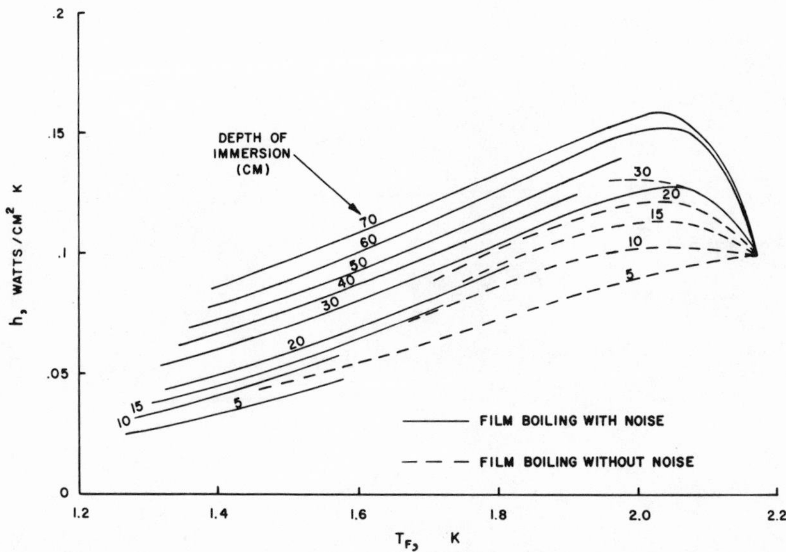

FIG. 15-10. Heat transfer coefficient as a function of bath temperature and depth of immersion (T_S = 400 K).[72]

consistent with data collected by other investigators[53,57,58,60,73] when one accounts for differences due to wire diameter (film boiling is inversely proportional to wire diameter with an exponential power of about $\frac{1}{4}$).

Noted also on the curves is the sharp difference between the data obtained with noise and without noise. First reported by Bussieres and Leonard,[76] a characteristic sound often described as similar to cavitation noise has been observed in many studies.[43,53,60,72–74] Motion pictures obtained both by backlighting[43,53,73,74] and shadowgraph procedures[77] reveal a large increase in amplitude and frequency when the film boiling process goes from no noise to noise. A photographic sequence of transition from silent to noisy film boiling, which has been edited from the motion pictures taken by Ebright and Irey,[74] is presented in Fig. 15-11. Frequency of oscillation gives quite dramatic evidence of the difference between the two modes. As reported by Ebright and Irey (Fig. 15-12), silent film boiling has a characteristic frequency of about 50 Hz. Noisy boiling has a characteristic frequency from about 500 to 900 Hz.* In neither regime was frequency found to be discernibly influenced by heat flux, depth, bath temperature, or wire diameter.† Film wavelength and amplitude have also been examined.[73,74]

* A few exceptional points had lower frequencies.

† It should be noted that only 0.005 and 0.008-in. wires were studied.

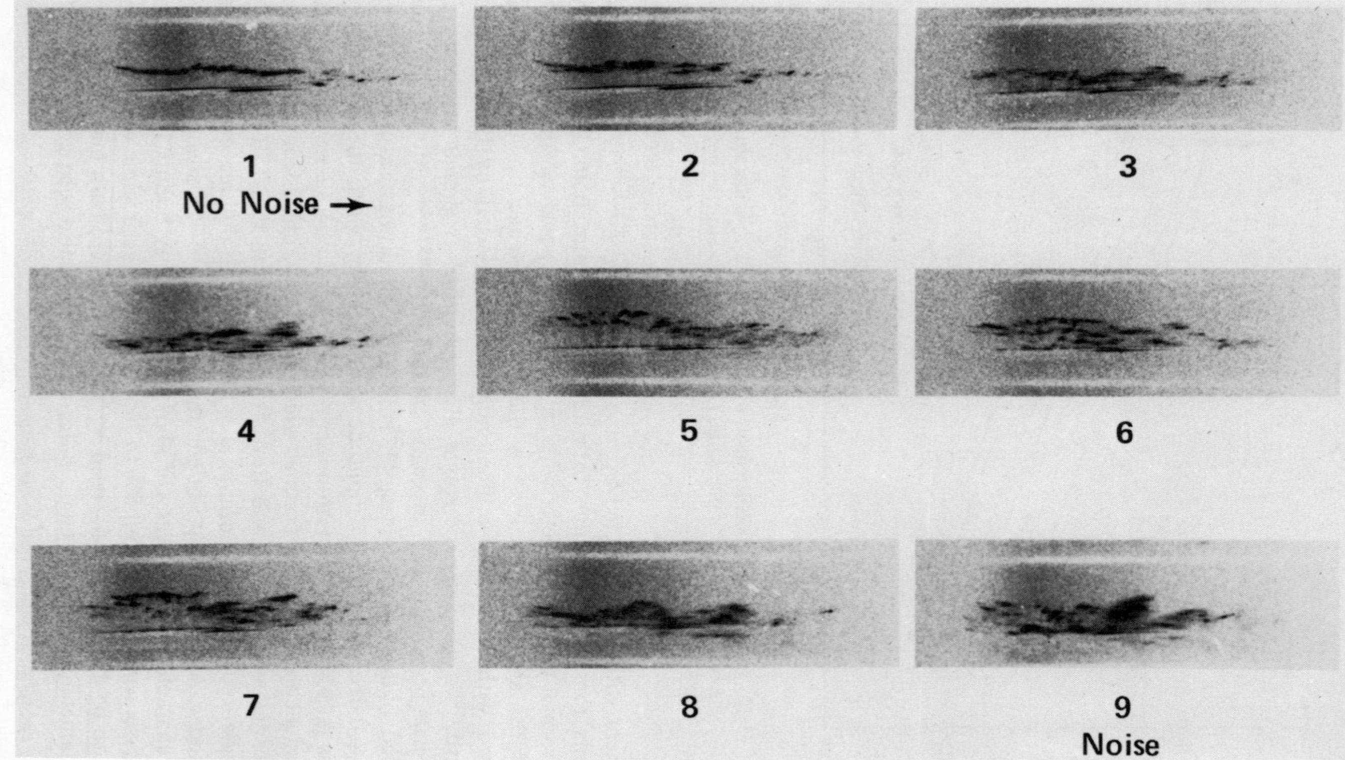

FIG. 15-11. Photographic sequence of spontaneous transition from silent to noisy film boiling.

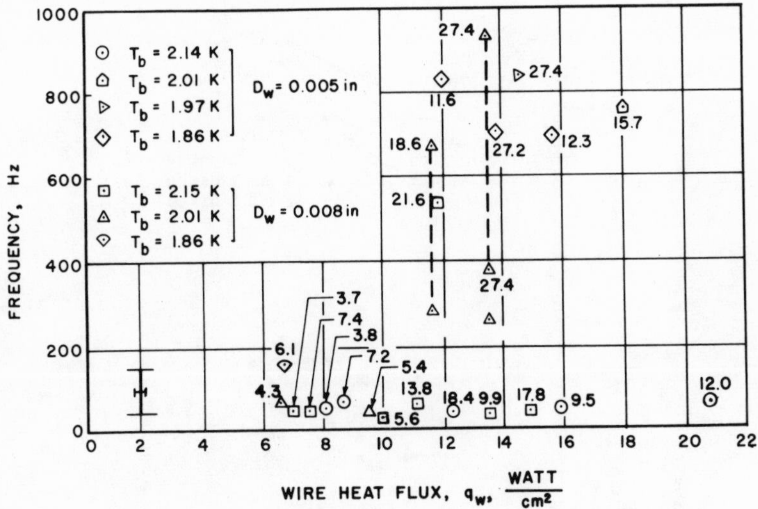

FIG. 15-12. Frequency vs. wire heat flux with indicated depths (cm).

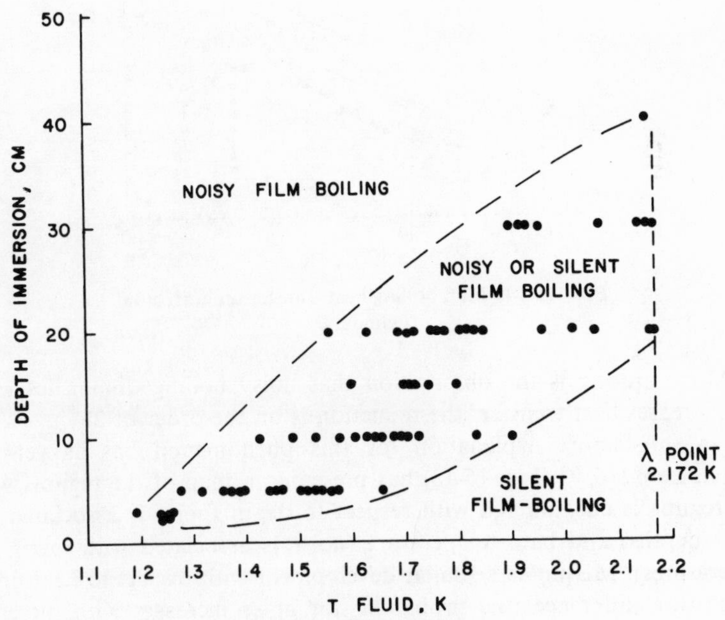

FIG. 15-13. Film boiling regimes.

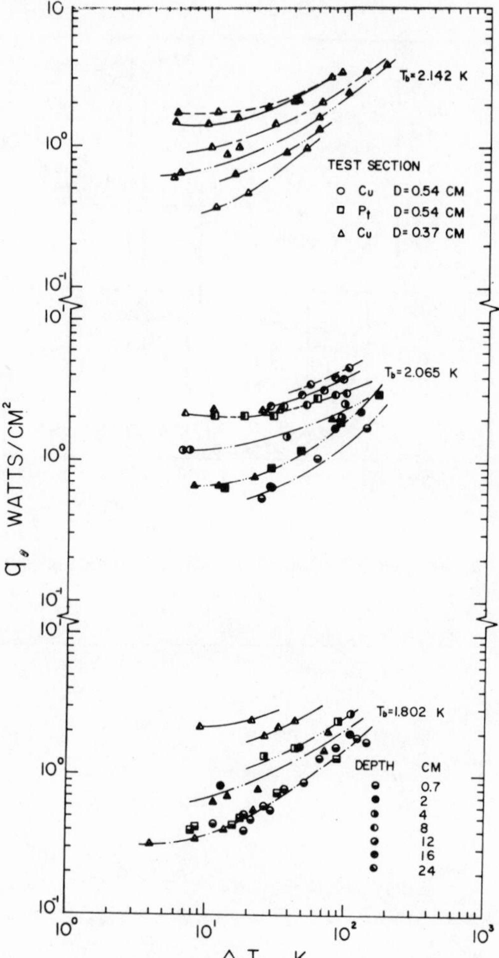

FIG. 15-14. Film boiling heat transfer on horizontal cylinders.

Most striking is the observation that noisy boiling diminishes rather than increases heat transfer; the reduction is on the order of 25%.

No satisfactory explanation for this phenomenon has as yet been devised. Leonard[78] (Fig. 15-13) has presented a map of the regions where each regime is encountered with respect to depth (noise is associated with deeper depths) and bath temperature (noise is associated with lower bath temperatures). Missing is a similar development with respect to heat flux or temperature difference (the probability of noise increases with increasing temperature difference).

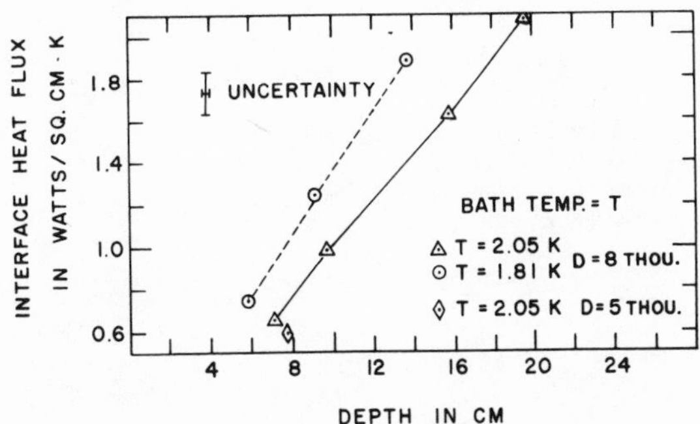

FIG. 15-15. Dependence of interface heat flux on depth.

The noise phenomenon is not limited to wire test surfaces. Noisy film boiling has been measured on large cylinders by Goodling and Irey[43] (Fig. 15-14).

Because of the total energy dissipation, film boiling measurements on larger composite test surfaces are limited to relatively low temperature differences.[43,44,47] This type of test surface does permit accurate data acquisition at temperature differences just above peak flux (Fig. 15-14).

The depth dependence evidenced in film boiling data appears to be explainable in terms of the same shift in local saturation temperature utilized in the peak flux correlation discussed above

$$T_i = T_b + \rho g H (dT/dP)_{sat} \tag{15-22}$$

Steed and Irey[73] have shown experimentally that the interface heat flux, defined by

$$q_i = qD/D_F \tag{15-23}$$

where D_F is the average film diameter, is a linear function of depth for each bath temperature and wire diameter (Fig. 15-15). Therefore, in terms of the local saturation temperature, Eq. (15-22), one can define an interface heat transfer coefficient

$$h_i = q_i/(T_i - T_b) \tag{15-24}$$

which is no longer depth dependent.

Steed and Irey also searched for an effect of the geometry of liquid helium II surrounding the test surface at fixed depth. They found no measurable change in the heat transfer rate under given conditions of test surface temperature, geometry, and depth for large percentage (up to 80) changes in evaporation surface area and volume of liquid between the test surface and the evaporation surface. This provided additional evidence that the effect is thermodynamically related to depth, as, for example, Eq. (15-22) implies.

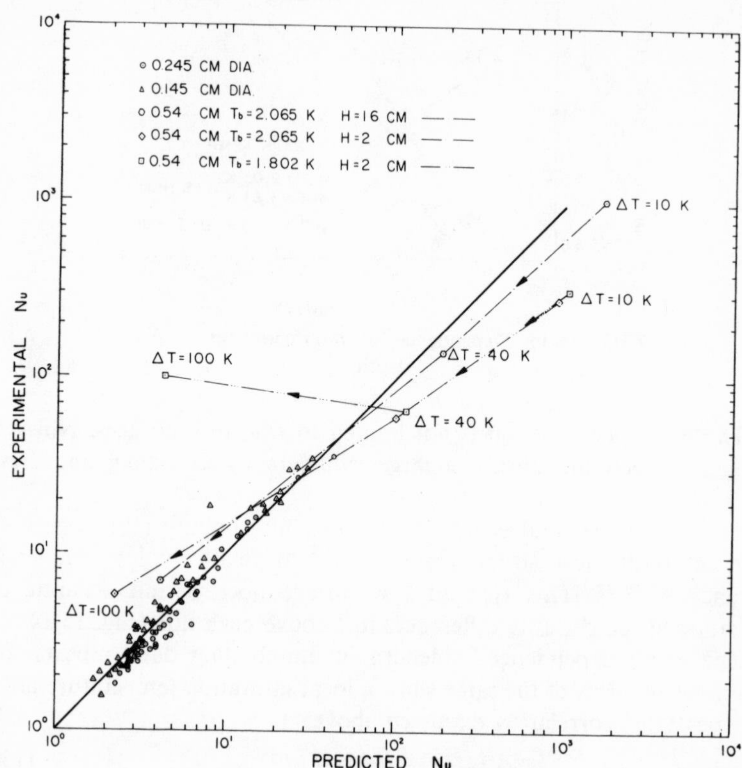

FIG. 15-16. Theoretical vs. experimental Nusselt numbers for Rivers and McFadden's boundary layer model.[79]

Rivers and McFadden[79] presented a boundary layer analysis of film boiling in liquid helium II. Their model accounts for liquid helium II's superfluidity in a boundary condition. They assign a heat flux q_i like that defined in Eq. (15-23) to the interface as an inhomogeneous boundary condition. Rivers and McFadden assumed that q_i was a function of the local thermodynamic equilibrium state. This assumption has been verified by Steed and Irey[73] (Fig. 15-15) and later by Ebright and Irey[74] with respect to bath temperature and depth. Neither of these investigators has satisfactorily resolved the question of the independence of q_i of geometry, since the variation of wire diameter in both studies was quite limited.

Holdredge and McFadden[80] have presented a comparison of the predictions of the Rivers and McFadden model with their film boiling data[46] (Fig. 15-16). Here q_i was assumed to be the peak flux q^* for the same test surface. Also shown are a few of the points collected by Goodling

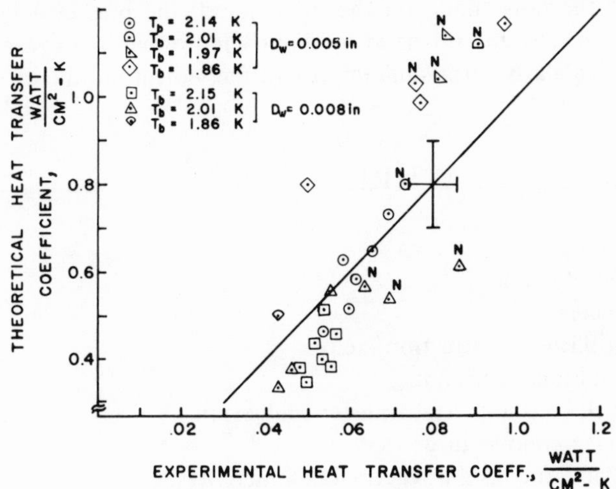

FIG. 15-17. Theoretical vs. experimental heat transfer coefficients for River's conduction model.

and Irey[43] for two bath temperatures and depths. Although the magnitude of the deviation of these data from the correlation is not much greater than that of Holdredge and McFadden, the dependence of the deviation on temperature difference gives an entirely different impression. The characteristic parameters on the Holdredge and McFadden points were not indicated.[79,80] In fact, in the region where q is below q^* (low ΔT), the assumption that $q_i = q^*$ cannot be correct. Therefore, the comparison made in Fig. 15-16 is not an adequate test of the theory.[79]

Ebright[60] found that peak flux was substantially greater than q_i on his wire studies. In addition, Ebright and Irey[74] found that the full Rivers and McFadden model did not correlate their test results. This is in no sense a failure of the theory since the analysis was based on the assumption that the boundary layer was thin. This assumption is essentially never satisfied by the wire test sections. Rivers[81] had in addition a very simple analysis applicable to cylindrical test surfaces in the limit of negligible convective transport. Here

$$(hD/k_v)C^{-(k_v/hD)} = q_i D/k_v(T_w - T_i) \tag{15-25}$$

where T_i is described by Eq. (15-22), k_v is the thermal conductivity of the vapor, and h is the heat transfer coefficient. The resulting comparison between the measured heat transfer coefficient and that calculated from Eq. (15-25), using the measured q_i, Eq. (15-23), is shown in Fig. 15-17. The correspondence between the analysis and experiment is within the experimental uncertainty for most of the silent, film boiling points. For a substantial portion of the noise data, on the other hand, the predicted heat transfer coefficient exceeds that observed by slightly more than the uncertainty. This result is particularly surprising, in view of the violent motion of the film.

15.5 NOMENCLATURE

C = constant
c_s = speed of sound
D = diameter
f = multiplicative correction factor
g = gravitational constant
$\hbar$ = Planck constant; heat transfer coefficient
h_K = Kapitza conductance
k = Boltzmann constant; thermal conductivity
$L\zeta$ = correlation length for the state of order
m = mass flow rate
N = dimensionless driving potential
n = number density of crystal
p = pressure
q = heat flux
R = universal gas constant
Re = Reynolds number
s = specific entropy
T = temperature
u = internal energy
$\bar{v}$ = local bulk velocity

Greek Letters

θ_D = Debye temperature
μ = viscosity
μ = chemical potential
ρ = density

Subscripts

b = liquid helium II bath
i = interface
n = normal fluid component
ph = phonons

S = surface
s = superfluid component
v = vapor

15.6 REFERENCES

1. R. London, *Superfluids*, Vols. I, II, John Wiley and Sons, New York (1950, 1954).
2. D. C. Freeman, in *Advances in Cryogenic Engineering*, Vol. 13, Plenum Press, New York (1968), p. 9
3. T. I. Smith, in *Advances in Cryogenic Engineering*, Vol. 13, Plenum Press, New York (1968), p. 102.
4. J. Wilks, *The Properties of Liquid and Solid Helium*, Oxford Univ. Press, Cambridge, England (1967).
5. K. R. Atkins, *Liquid Helium*, Cambridge Univ. Press, Cambridge, England (1959).
6. C. T. Lane, *Superfluid Physics*, McGraw-Hill Book Co., New York (1962).
7. V. Arp, *Cryogenics* **10**, 96 (1970).
8. I. M. Khalatnikov, *Introduction to the Theory of Superfluidity*, W. A. Benjamin, (1965).
9. E. M. Lifshitz and E. L. Andronikashvili, *A Supplement to Helium*, Consultants Bureau, New York (1959).
10. W. E. Keller, *Helium-3 and Helium-4*, Plenum Press, New York (1969).
11. P. L. Kapitza, *J. Physics* (*Moscow*) **4**(3), 181 (1941).
12. R. Kronig and A. Thellung, *Physica* **16**, 678 (1950).
13. R. A. Kronig, A. Thellung, and H. H. Woldringh, *Physica* **18**, 21 (1952).
14. C. J. Gorter, K. W. Taconis, and J. J. M. Beenakker, *Physica* **17**, 841 (1951).
15. D. M. Lee and H. A. Fairbank, *Phys. Rev.* **116**, 1359 (1959).
16. C. F. Mate and S. P. Sawyer, in *Proc. 11th Intern. Conf. on Low Temp. Phys.*, Vol. **I**, Univ. of St. Andrews, St. Andrews, Scotland (1969), p. 579.
17. T. H. K. Frederking, *Chem. Eng. Progr. Symp. Series* **64**(87), 21 (1968).
18. C. F. Mate and S. P. Sawyer, *Phys. Rev. Lett.* **20**(16), 834 (1968).
19. N. S. Snyder, NBS Tech. Note No. 385 (1969); also *Cryogenics* **10**, 89 (1970).
20. R. E. Jones and W. B. Pennebaker, *Cryogenics* **3**(4), 215 (1963).
21. T. J. Love, *Radiative Heat Transfer*, Merrill (1968).
22. A. J. Dekker, *Solid State Physics*, Prentice-Hall, New Jersey (1957).
23. R. C. Johnson and W. A. Little, *Phys. Rev.* **130**(2), 596 (1963).
24. W. A. Little, in *Proc. 7th Intern. Conf. Low Temp. Phys.*, Univ. of Toronto Press (1960), p. 482.
25. J. I. Gittleman and S. Bozowski, *Phys. Rev.* **128**(2) 646 (1962).
26. D. A. Neeper, D. C. Pearce, and R. M. Wasilik, *Phys. Rev.* **156**(3), 764 (1967).
27. L. J. Challis, *Proc. Phys. Soc.* **80**(Part 3), 759 (1962).
28. L. J. Barnes and J. R. Dillinger, *Phys. Rev. Lett.* **10**(7), 287 (1963).
29. L. J. Challis and D. N. Cheeke, *Progr. in Refr. Sci. Tech.* **1**, 227 (1965).
30. R. J. Hesser, R. C. Chapman, Y. W. Chang, and T. H. K. Frederking, AIChE Preprint 10, Eleventh National Heat Transfer Conference, AIChE, New York (1969).
31. J. D. N. Cheeke, in *Proc. 11th Intern. Conf. on Low Temp Phys.*, Vol. 1 (1969), p. 567.
32. K. Wey-Yen, *Soviet Phys.—JETP* **15**(4), 635 (1962).
33. H. M. Rosenburg, *Low Temperature Solid State Physics*, Clarendon Press, Oxford, England (1963).

34. I. M. Khalatnikov, *Zh. Eksp. Teor. Fiz.* **22**(6) 687 (1952).
35. L. J. Challis, K. Dransfeld, and J. Wilks, *Proc. Roy. Soc.* **A260**(1300), 31 (1961).
36. W. A. Little, *Phys. Rev.* **123**(6), 1909 (1961).
37. G. L. Pollack, *Rev. Mod. Phys.* **41**(1), 48 (1969).
38. D. White, O. D. Gonzales, and H. L. Johnston, *Phys. Rev.* **89**(3), 593 (1953).
39. K. N. Zinoveva, *Zh Eksp. Teor. Fiz.* **25**, 235 (1953).
40. H. A. Fairbank and J. Wilks, *Proc. Roy. Soc.* **A231**(1187), 545 (1955).
41. N. J. Brow and D. V. Osborne, *Phil. Mag.* **3**(36), 1463 (1958).
42. A. C. Anderson, J. I. Connolly, and J. C. Wheatley, *Phys. Rev.* **135**(4A), A910 (1964).
43. J. S. Goodling and R. K. Irey, in *Advances in Cryogenic Engineering*, Vol. 14, Plenum Press, New York (1969), p. 159.
44. B. W. Clement and T. H. K. Frederking, in *Bull. Intern. Inst. Refr., Annex*, **1966**, 49.
45. R. A. Madsen and P. W. McFadden, in *Advances in Cryogenic Engineering*, Plenum Press, New York (1968), p. 617.
46. R. M. Holdredge and P. W. McFadden, in *Advances in Cryogenic Engineering*, Vol. 11, Plenum Press, New York (1966), p. 507.
47. D. N. Lyon, *Chem. Eng. Progr. Symp. Series* **64**(87), 82 (1968).
48. V. P. Peshkov, *Soviet Phys.—JETP* **35**(6), 443 (1959).
49. V. P. Peshkov, *Soviet Phys.—JETP* **3**(4), 628 (1956).
50. E. L. Andronikashvili and G. G. Mirskaia, *Soviet Phys.—JETP* **2**(3) 406 (1956).
51. J. S. Goodling, Ph.D. Dissertation, Univ. of Florida (1968) (available from University Microfilms, Ann Arbor, Michigan).
52. R. K. Irey, P. W. McFadden, and R. A. Madsen, in *Advances in Cryogenic Engineering*, Vol. 10, Plenum Press, New York (1965), p. 361.
53. R. C. Steed, Ph.D. Dissertation, Univ. of Florida (1968) (available through University Microfilms, Ann Arbor, Michigan).
54. A. C. Leonard and E. R. Lady, in *Advances in Cryogenic Engineering*, Vol. 16, Plenum Press, New York (1961), p. 378.
55. T. H. K. Frederking and R. L. Haben, *Cryogenics* **8**(1), 32 (1968).
56. D. N. Lyon, in *Advances in Cryogenic Engineering*, Vol. 10, Plenum Press, New York (1965), p. 371.
57. L. Rinderer and F. Haenseler, *Helv. Phys. Acta* **2**(4), 322 (1959).
58. T. H. K. Frederking, *Forschung* **27**, 17 (1961).
59. G. P. Lemieux and A. C. Leonard, in *Advances in Cryogenic Engineering*, Vol. 13, Plenum Press, New York (1968), p. 624.
60. R. L. Ebright, Ph.D. Dissertation, Univ. of Florida (1969) (available from University Microfilms, Ann Arbor, Michigan).
61. J. S. Vinson, F. J. Agee, R. J. Manning, and F. L. Hereford, *Phys. Rev.* **168**(1), 180 (1968).
62. P. W. McFadden and R. M. Holdredge, *Bull. Intern. Inst. Refr., Annexe*, **1966**, 259.
63. T. H. K. Frederking, R. L. Haben, and R. H. Madsen, in *Advances in Cryogenic Engineering*, Vol. 17, Plenum Press, New York (1972), p. 323.
64. J. R. Clow and J. D. Reppy, *Phys. Rev. Lett.* **19**, 291 (1967).
65. J. R. Clow and J. D. Reppy, *Phys. Rev. Lett.* **16**(20) 887 (1966).
66. R. P. Henkel, E. N. Smith, and J. D. Reppy, *Phys. Rev. Lett.* **23**, 1276 (1969).
67. H. E. Rorschach and F. A. Romberg, in *Proc. Intern. Conf. Low. Temp. Phys. Chem.*, Madison, Wisconsin (1957), p. 35.
68. S. G. Sydoriak and T. R. Roberts, *Bull. Intern. Inst. Refrig., Annex*, **1966**, 115.
69. V. Purdy, T. H. K. Frederking, R. W. Boom, C. A. Guderjahn, G. A. Domoto, and

C. L. Tien, in *Advances in Cryogenic Engineering*, Vol. 14, Plenum Press, New York (1969), p. 146.
70. C. N. Whetstone and R. W. Boom, in *Advances in Cryogenic Engineering*, Vol. 13, Plenum Press, New York (1968), p. 68.
71. T. H. K. Frederking and C. Linnet, in *Advances in Cryogenic Engineering*, Vol. 13, Plenum Press, New York (1968), p. 80.
72. A. C. Leonard and G. P. Lemieux, ASME Paper No. 67-WA/Ht-37 (1967).
73. R. C. Steed and R. K. Irey, in *Advances in Cryogenic Engineering*, Vol. 15, Plenum Press, New York (1970), p. 299.
74. F. L. Ebright and R. K. Irey, in *Advances in Cryogenic Engineering*, Vol. 16, Plenum Press, New York (1961).
75. R. B. Steward and V. J. Johnson, (editors), "A Compendium of Properties of Materials at Low Temperatures," Armed Services Tech. Info. Agency, Arlington, Virginia (1961).
76. P. Bussieres and A. C. Leonard, *Cryogenic News* **2**, 4 (1967).
77. D. M. Coulter, A. C. Leonard, and J. G. Pike, in *Advances in Cryogenic Engineering*, Vol. 13, Plenum Press, New York (1968), p. 640.
78. A. C. Leonard, "Helium II Noise Film-Boiling and Silent Film-Boiling Heat Transfer Coefficient Values," presented at the 3rd Intern. Cryo. Engr. Conf., West Berlin, May 1970.
79. W. J. Rivers and P. W. McFadden, *Trans. ASME, J. Heat Transfer* **88C**(4), 343 (1966).
80. R. M. Holdredge and P. W. McFadden, "Film Boiling on Horizontal Cylinders in Helium II," ASME Paper No. 70-HT-3 (1970).
81. W. J. Rivers, Ph.D. Dissertation, Purdue Univ. (1964) (available from University Microfilms, Ann Arbor, Michigan).

INDEX